D0867231

TEMPO AND MODE IN EVOLUTION

GENETICS AND PALEONTOLOGY 50 YEARS AFTER SIMPSON

WALTER M. FITCH
FRANCISCO J. AYALA
Editors

NATIONAL ACADEMY OF SCIENCES

NATIONAL ACADEMY PRESS
Washington, D.C. 1995

NATIONAL ACADEMY PRESS • 2101 CONSTITUTION AVENUE, N.W. • WASHINGTON, D.C. 20418

This volume is based on the National Academy of Sciences' Colloquium on the Tempo and Mode of Evolution. The articles appearing in these pages were contributed by speakers at the colloquium and have not been independently reviewed. Any opinions, findings, conclusions, or recommendations expressed in this volume are those of the authors and do not necessarily reflect the views of the National Academy of Sciences.

The National Academy of Sciences is a private, nonprofit, self-perpetuating society of distinguished scholars engaged in scientific and engineering research, dedicated to the furtherance of science and technology and to their use for the general welfare. Upon the authority of the charter granted to it by the Congress in 1863, the Academy has a mandate that requires it to advise the federal government on scientific and technical matters. Dr. Bruce M. Alberts is president of the National Academy of Sciences.

Library of Congress Cataloging-in-Publication Data

Tempo and mode in evolution: genetics and paleontology 50 years after Simpson/ Walter M. Fitch and Francisco J. Ayala, editors, for the National Academy of Sciences.
p. cm.
Includes bibliographical references and index.
ISBN 0-309-05191-6
1. Evolution (Biology) 2. Genetics. 3. Paleontology. I. Fitch, Walter M., 1929– . II. Ayala, Francisco J., 1934– .
III. National Academy of Sciences (U.S.)
QH366.2.T46 1995 94-37675
575--dc20 CIP

Printed in the United States of America

Preface

WALTER M. FITCH AND FRANCISCO J. AYALA

George Gaylord Simpson said in his classic *Tempo and Mode in Evolution* (1944) that paleontologists enjoy special advantages over geneticists on two evolutionary topics. One general topic, suggested by the word "tempo," has to do with "evolutionary rates. . ., their acceleration and deceleration, the conditions of exceptionally slow or rapid evolutions, and phenomena suggestive of inertia and momentum." A group of related problems, implied by the word "mode," involves "the study of the way, manner, or pattern of evolution, a study in which tempo is a basic factor, but which embraces considerably more than tempo" (pp. xvii–xviii).

Simpson's book was self-consciously written in the wake of Theodosius Dobzhansky's *Genetics and the Origin of Species* (1937). The title of Dobzhansky's book suggested its theme: the role of genetics in explaining "the origin of species"—i.e., a synthesis of Darwin's theory of evolution by natural selection and the maturing science of genetics. In the introduction to his book, Simpson averred that an essential part of his study was an "attempted synthesis of paleontology and genetics," an effort that pervaded the whole book, but was particularly the subject of the first two chapters, which accounted for nearly half the book's pages.

Darwin believed that evolutionary change occurs by natural selection of small individual differences appearing every generation within any species. Singly the changes effected by selection are small but, given enough time, great changes can take place. Two of Darwin's most dedicated supporters, Thomas Huxley and Francis Galton, argued instead that evolution occurs by selection of discontinuous variations, or sports; evolution proceeds rapidly by discrete leaps. According to Huxley, if natural selection operates only upon gradual differences among individuals, the gaps between existing species and in the paleontological record could not be explained. For Galton, evolution was not a smooth and uniform process, but proceeded by "jerks," some of which imply considerable organic change.

This controversy was continued in the latter part of the nineteenth century by the biometricians Karl Pearson and W. F. R. Weldon, who believed, like Darwin, in the primary importance of common individual differences, and by the geneticist William Bateson, who maintained the primary importance of discontinuous variations. The rediscovery of Mendelian inheritance in 1900 provided what might have been common grounds to resolve the conflict. Instead, the dispute between biometricians and geneticists extended to continental Europe and to the United States. Bateson was the champion of the Mendelians, many of whom accepted the mutation theory proposed by De Vries, and denied that natural selection played a major role in evolution. The biometricians for their part argued that Mendelian characters were sports of little importance to the evolutionary process.

The conflict between Mendelians and biometricians was resolved between 1918 and 1931 by the work of R. A. Fisher and J. B. S. Haldane in England, Sewall Wright in the United States, and S. S. Chetverikov in Russia. Independently of each other, these authors proposed theoretical models of evolutionary processes which integrate Mendelian inheritance, natural selection, and biometrical knowledge.

The work of these authors, however, had a limited impact on the biology of the time, because it was formulated in difficult mathematical language and because it was largely theoretical with little empirical support. Dobzhansky's *Genetics and the Origin of Species* completed the integration of Darwinism and Mendelism in two ways. First, he gathered the empirical evidence that corroborated the mathematico-theoretical framework. Second, he extended the integration of genetics with Darwinism beyond the range of issues treated by the mathematicians, into critical evolutionary issues, such as the problem of speciation, not readily amenable to mathematical treatment. Moreover, Dobzhansky's book was written in prose that biologists could understand.

The formulation of the modern theory of evolution may well be traced to the publication in 1937 of *Genetics and the Origin of Species* (second edition in 1941; third in 1951). Several important books appeared in the ensuing years that deepened the synthesis and extended it to subjects, such as systematics (Ernst Mayr, *Systematics and the Origin of Species*, 1942), zoology (Julian S. Huxley, *Evolution: The Modern Synthesis*, 1942), paleontology (Simpson's *Tempo and Mode,* 1944), and botany (G. Ledyard Stebbins, *Variation and Evolution in Plants,* 1950).

The first two chapters of *Tempo and Mode* embraced a subject totally innovative at the time. Simpson first defined rates of evolution as resultants of genetic change. Then, he elucidated the contribution of various population genetic "determinants" to evolutionary rates. Simpson's determinants are variability, rate and effect of mutations, generation length, population size, and natural selection. Many of his conclusions have resisted the test of time: variability within populations and not between populations is of consequence in determining evolutionary rates; high variability is "a sort of bank in which mutations are on deposit, available when needed" so that populations need not wait for mutation to occur; mutations per gene are rare, but per population they are not; mutations with small effects are much more frequent than those with relatively large effects; rates of evolution are time dependent, rather than generation dependent; selection is more effective in large than in very small populations; and a myriad other insights and clever calculations that illustrate conceptual points.

Simpson develops a wonderful argument against the misconception that selection is purely a negative process. Selection controls the frequencies of different genes in a population and thus it determines, within limits, which combinations will be realized and in what proportions. Thus selection does not "simply kill off or permit to live fixed types of organisms delivered to it . . . Selection also determines which among the millions of possible types of organisms will actually arise, and it is therefore a truly creative factor in evolution" (p. 80).

Many current motifs in paleontology appeared first in *Tempo and Mode,* at least in their modern definition. Simpson had much of interest to say about the differentiating characteristics of microevolution and macroevolution and about the troubling (at the time) issue of discontinuities in the fossil record. Discontinuities result, on occasion, from deficiencies of the fossil record; but often, as in the origin of mammalian orders and other major transitions, because animals were evolving at unusually high rates. Increased rates of evolution ensue, in part, from reduced population numbers, a state of affairs likely when new niches or adaptive zones are invaded.

Simpson saw that low rates as well as high rates of evolution occur in the record. He called them "bradytelic" and "tachytelic" rates, and "horotelic" those that evolve at the standard rate of the group. The stasis that proponents of punctuated equilibrium would stress three decades later, Simpson recognized in the "so-called living fossils, groups that survive today and that show relatively little change since the very remote time when they first appeared in the fossil record" (p. 124).

Similarly, "major transitions do take place at relatively great rates over short periods of times and in special circumstances" (p. 207).

Simpson failed to anticipate that molecular biology would make it possible to measure rates of evolution "most desirably" as "amount of genetic change in a population per year, century, or other unit of absolute time" (p. 3). He was defeatist on this subject. Immediately after the phrase just quoted, he wrote: "This definition is, however, unusable in practice." It would be unfair to claim lack of vision here, where nothing short of a sorcerer's clairvoyance could have anticipated the magician's tricks of molecular biology and its boundless contributions to elucidating the history of evolution. It is in this respect that the papers that follow depart the most from the themes of *Tempo and Mode in Evolution*. Indeed, most of what is now known about phylogenetic relationships among organisms that lived during the first three billion years of life on earth is an outcome of molecular biology; and three billion years amounts to most of the history of life.

The sixteen papers that follow were presented and discussed at a colloquium sponsored by the National Academy of Sciences to celebrate the 50th anniversary of the publication of Simpson's *Tempo and Mode in Evolution*. The Colloquium was held on January 27–29, 1994, at full capacity in the Academy's splendid facility in Irvine, California, the Arnold and Mabel Beckman Center. We are grateful to the NAS for the generous grant that financed the Colloquium, and to the staff of the Beckman Center for their skill and generous assistance.

REFERENCES

Dobzhansky, Th. (1937) *Genetics and the Origin of Species* (Columbia Univ. Press, New York).
Dobzhansky, Th. (1941) *Genetics and the Origin of Species*, 2nd ed. (Columbia Univ. Press, New York).
Dobzhansky, T. (1951) *Genetics and the Origin of Species*, 3rd ed. (Columbia Univ. Press, New York).
Huxley, J. S. (1942) *Evolution: The Modern Synthesis* (Harper, New York).
Mayr, E. (1942) *Systematics and the Origin of Species* (Columbia Univ. Press, New York).
Simpson, G. G. (1944) *Tempo and Mode in Evolution* (Columbia Univ. Press, New York).
Stebbins, G. L. (1950) *Variation and Evolution in Plants* (Columbia Univ. Press, New York).

Contents

TEMPO AND MODE IN EVOLUTION

Part I

EARLY LIFE

There is no fossil record of the evolution that preceded cellular life, but the process can be elucidated by molecular investigation of modern organisms. One issue is the differentiation between genotype and phenotype, mediated by the mechanisms of transcription and translation that express genes as proteins. In Chapter 1, W. Ford Doolittle and James R. Brown explore the question whether the last common ancestor of all life was a "progenote" endowed with a genetic information transfer system that was much more rudimentary than at present.

Nancy Maizels and Alan M. Weiner, in Chapter 2, conclude that transfer RNA-like molecules predate the progenote. They evolved, *before* the advent of messenger RNA and templated protein synthesis, as regulatory elements of replication. These molecules were later hijacked for their present function in amino acid transfer, which required the evolution of the bottom half of modern tRNA molecules, which includes the anticodon.

In 1944, when Simpson's *Tempo and Mode* was published, the only fossils known were animals and plants that had lived since the Cambrian Period, some 550 million years ago. Since the 1960s, numerous fossil remains have been discovered of prokaryotic microbes, some of which are more than 3 billion years old. These simple asexual organisms were ecological generalists that evolved at astonishingly slow rates. As J. William Schopf, in Chapter 3, puts it: "In both tempo and mode of evolution, much of the Precambrian history of life . . . appears to have differed markedly from the more recent

Phanerozoic evolution of megascopic, horotelic, adaptationally specialized eukaryotes."

Eukaryotic microfossils appeared in the late Paleoproterozoic, some 1700 million years ago. Their evolution was at first slow, but diversity and turnover rates greatly increased around 1000 million years ago. Evolutionary rates accelerated again during the Cambrian, as Andrew H. Knoll shows in Chapter 4: protistan diversity increases by a factor of two and turnover rates by a factor of ten.

1

Tempo, Mode, the Progenote, and the Universal Root

W. FORD DOOLITTLE AND JAMES R. BROWN

Simpson sought in *Tempo and Mode in Evolution* to explain large-scale variations in evolutionary rate and pattern apparent in the fossil record (Simpson, 1944). By *tempo* he meant "rate of evolution . . . practically defined as amount of morphological change relative to a standard," and by *mode* he meant "the way, manner, or pattern of evolution." For those of us concerned with the evolution of molecules rather than organisms, issues of tempo mostly have to do with the molecular clock, while questions about mode address mutational mechanisms and forces driving changes in gene and genome structure. In this article, we focus on the period of early cellular evolution, between the appearance of the first self-replicating informational macromolecule and the deposition of the first microfossils, by all accounts already modern cells (Schopf, 1994). We ask whether major shifts in predominant mode occurred during this period, and (since the answer is of course yes) whether we might actually come to know anything other than the vaguest generalities about these shifts.

W. Ford Doolittle is a Fellow of the Canadian Institute for Advanced Research and professor of biochemistry at Dalhousie University, Halifax, Nova Scotia. James R. Brown is a postdoctoral fellow in biochemistry at Dalhousie University, whose work is supported by the Medical Research Council of Canada.

Stages in the Evolution of the Cellular Information-Processing System

In Figure 1 we present a fanciful representation of the evolution of the information transfer system of modern cells and propose that it be seen as divisible into three phases, differing profoundly in *both* tempo and mode. The first (Figure 1 *Bottom*) would be accepted by all who speculate on the origin of Life as a period of **preDarwinian evolution**: without replication there are no entities to evolve through the agency of natural selection. We call the second period, between the appearance of the first self-replicating informational molecule and the appearance of the first "modern" cell, the period of **progressive Darwinian evolution** (Figure 1 *Middle*). "Progress" is of course an onerous concept in evolutionary theory (Ayala, 1988). Nevertheless, we submit that, as its uniquely defining feature or mode, this second phase witnessed the fixation of many mutations improving the accuracy, speed, and efficiency of information transfer overall and, thus, the adaptedness of cells (or simpler precellular units of selection) under almost any imaginable conditions. Nowadays (in the third period, that of **postprogressive Darwinian evolution**; Figure 1 *Top*), most mutations that are fixed by selection improve fitness only for specific environmental regimes. But earlier, when evolution did exhibit progress, selection forged successive generations of organisms (or simpler units) in which phenotype was more reliably coupled to genotype. Individuals from later in this period would have almost always outperformed their ancestors if placed in direct competition with them.

How could we hope to know anything about this ancient era of radically different tempo and mode? *If* divergences that established the major lineages of contemporary living things occurred before completion of the period of progressive Darwinian evolution, then we would expect that the information processing systems of these lineages would differ from each other—the earlier the divergence, the more profound the difference. That is, components of the replication, transcription, and translation machineries that were still experiencing progressive Darwinian evolution at the time of divergence should be differently refined or altogether separately fashioned (nonhomologous) in major lineages. Thus, comparisons between modern major groups (such as prokaryotes and eukaryotes) might lead to informed guesses about primitive ancestral states.

As an exemplary exercise, Benner and colleagues (Benner *et al.*, 1993) inferred from the fact that archaebacteria, eubacteria, and eukaryotes produce ribonucleotide reductases that are not demonstrably homologous that their last common ancestor used a ribozyme for the reduction

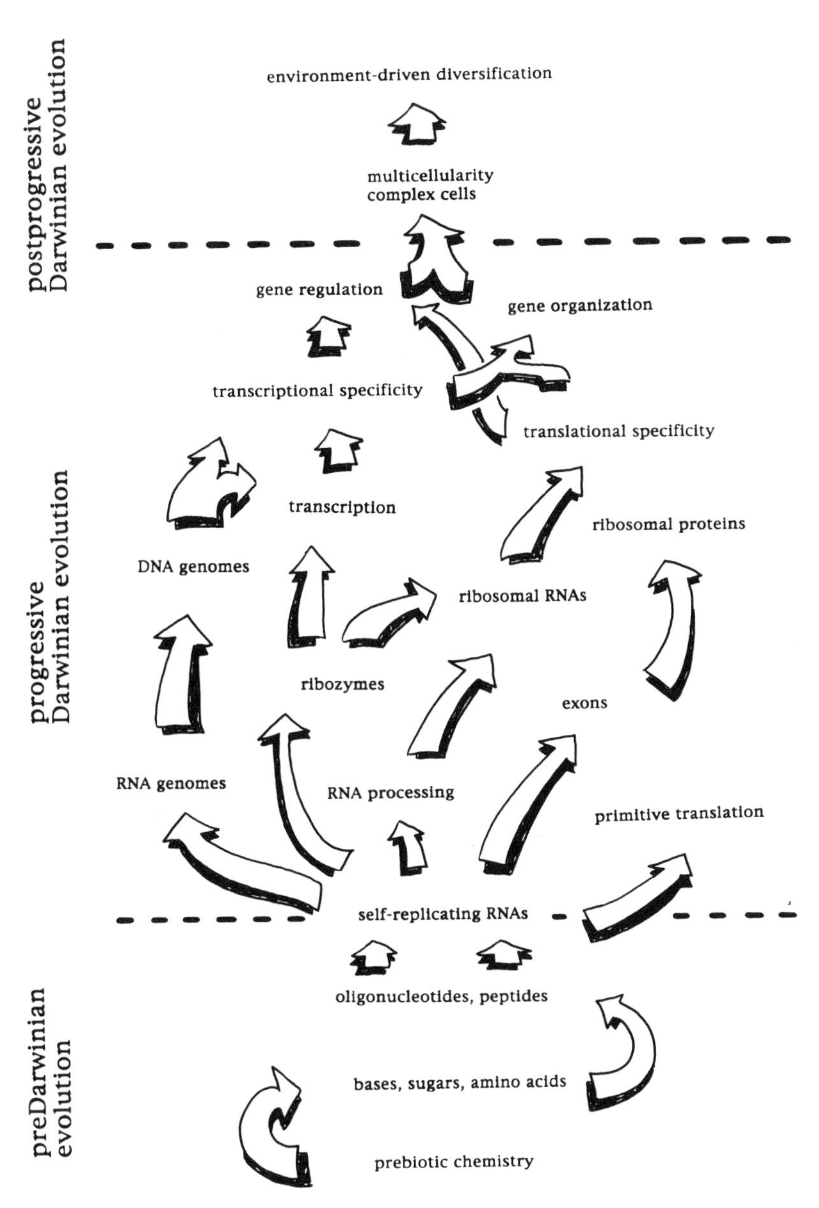

FIGURE 1 Fanciful interpretation of early evolution. The underlying assumption is that some contemporary processes and molecules had to appear before others and that the evolution of the information processing system involved interactions between separately evolving components.

of ribonucleotides. Benner's group also (and much more persuasively) concluded, from the evident homology of DNA (or RNA) polymerases in the three domains, that the transition from RNA to DNA genomes had itself already been made by that last common ancestor, whatever its residual reliance on ribozymology (Benner *et al.*, 1989).

To make such inferences about past events through use of parsimony arguments to reconstruct common ancestors from knowledge of the different paths taken by descendants, we must know that the contemporary groups compared really did begin to diverge at an appropriately ancient date. In the rest of this review, we consider developments in our thinking about the relationships between basic kinds of living things primarily as they bear on this issue, asking *if there is any reason to believe that the cenancestor was a progenote.*

Of the two new terms introduced here, **cenancestor** is Walter Fitch's for "the most recent common ancestor to all the organisms that are alive today (Fitch and Upper, 1987)". **Progenote** is George Fox and Carl Woese's descriptor for "a theoretical construct, an entity that, by definition, has a rudimentary, imprecise linkage between its genotype and phenotype (Woese, 1987)"—a creature still experiencing progressive Darwinian evolution, in other words.

The Basic Kinds of Living Things

For more than a century, microbiologists suspected that bacteria, because of their small size and seemingly primitive structure, might differ fundamentally from animals, plants, and even fungi. The blue–green algae (now "cyanobacteria") might be intermediate, looking like bacteria but acting like plants. Chatton in 1937 (Chatton, 1937) and Stanier and van Niel in 1941 (Stanier and van Niel, 1941) proposed that these two groups share a common *cellular organization* distinguishing them as **prokaryotes** from the rest of the living world, or **eukaryotes**. A clear statement of the differences, however, required further work in biochemistry, genetics, and cellular ultrastructure. By 1962, Stanier and van Niel (Stanier and van Niel, 1962) were prepared to define prokaryotes in terms of the specific features they shared as well as the eukaryotic characteristics they lacked. They wrote that:

> . . . the principal distinguishing features of the prokaryotic cell are:
> 1. absence of internal membranes which separate the resting nucleus from the cytoplasm, and isolate the enzymatic machinery of photosynthesis and respiration in specific organelles;
> 2. nuclear division by fission, not by mitosis, a character possibly related to the presence of a single structure which carries all the genetic information of the cell; and

3. the presence of a cell wall which contains a specific mucopeptide as its strengthening element.

By 1970, Stanier could confidently state that

> . . . advancing knowledge in the domain of cell biology has done nothing to diminish the magnitude of the differences between eukaryotic and prokaryotic cells that could be described some ten years ago: if anything, the differences now seem greater (Stanier, 1970).

But, cautiously endorsing Lynn Margulis' assertion that eukaryotic cells are themselves the result of the fusion of separate (prokaryotic) evolutionary lineages (Margulis, 1970), he went on to note that

> . . . the only major links [between the two cell types] which have emerged from recent work are the many significant parallelisms between the entire prokaryotic cell and two component parts of the eukaryotic cell, its mitochondria and chloroplasts.

This linkage has since been amply supported by molecular sequence data (Gray and Doolittle, 1982), and the endosymbiont hypothesis for the origin of eukaryotic organelles of photosynthesis and respiration has become a basic tenet of the contemporary evolutionary consensus.

Together, the prokaryote/eukaryote dichotomy and the endosymbiont hypothesis for the origin of mitochondria and chloroplasts informed and (no doubt) constrained the biology and molecular biology of the 1960s, 1970s, and early 1980s, providing the framework within which all of the results of biochemists, geneticists, and evolutionists were interpreted (Figure 2). In typical text books from this era, genes in *Escherichia coli* are compared and contrasted to their counterparts in yeast, mouse, and man, with differences interpreted either in terms of the relatively advanced and complex state of the latter or the admirably streamlined features of the former. The paradigm has been extraordinarily fruitful: without such a grand scheme for organizing our knowledge of cell and molecular biology, we would have become lost in the details. It also seems safe to say that, *for the organisms studied by most molecular biologists* in those decades, this view of things is substantially correct and invaluable in interpreting the differences in the information-transfer systems of prokaryotes and eukaryotic nuclei, chloroplasts, and mitochondria.

As well, this view was easily consistent with the most straightforward interpretation of the fossil record. As reviewed by Schopf and Knoll elsewhere in this volume, unquestionable prokaryotes, by all available measures indistinguishable from modern cyanobacteria, appeared more than 3.5 billion years ago (Schopf, 1994; Knoll, 1994). Fossils that are undeniably eukaryotic are not seen for another 1 to 1.5 billion years, ample time for the symbioses required by Margulis.

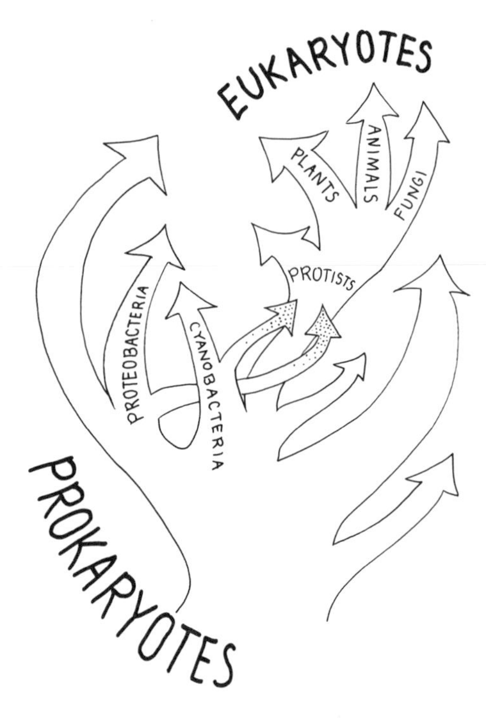

FIGURE 2 Prevalent evolutionary view between 1970 and 1977. The eukaryotic nuclear lineage arose from within the already characterized prokaryotes (eubacteria, perhaps a mycoplasma). Mitochondria descend from endosymbiotic proteobacteria, and plastids descend from endosymbiotic cyanobacteria.

The Woesian Revolution

The consensus represented by Figure 2 rested on comparative ultrastructural, biochemical, and physiological data and on a modest accumulation of primary (protein) sequence information, mostly from cytochromes and ferredoxins. In 1978, Schwartz and Dayhoff summarized this information and the then even-more-limited data from ribosomal RNA (rRNA)—in particular, 5S rRNA (Schwartz and Dayhoff, 1978). The endosymbiotic nature of organelles was well supported, but the origin of the nuclear genome (that is, the genome of the host for these endosymbioses) remained a mystery. A grand reconstruction of all of the main events of evolution with a single molecular chronometer was called for.

Such a grand reconstruction was the goal of Woese, who had begun, in the late 1960s, to assemble catalogs of the sequences of the oligonucleotides released by digestion of *in vivo*-labeled 16S rRNA with T1 ribonuclease. Comparing catalogs from different bacteria (scoring for presence or absence of identical oligonucleotides) by methods of numerical taxonomy allowed the construction of dendrograms showing relationships between them (Woese and Fox, 1977). Methods have been updated, cataloging giving way to reverse-transcriptase sequencing of rRNA, and this in turn to cloning (and now PCR cloning) of DNAs encoding rRNA (rDNAs). The data bases presently contain partial or complete sequences for some 1500 small-subunit rRNAs from prokaryotes and a rapidly growing collection of eukaryotic cytoplasmic small-subunit sequences, which track the evolutionary history of the nucleus (Olsen *et al.*, 1994).

The rRNA data support the consensual picture represented in Figure 2 in many important ways. Such data not only confirm that chloroplasts and mitochondria descend from free-living prokaryotes but also show that the former belong close to (perhaps within) the cyanobacteria, while the latter derive from the alpha subdivision of the purple bacteria (proteobacteria). These data also establish relationships within the bacteria that are sensible in terms of advancing knowledge of prokaryotic biochemical and ecological diversity and often congruent with more traditional classification schemes, at least at lower taxonomic rank. However, there were two major surprises, both announced by Woese and colleagues in 1977 (Woese and Fox, 1977; Fox *et al.*, 1977).

The first was that the eukaryotic nuclear lineage, as tracked by (18S) cytoplasmic small-subunit rRNA, was not demonstrably related to any specific, previously characterized prokaryotic lineage (Figure 3). This was not expected: the endosymbiont hypothesis saw the endosymbiotic host arising *within* the bacteria, the descendant of some otherwise typical prokaryote that had lost its cell wall and acquired the ability to engulf other cells. Differences in primary sequence between eukaryotic and prokaryotic small-subunit rRNAs also bespoke differences in secondary structure, consonant with the known differences in size (80S versus 70S), ribosomal protein content (75–90 polypeptides rather than 50–60), and function (initiation through "scanning" rather than base-pairing *via* the Shine–Dalgarno sequence, unformylated rather than formylated initiator tRNA).

Because of these differences, Woese argued that the ribosome of the last common ancestor of bacteria and eukaryotes (their nuclear-cytoplasmic part, that is) was itself a *primitive* ribosome, a structure still experiencing progressive Darwinian evolution. He ventured (Woese, 1982) that the same might be said for other components of this cenancestral information processing system and that:

> . . . in such a progenote, molecular functions would not be of the complex, refined nature we associate with functions today. Thus subsequent evolution would alter functions mainly in the sense of refining them. In this way, the

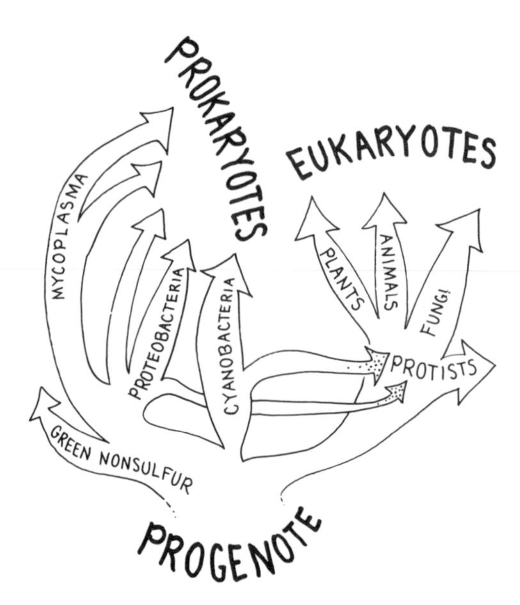

FIGURE 3 Implications of the rRNA data of Woese and Fox (1977). Prokaryotic (eubacterial) lineages from which the eukaryotic nuclear lineage was thought to have evolved were entirely separate from that lineage. Distinct properties of rRNAs suggested that the ribosome of the last common ancestor was a primitive ribosome and that the last common ancestral cell was a primitive "progenote," still experiencing progressive Darwinian evolution.

molecular differences among the three major groups would be in refinements of functions that occurred separately in the primary lines of descent, after they diverged from the universal ancestor.

In other words, the cenancestor *was* a progenote—one of the series of ancestral forms in which the phenotype–genotype coupling was actively evolving, and we might learn about progressive Darwinian evolution by comparing prokaryotic and eukaryotic (nuclear) molecular biology. Woese went on:

> . . . it is hard to avoid concluding that the universal ancestor was a very different entity from its descendants. If it were a more rudimentary sort of organism, then the tempo of its evolution would have been higher and the mode of its evolution highly varied, greatly expanded (Woese, 1987).

This view came to play a dominant role in the molecular biology and evolutionary microbiology of the 1980s and early 1990s. The prokaryote/eukaryote dichotomy remained, but as a vertical split, separating living things into two camps from the very beginning rather than marking a more recent but crucial transition in the grade of cellular organization. The inference that the cenancestor was a rudimentary being gave aid and comfort to those of us who had always doubted that the profound differences in gene and genome structure between eukaryotic nuclei and prokaryotes were improvements or advancements wrought in the former after their emergence from among the latter. Eukaryotic nuclear genomes are after all very messy structures, with vast amounts of seemingly unneccesary "junk" DNA, difficult-to-rationalize complexities in mechanisms of transcription and mRNA modification and processing, and needless scattering of genes that often in prokaryotes would be neatly arranged into operons. It might be easiest to see nuclear genomes as in a primitive state of organization, which prokaryotes, by dint of vigorous selection for economy and efficiency ("streamlining"), have managed to outgrow.

Such a view gained credence from and lent credence to the still popular although increasingly untenable "introns early" hypothesis or "exon theory of genes" (Doolittle, 1991). In brief, the notion here is that (*i*) the first self-replicators were small RNAs, which became translatable into small peptides; (*ii*) such "minigenes" came together to form the (RNA) ancestors of modern genes, introns marking the sutures; and (*iii*) the subsequent history of introns has been one of loss: streamlining has removed them entirely from the genes of prokaryotes but has been less effective in eukaryotes for a variety of reasons (less intense selection, lack of transcription–translation coupling as a driving force).

The second surprise from the rRNA data is depicted in Figure 4. In addition to showing the profound division between eukaryotes (their nuclei) and prokaryotes just discussed, these data identified two deeply diverging groups, two "primary kingdoms" within the prokaryotes. Woese and Fox called the first, which included *E. coli* and other proteobacteria, *Bacillus subtilis*, mycoplasma, the cyanobacteria, and indeed all prokaryotes about which we had accumulated any extensive biochemical or molecular genetic information, the "**eubacteria**" (Woese and Fox, 1977). It was these organisms that Stanier and van Niel had in mind when defining the prokaryote–eukaryote dichotomy in the 1950s and 1960s and on which most of us still fashion our beliefs about prokaryotes. The second primary kingdom, the "**archaebacteria**," included organisms that, although certainly not unknown to microbiologists, had been little studied at the cellular and molecular level, and whose inclusion within the prokaryotes therefore rested at that time on only the most basic of criteria (absence of a nucleus).

Archaebacteria are organisms of diverse morphology and radically different phenotypes, including the obligately anaerobic mesophilic methanogens, the aerobic and highly salt-dependent extreme halophiles, the amazing (because capable of growth up to at least 110°C)

FIGURE 4 Further implications of the rRNA data (Woese and Fox, 1977). A third group, the archaebacteria, seemed as distant evolutionarily from eubacteria and eukaryotes as these were from each other.

extreme thermophiles, and still completely uncharacterized and unseen meso- or psychrophiles, which are related to the extreme thermophiles and known only from PCR products amplified from the open ocean (DeLong, 1992). Uniting them are a number of basic characters unrelated to rRNA sequence and more than adequate to support their taxonomic and phylogenetic unity in spite of this diversity. These include unique isopranyl ether lipids (and the absence of acyl ester lipids found in eubacteria and eukaryotes); characteristic genetic organization, sequence, and function of RNA polymerase subunits; structural and functional characteristics of ribosomes and modification patterns of tRNAs; varied but unique cell-envelope polymers; and distinctive antibiotic sensitivities and insensitivities (Zillig *et al.*, 1993).

Rooting the Universal Tree

Woese felt that the differences between archaebacteria and either eubacteria or eukaryotes were of a sufficiently fundamental nature to indicate that *all three* primary kingdoms must have begun to diverge during the period of progressive evolution from a progenote. But there was no way to decide the order of branching—whether the first divergence in the universal tree separated (*i*) eubacteria from a line that was to produce archaebacteria and eukaryotes, or (*ii*) a proto-eukaryotic lineage from a fully prokaryotic (eubacterial and archaebacterial) clade, or (*iii*) the (the third and least popular possibility) archaebacteria from eukaryotes and eubacteria.

There is in fact *in principle* no way to decide this or to root such a universal tree based only on a collection of homologous sequences. We can root any sequence-based tree relating a restricted group of organisms (all animals, say) by determining which point on it is closest to an "outgroup" (plants, for example). But there can be no such organismal outgroup for a tree relating all organisms, and the designation of an outgroup for any less-embracing tree involves an *assumption*, justifiable only by other unrelated data or argument. Alternatively, we might root a universal tree by assuming something about the direction of evolution itself: Figure 2 for instance is rooted in the belief that prokaryotic cellular organization preceded eukaryotic cellular organization. But in fact the progenote hypothesis itself is such an assumption about the direction of evolution: we cannot use it to prove its own truth. We must establish which of the three domains diverged first by some other method—unrelated to either outgroup organisms or theories about primitive and advanced states—before we can start to use three-way comparative studies to make guesses about the common ancestor.

A solution to this problem was proposed and implemented by Iwabe and colleagues (Iwabe *et al.*, 1989), in 1989. Although there can be no *organism* that is an outgroup for a tree relating all organisms, we can root an all-organism tree based on the sequences of outgroup *genes* produced by gene duplication prior to the time of the cenancestor. The reasoning is as follows. Imagine such an ancient gene duplication producing genes *A* and *A′*, both retained in the genome of the cenancestor and all descendant lineages (Figure 5). Then either *A* or *A′* sequences can be used to construct unrooted all-organism trees, and the *A* tree can be rooted with any *A′* sequence, and the *A′* tree can be rooted with any *A* sequence. As well, there is a built-in internal check, because both trees should have the same topology!

What Iwabe *et al.* needed, then, were sequences of gene pairs that (because all organisms have two copies) must be the product of a precenancestral gene duplication and for which eubacterial, archaebacterial, and eukaryotic versions were known. Two data sets met their criteria—the α and β subunits of F_1 ATPases and the translation elongation factors EF-1α (Tu) and EF-2 (G). With either data set, rooted trees showing archaebacteria and eukaryotic nuclear genomes to be sister groups were obtained; eubacteria represented the earliest divergence from the universal tree (Figure 6).

The archaebacteriological community was already primed to accept this conclusion. At the very first meeting of archaebacterial molecular

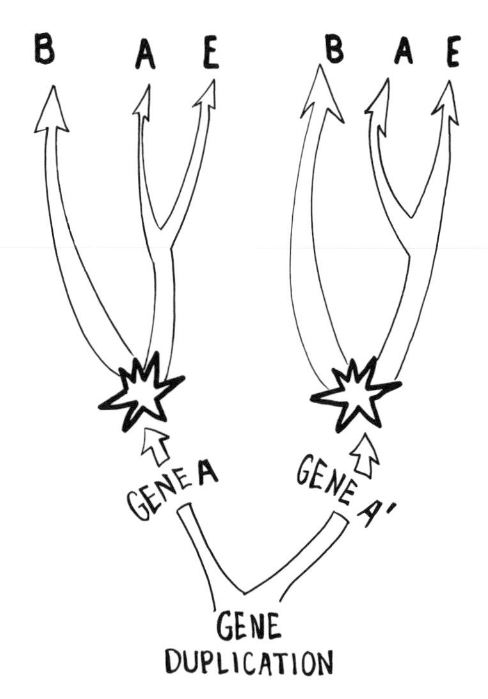

FIGURE 5 Rooting method of Iwabe and coworkers (1989). Two unrooted trees are constructed for eubacteria (B), archaebacteria (A), and eukaryotes (E), one with sequences of *A* genes and one with sequences of *A'* genes. *A* and *A'* genes are products of a gene duplication that must have predated the time of the last common ancestor because both are found in B, A, and E. The *A* tree can be rooted with *A'* sequences and vice versa, so that the cenacestor (the universal root) is in the position shown by the star for either tree.

biologists in Munich in 1981, one could sense a general feeling that archaebacteria were somehow "missing links" between eubacteria and eukaryotes. Zillig in particular stressed the (still-supported) eukaryote-like structural and functional characteristics of archaebacterial RNA polymerases (Zillig *et al.*, 1982). In the subsequent 7 or 8 years, further gene sequences for proteins of the information-transfer system (ribosomal proteins, DNA polymerase) that looked strongly eukaryote-like had appeared. Although not rootable, these data too seemed to support a specific archaebacterial/eukaryotic affinity (Ramirez *et al.*, 1993).

In 1990, Woese, Kandler and Wheelis incorporated the Iwabe rooting in a new and broader exegesis on the significance of the tripartite division of the living world (Woese *et al.*, 1990). This treatment elevated the rank of the three primary kingdoms to "domains" (since kingdom status was already well accepted for animals, plants, and fungi within the eukaryotes) and renamed them Bacteria, Archaea, and Eucarya. There were immediate and strong complaints from key figures in the evolutionary community, principally Lynn Margulis, Ernst Mayr and Tom Cavalier-Smith (Margulis and Guerrero, 1991; Mayr, 1990; Cavalier-Smith, 1992).

The objections touch many of the usual bases in evolutionary debates. Strict cladists would applaud the removal of "bacteria" from the name of the archaea, for instance, and would agree that the term "prokaryote" should not be used as a clade name because it describes a paraphyletic group. However, Woese and colleagues proposed the renaming not from cladist scruples but because of their belief in the profound nature of the phenotypic differences between archaebacteria and eubacteria. Mayr is no cladist either, but as a "gradist" he sees the change in cellular

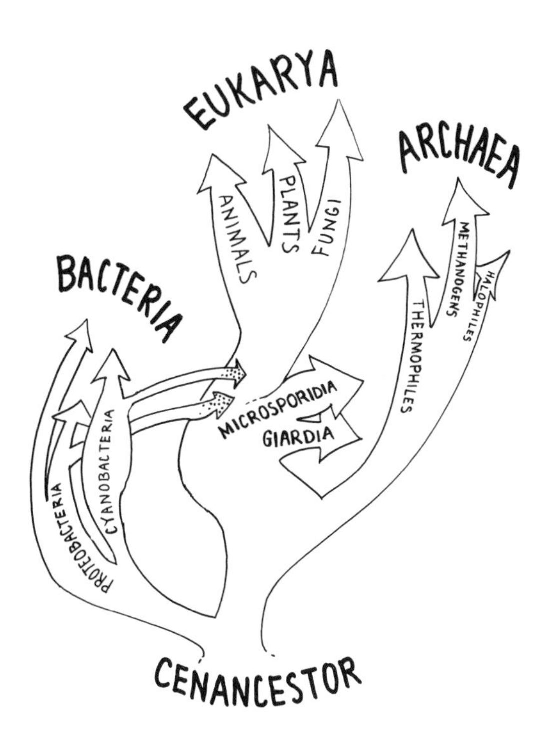

FIGURE 6 Currently accepted (Iwabe) rooting and renaming by Woese and collaborators of three primary kingdoms as domains (1990). Microsporidia and *Giardia* are archezoans thought to have diverged from the rest of the eukaryotic nuclear lineage before the acquisition of mitochondria or plastids through endosymbiosis.

grade represented by the prokaryote → eukaryote cellular transition as *the* major event in cell evolution. In lodging his objections to the paper of Woese, Kandler and Wheelis, he writes:

> . . . as important as the molecular distance between the Archaebacteria and Eubacteria may seem to a specialist, as far as their general organization is concerned, the two kinds of prokaryotes are very much the same. By contrast, the series of evolutionary steps in cellular organization leading from the prokaryotes to the eukaryotes, including the acquisition of a nucleus, a set of chromosomes and the acquisition, presumably through symbiosis, of various cellular organelles (chloroplasts, mitochondria and so on) results in the eukaryotes in an entirely new level of organization . . . (Mayr, 1990).

Tom Cavalier-Smith (Cavalier-Smith, 1992) echoes this view:

> Woese has repeatedly and mistakenly asserted that his recognition and firm establishment of the kingdom Archaebacteria (certainly a great and important breakthrough) invalidates the classical distinction between prokaryotes and eukaryotes. But as archaebacteria fall well within the scope of prokaryotes and bacteria as classically defined, it does nothing of the kind.

The questions that have to do with data rather than philosophy are: (*i*) what and how many traits distinguish the domains from each other (or betray a closer affinity between any two), (*ii*) how "fundamental" are these traits, and (*iii*) are such traits universally present within one (or two) groups and universally absent from the other(s) or is there in reality more of a mixing. For all the richness of our understanding of individual aspects of the biology of individual organisms, we are still very much in the dark, especially for answers to the second and third questions. Only recently, for instance (Cavalier-Smith, 1993), have we come to realize that archezoa (primitively amitochondrial eukaryotes) have 70S ribosomes, with rRNAs of the sizes and classes found in prokaryotes (archaebacteria or eubacteria). We know very little about possible forerunners of cytoskeletal proteins and functions in archaebacteria, although there have long been hints of such (Stein and Searcy, 1981). Even the eubacteria have not been plumbed in depth—newly discovered deeply branching lineages like *Aquifex* and the Thermotogales remain almost completely unknown in molecular or biochemical terms.

Implications of the Rooting for an Understanding of Tempo and Mode in Early Cellular Evolution

The Iwabe rooting and the renaming of the three domains as Bacteria, Archaea, and Eukarya (Figure 6) have found, in spite of these philosophical concerns, wide acceptance in the last 3 or 4 years. Together with increasing general understanding of gene and genome structure and function in the archaebacteria, the rooting has unavoidable impli-

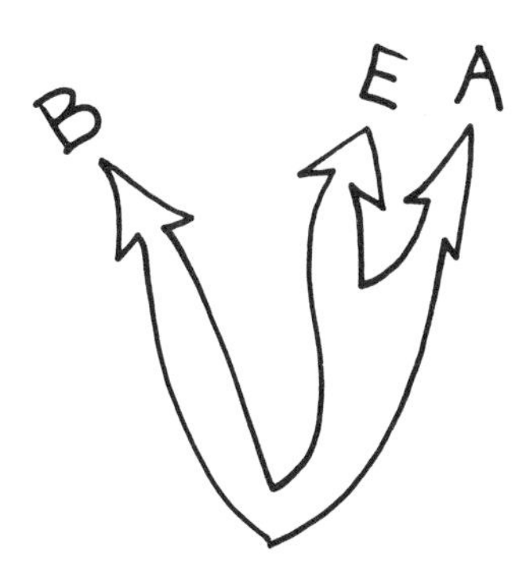

Argininosuccinate synthetase
(ATPase α subunit)
(ATPase β subunit)
DNA polymerase B
Ef-1α/Tu
Ef-G/2
HMG-CoA reductase
Isoleucyl-tRNA synthetase
Ribosomal proteins L2, L3
L6, L10, L22, L23, S9, S10
RNA polymerase II

(Acetyl-coenzyme A synthetase)
(Aspartate aminotransferase)
(Citrate synthase)
(Glutamate dehydrogenase II)
(Glutamine synthetase)
Gyrase B
(HSP70)
Ribosomal proteins L11, (S17)

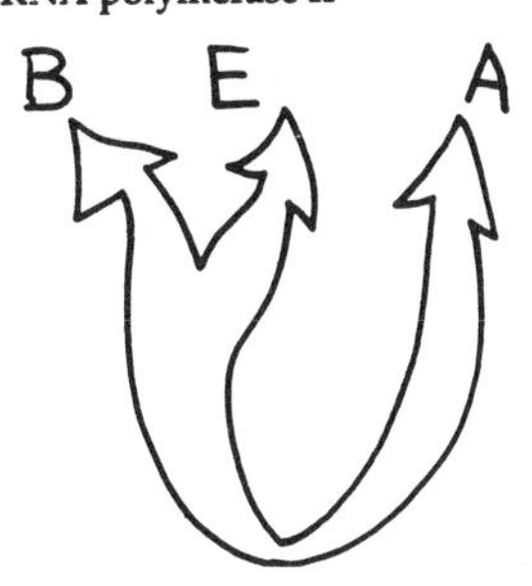

Enolase
(FeMn superoxide dismutase)
(*his* C)
(GAPDH)
MDH
Phosphoglycerate kinase
***trp* C**

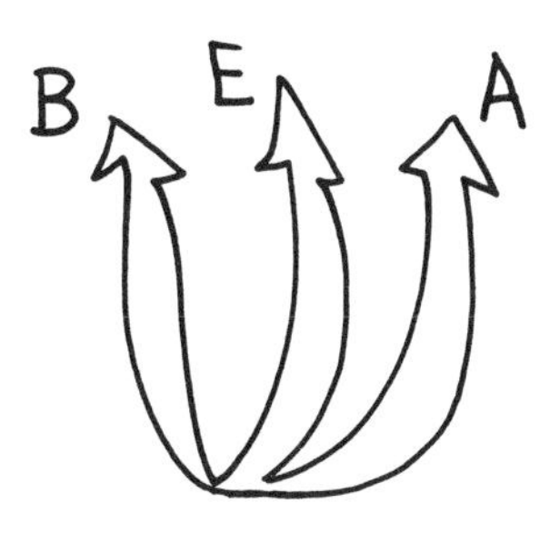

(Carbamoyl-phosphate synthetase)
(Porphobilinogen synthase)
Ribosomal protein S15
(*trp* A)
(*trp* B)
(*trp* D)
(*trp* E)

FIGURE 7 Categorization of proteins according to support for four possible rooted universal-tree topologies. Proteins were assigned to a particular topology based on the two domains determined to be closest after calculation of interdomain distances. Interdomain distances were estimated from the means of multiple pairwise comparisons between different species of each domain by using the program PROTDIST of the PHYLIP version 3.5 package (James R. Brown and W. Ford Doolittle, in preparation). For protein data sets shown in parentheses, at least one sequence did not support the monophyletic groups (B, A, and E) expected from rRNA data, according to the neighbor-joining method.

cations concerning the nature of the cenancestor and the possibility of learning about the period of progressive Darwinian evolution.

For instance, a specific lesson can be drawn from work in our laboratory (Cohen *et al.*, 1992; Lam *et al.*, 1990). The halophilic archaebacterium *Haloferax volcanii* was shown, by a variety of physical and genetic techniques, to have a genome made up of a large circular DNA of 2.92 million base pairs (Mbp) and several smaller but still sizeable molecules at 690, 442, 86, and 6 kbp. Of 60 or 70 genes known from cloned and sequenced fragments or through mutants, all but a doubtful 1 mapped to the 2.9 Mbp circle, which we thus called the chromosome, considering it similar to eubacterial chromosomes. There may of course be only so many ways to assemble a small genome: more telling is the fact that genes on this chromosome, and in thermophilic and methanogenic archaebacteria as well, are often organized into operons—cotranscribed and coordinately regulated clusters of overlapping genes controlling biochemically related functions. This too might be dismissed as a convergent or coincidental "eubacterial" feature (operons being unknown in eukaryotes), but the finding of tryptophan operons in *Haloferax* and in a methanogen and in the thermophile *Sulfolobus* (Meile *et al.*, 1991; Tutino *et al.*, 1993) seems more than coincidental, since clustering of tryptophan bioynthetic genes is almost universal among eubacteria. Most compelling of all are ribosomal protein gene clusters. In the L11-L10 clusters and the spectinomycin, S10, and streptomycin operons, 4 of 4, 11 of 11, 8 of 8, and 3 of 3 ribosomal protein genes are linked *in the very same order* (Ramirez *et al.*, 1993) in *E. coli* and in the archaebacteria that have been looked at (often including a halophile, a methanogen, and a thermophile). These remarkable organizational similarities cannot be mere coincidence and are most unlikely to reflect convergence, since there is no clear reason why the genes *must* be linked in these precise orders. In fact, gene order is conserved even when positions of promoters (and hence units of coordinate regulation) are not. The last common archaebacterial/eubacterial ancestral genome must have had operons just like this and likely was very much like the present *E. coli* or *Haloferax* genomes in other specific and general respects (including origins and mechanism of replication, and so forth). *If the Iwabe rooting is right,* the last common archaebacterial/eubacterial ancestor is the last common ancestor of all Life. The genome of this cell, the cenancestor, would have been—as far as its organization is concerned—remarkably like that of a modern eubacterium, and we would have no hope of recreating the period of progressive Darwinian evolution by the comparative method.

There is a consolation, however, if this is true. We can then more surely say that the eukaryotic nuclear genome has become drastically disorganized (or reorganized) since its divergence from its more immediate archaebacterial ancestor. As well, other characteristic features of eukaryotic nuclear molecular biology, such as multiple RNA polymerases and complex mRNA processing and intron splicing, must have appeared since this divergence.

We simply do not know how soon after the nuclear divergence these changes were wrought. The eukaryotes whose molecular biology we understand well—animals and fungi—are part of what has been called "the crown" (Sogin, 1991) of the eukaryotic subtree (Figure 6). Very few genes have been cloned from protists diverging below the trypanosomes, and virtually nothing is known about their expression. It would not be foolish, if the Iwabe rooting holds, to anticipate that some diplomonads or microsporidia, which are thought to have diverged from the rest of the eukaryotes before the mitochondrial invasion, will turn out to have operons. Hopes of finding out just how archaebacteria-like such archezoal eukaryotic genomes are have now captured the interests and energies of several laboratories.

But Is the Rooting Right?

In a sense this new direction is an old one. Once again, we are examining the prokaryote → eukaryote transition. Once again, as in Figure 2, we see the eukaryotic nuclear genome as the highly modified descendant of an already well-formed prokaryotic genome. The difference is that the immediate prokaryotic ancestors of the eukaryotic nuclear-cytoplasmic component are cells of a type we did not know when we first adopted the view shown in Figure 2. How we feel about the importance and novelty of this Hegelian outcome may depend on the side we take in the clade *versus* grade (Woese *versus* Mayr and Cavalier-Smith) debate discussed above. More to the point, however, is the possibility that we have accepted the Iwabe rooting, and consequently its implications for the modernity of the cenancestor and the radical remaking of the nuclear genome, too quickly and too uncritically. Iwabe and colleagues' data set included only one archaebacterial ATPase subunit pair (from *Sulfolobus*), only one elongation factor pair (*Methanococcus*), and a very limited representation of eubacterial sequences. As Gogarten (Hilario and Gogarten, 1993), and Forterre and his coworkers have recently and persuasively argued, both data sets can be questioned (Forterre *et al.*, 1993). There is increasing evidence for multiple gene duplication events in the history of the ATPase genes, and it is difficult to distinguish orthologues (descendant from the same cenancestral α or β subunit gene) from paralogs (descendants of more distant homologs produced by gene duplication before the cenancestor). For the elongation factors, the alignment between EF1-α/Tu types and EF-2/G species, on the correctness of which the accuracy of the rooting absolutely depends, is highly problematic. More data for precenancestral gene duplications are sorely needed.

Along with the ATPase and elongation factor gene duplication analyses, it has become common to stress the similarity in sequence of archaebacterial and eukaryotic RNA polymerase subunits or (certain) ribosomal proteins. These indeed have shown a close archaebacterial/eukaryotic relationship by a variety of measures (Zillig *et al.*, 1993). A broader survey of homologous genes for which readily alignable sequences are available for at least one species of each of the three domains is presented as Figure 7. (Eukaryotic nuclear genes suspected of being more recent acquisitions from bacterial endosymbiosis and extensively polyphyletic gene data sets are not shown.) In this figure, mean interdomain distances were used to construct midpoint rooted trees. The Iwabe tree *is* the most frequent among them, but not significantly so, and of course midpoint rootings can be correct only with constant molecular clocks.

So we must continue to remain open. If the currently accepted rooting were wrong, then an archaebacterial/eubacterial sisterhood seems the next most likely possibility, given the remarkable similarity in genetic organization between these two prokaryotic domains. The cenancestor could (again) be seen as a more primitive cell. Although it would have to possess all of those biochemical features known to be homologous in archaebacteria, eubacteria, and eukaryotes now (DNA genome, DNA polymerases, RNA polymerases, two-subunit ribosomes, the "universal code," most of metabolism, and many features of cell-cycle and growth regulation), we are free to see its genome as eubacteria-like, eukaryote-like, or something altogether different still (Woese, 1982; Woese, 1987). The fluid exchange of genes between lineages imagined by Woese in his early descriptions of the progenote remains possible.

Where Next?

The root of the universal tree is still "up in the air," and we don't know as much about the cenancestor as we had hoped. Why is this? One possibility is that we are pushing molecular phylogenetic methods to their limits: although we have reasonable ways of assessing how well any given tree is supported by the data on which it is based, methods for determining the likelihood that this is the "true tree" are poorly developed. Another is hidden paralogy—gene duplication events (of which there are only scattered detected survivors, different in different lineages) are fatal to the enterprise of phylogenetic reconstruction.

A mammalian tree drawn on the basis of myoglobin sequences from some species and hemoglobin sequences from others would be accurate as far as the molecules (which are all homologues) are concerned, but would be seriously wrong for the organisms. A third possibility, formally identical to paralogy in its baleful consequence for tree construction, is lateral (horizontal) gene transfer. Certainly such transfer has occurred within and between domains, early and late in their evolution (Smith *et al.*, 1992). Zillig and Sogin (Zillig *et al.*, 1993; Sogin, 1991) have drawn (quite different) scenarios in which extensive lateral transfer is invoked to explain the multiplicity of trees shown in Figure 7, each of which can then be taken at face value.

What renders all such attempts to resolve the current dilemma unnecessary and dangerously premature is the certainty that we will soon have enormously many more data. Total genome sequencing projects are under way for several eubacteria (*E. coli, B. subtilis*, a mycoplasma, and two mycobacteria), several archaebacteria (including *Sulfolobus solfataricus*), and, of course, a number of "crown" eukaryotes seen as more direct models for the human genome. Instead of at most three dozen data sets with representative gene sequences from all three domains, we should have 3000. If the data in aggregate favor a single tree, this should be apparent. If there have been lateral transfers of related or physically linked genes, then we might be able to see them. If transfer has so scrambled genomes that we can no longer talk sensibly about the early evolution of cellular lineages but only of lineages of genes, then that too should be apparent, as would the need to change the very language with which we address an evolutionary process so radically different in both tempo and mode.

We should not allow our current confusion about the root to discourage us, and it is heartening to remember how far we have come. The prokaryote–eukaryote distinction has replaced that between animals and plants, and although we may no longer see that distinction as clearly as Stanier and van Niel thought they did, it is because we know more about the diversity of microbes; we will never go back to a world of just animals and plants. Similarly, the endosymbiont hypothesis for the origin of mitochondria and chloroplasts is as firmly established as any fact in biology; we will not return to the belief in direct filiation (bacteria → cyanobacteria → algae → all other eukaryotes) which preceded it. As for the archaebacteria, although there remains some doubt as to their "holophyly" (thermophiles may be especially close to eukaryotes) and legitimate debate over the philosophical and biological implications of their existence for the meaning of the word "prokaryote," we will never again see these fascinating creatures scattered taxonomically among the bacteria, as uncertain relatives of known eubacterial groups.

Methodologically, rRNA seems unlikely ever to lose pride of place as *the* most reliable molecular chronometer: Woese's original choice of this universally essential, functionally conservative and slowly evolving species was well justified. At the same time, protein data will increasingly supplement rRNA sequences—rRNAs may mislead us when they show base compositional biases, and there is anyway no single molecule which defines a cellular lineage, once lateral transfer is admitted. Molecular evolution is maturing, which means that the arguments of molecular evolutionists are becoming more pluralistic and subtler. We should welcome this, and the dialectic which assures that evolutionary theories are rarely wholely overthrown but instead are incorporated in unexpected ways and with unanticipated benefits into succeeding generations of biological thinking.

SUMMARY

Early cellular evolution differed in both mode and tempo from the contemporary process. If modern lineages first began to diverge when the phenotype–genotype coupling was still poorly articulated, then we might be able to learn something about the evolution of that coupling through comparing the molecular biologies of living organisms. The issue is whether the last common ancestor of all life, the *cenancestor*, was a primitive entity, a *progenote*, with a more rudimentary genetic information-transfer system. Thinking on this issue is still unsettled. Much depends on the placement of the root of the universal tree and on whether or not lateral transfer renders such rooting meaningless.

Work in this laboratory described in this manuscript is supported by the Medical Research Council of Canada, of which agency J.R.B. is also a Postdoctoral Fellow. W.F.D. is a Fellow of the Canadian Institute for Advanced Research.

REFERENCES

Ayala, F. J. (1988) Can "Progress" be defined as a biological concept? In *Evolutionary Progress*, ed. Nitecki, M. H. (Univ. Chicago Press, Chicago), pp. 75–96.

Benner, S. A., Cohen, M. A., Gonnet, G. H., Berkowitz, D. B. & Johnsson, K. P. (1993) Reading the palimpsest: contemporary biochemical data and the RNA world. In *The RNA World*, eds. Gesteland, R. F. & Atkins, J. F. (Cold Spring Harbor Lab. Press, Plainview, NY), pp. 27–70.

Benner, S. A., Ellington, A. D. & Tauer, A. (1989) Modern metabolism as a palimpsest of the RNA world. *Proc. Natl. Acad. Sci. USA* **86,** 7054–7058.

Cavalier-Smith, T. (1992) Bacteria and eukaryotes. *Nature (London)* **356,** 570.
Cavalier-Smith, T. (1993) Kingdom Protozoa and its 18 phyla. *Microbiol. Rev.* **57,** 953–994.
Chatton, E. (1937) *Titres et Travauz Scientifiques* (Setes, Sottano, Italy).
Cohen, A., Lam, W. C., Charlebois, R. L., Doolittle, W. F. & Schalkwyk, L. C. (1992) Localizing genes on the map of the genome of *Haloferx volcanii*, one of the archaea. *Proc. Natl. Acad. Sci. USA* **89,** 1602–1606.
DeLong, E. F. (1992) Archaea in coastal marine environments. *Proc. Natl. Acad. Sci. USA* **89,** 5685–5689.
Doolittle, W. F. (1991) The origins of introns. *Curr. Biol.* **1,** 145–146.
Fitch, W. M. & Upper, K. (1987) The phylogeny of tRNA sequences provides evidence for ambiguity reduction in the origin of the genetic code. *Cold Spring Harbor Symp. Quant. Biol.* **52,** 759–767.
Forterre, P., Benanchenhou-Lahfa, N., Canfalonieri, F., Duguet, M., Elie, C. & Labedan, B. (1993) The nature of the last universal ancestor and the root of the tree of life: still open questions. *BioSystems* **28,** 15–32.
Fox, G. E., Magrum, L. J., Batch, W. E., Wolfe, R. S. & Woese, C. R. (1977) Classification of methanogenic bacteria by 16S ribosomal RNA characterization. *Proc. Natl. Acad. Sci. USA* **74,** 4537–4541.
Gray, M. W. & Doolittle, W. F. (1982) Has the endosymbiont hypothesis been proven? *Microbiol. Rev.* **46,** 1–42.
Hilario, E. & Gogarten, J. P. (1993) Horizontal transfer of ATPase genes—the tree of life becomes a net of life. *BioSystems* **31,** 111–119.
Iwabe, N., Kuma, K., Hasegawa, M., Osawa, S. & Miyata, T. (1989) Evolutionary relationship of archaebacteria, eubacteria, and eukaryotes inferred from phylogenetic trees of duplicated genes. *Proc. Natl. Acad. Sci. USA* **86,** 9355–9359.
Knoll, A. H. (1994) *Proc. Natl. Acad. Sci. USA* **91,** 6743–6750.
Lam, W. C., Cohen, A., Tsouluhas, D. & Doolittle, W. F. (1990) Genes for tryptophan biosynthesis in the archaebacterium *Haloferax (Halobacterium) volcanii*. *Proc. Natl. Acad. Sci. USA* **87,** 6614–6618.
Maizels, N. (1994) *Proc. Natl. Acad. Sci. USA* **91,** 6729–6734.
Margulis, L. & Guerrero, R. (1991) Kingdoms in turmoil. *New Sci.* **129,** 46–50.
Margulis, L. (1970) *Origin of Eukaryotic Cells* (Yale Univ. Press, New Haven, CT).
Mayr, E. (1990) A natural system of organisms. *Nature (London)* **348,** 491.
Meile, L., Stettler, R., Banholzer, R., Kotik, M. & Leisinger, T. (1991) Tryptophan gene cluster of *Methanobacterium thermoautotrophicum* Marburg: Molecular cloning and nucleotide sequence of a putative *trpEGCBAD* operon. *J. Bacteriol.* **173,** 5017–5023.
Olsen, G. J., Woese, C. R. & Overbeek, R. (1994) The winds of (evolutionary) change: Breathing new life into microbiology. *J. Bacteriol.* **176,** 1–6.
Ramirez, C., Kopke, A. K. E., Yang, C.-C., Boeckh, T. & Matheson, A. T. (1993) Chapter 14: The structure, function and evolution of archaeal ribosomes. In *The Biochemistry of Archaea (Archaebacteria)*, eds. Kales, M., Kushner, D. J. & Matheson, A. T. (Elsevier, Amsterdam), pp. 439–466.
Schopf, J. W. (1994) *Proc. Natl. Acad. Sci. USA* **91,** 6735–6742.
Schwartz, R. M. & Dayhoff, M. O. (1978) Origin of prokaryotes, eukaryotes, mitochondria and chloroplasts. *Science* **199,** 395–403.
Simpson, G. G. (1944) *Tempo and Mode in Evolution* (Columbia Univ. Press, New York), pp. xxix-xxx.
Smith, M. W., Feng, D.-F. & Doolittle, R. F. (1992) Evolution by acquisition: the case for horizontal gene transfer. *Trends Biochem. Sci.* **17,** 489–493.

Sogin, M. C. (1991) Early evolution and the origin of eukaryotes. *Curr. Opin. Genet. Dev.* **1,** 457–463.

Stanier, R. Y. & van Niel, C. B. (1941) The main outlines of bacterial classification. *J. Bacteriol.* **42,** 437–466.

Stanier, R. Y. & van Niel, C. B. (1962) The concept of a bacterium. *Arch. Microbiol.* **42,** 17–35.

Stanier, R. Y. (1970) Some aspects of the biology of cells and their possible evolutionary significance. *Symp. Soc. Gen. Microbiol.***20,** 1–38.

Stein, D. B. & Searcy, K. B. (1981) A microplasma-like archaebacterium possibly related to the nucleus and cytoplasm of eukaryotic cells. *Ann. N.Y. Acad. Sci.* **361,** 312–323.

Tutino, M. L., Scarano, G., Marino, G., Sannia, G. & Cubellis, M. V. (1993) Tryptophan biosynthesis genes *trpEGC* in the thermoacidophilic archaebacterium *Sulfolobus solfataricus*. *J. Bacteriol.* **175,** 299–302.

Woese, C. R. & Fox, G. E. (1977) Phylogenetic structure of the prokaryotic domain: the primary kingdoms. *Proc. Natl. Acad. Sci.USA* **74,** 5088–5090.

Woese, C. R. (1982) Archaebacteria and cellular origins: an overview. *Zbl. Bakt. Hyg. I. Abt. Urig.* **C3,** 1–17.

Woese, C. R. (1987) Bacterial evolution. *Microbiol. Rev.* **51,** 221–271.

Woese, C. R., Kandler, O. & Wheelis, M. L. (1990) Towards a natural system of organisms: Proposal for the domains Archaea, Bacteria, and Eucarya. *Proc. Natl. Acad. Sci. USA* **87,** 4576–4579.

Zillig, W., Palm, P., Klenk, H.-P., Langer, D., Hüdepohl, U., Hain, J., Lanzendörfer, M. & Holz, I. (1993) Chapter 12: Transcription in archaea. In *The Biochemistry of Archaea (Archaebacteria),* eds. Kales, M., Kushner, D. J. & Matheson, A. T. (Elsevier, Amsterdam), pp. 367–391.

Zillig, W., Stetter, K. O., Schnabel, R., Madon, J. & Gierl, A. (1982) Transcription in archaebacteria. *Zbl. Bakt. Hyg. I. Abt. Orig.* **C3,** 218–227.

2

Phylogeny from Function: The Origin of tRNA Is in Replication, not Translation

NANCY MAIZELS AND ALAN M. WEINER

The diversity evident in contemporary organisms and in the fossil record is remarkable. But perhaps even more remarkable is the fact that all living species, and perhaps all species that have ever lived on earth, share a common biochemistry. In particular, the central pathways of macromolecular synthesis—DNA replication, transcription, and translation—are conserved in all three contemporary kingdoms: Eucarya, Eubacteria, and Archaea (Woese *et al.*, 1990).

The challenge for those interested in biochemical evolution is to deduce, from contemporary molecules and organisms, how these pathways arose. Replication, transcription, and translation all require a multiplicity of interactions among a large cast of macromolecules. It is inconceivable that any of these complex processes arose full blown. A plausible scenario for the evolution of any pathway must therefore provide a persuasive explanation for the selective advantage conferred by each new component as it was added to the pathway.

The discovery of the catalytic properties of RNA (Cech *et al.*, 1981; Guerrier-Takada *et al.*, 1983) showed that a single chemical species could function as both genome and enzyme. This suggested an evolutionary scheme, simple in outline, which is shown in Figure 1. Prebiotic reactions on the primordial earth generated RNA, RNA became the first

Nancy Maizels is associate professor and Alan M. Weiner is professor of molecular biophysics and biochemistry at Yale University School of Medicine, New Haven, Connecticut.

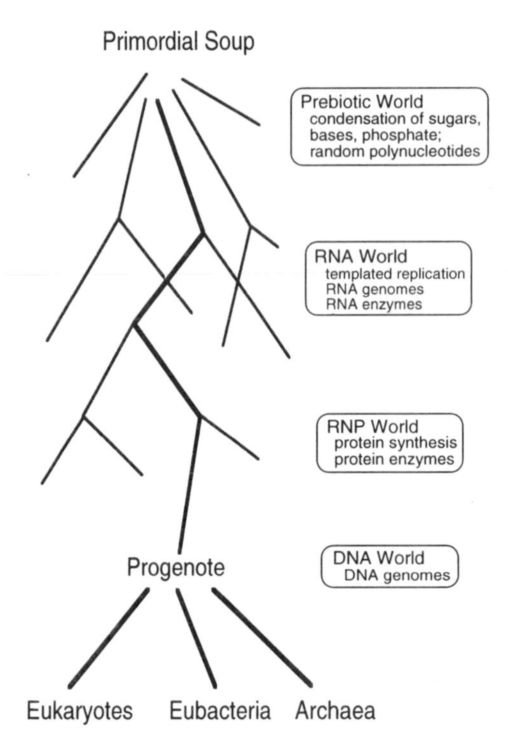

FIGURE 1 From the primordial soup to the three contemporary kingdoms. Woese (1990) originally used the term "progenote" to describe a common ancestor with an inaccurate translation system, but we make no assumption about the accuracy of translation in the progenote. Accuracy is a matter of degree, and the important point is that translation in this ancestor must have been substantially similar to our own. For clarity, we show DNA as evolving after templated protein synthesis. An interesting alternative, suggested by J. Szostak (Harvard Medical School, personal communication), is that DNA evolved much earlier, perhaps even in the RNA world, as a catalytically inactive storage form of RNA.

self-replicating molecule, early living systems arose with RNA genomes and relatively simple biochemistries, and these evolved into cells with DNA genomes and complex biochemistries. Indeed, the biochemistry of all living organisms is so similar that there is little doubt all life on earth can be traced back to a single common ancestor, which Woese named the "progenote" (Woese, 1990). Moreover, because the translation apparatus is fundamentally similar in all three contemporary kingdoms, this progenote must have been capable of templated protein synthesis substantially like our own.

Here we propose a phylogeny for the origin of tRNA based on the ubiquity and conservation of tRNA-like structures in the *replication* of contemporary genomes, and we discuss the evidence in contemporary molecules that leads to and supports this phylogeny. The unique aspect of this phylogeny is that it places the origin of tRNA in replication, *before* the advent of templated protein synthesis. This implies that tRNAs arose before the other components of the translation apparatus, that aminoacyl-tRNA synthetases arose next, and that both tRNA and aminoacyl-tRNA synthetases predated the anticodon and mRNA.

METHODS

Molecular Fossils. In the classical terms of paleobiology, there is no fossil record of the most ancient forms of life or the molecules from which they were made. However, just as mineralized bones, shells, or cell walls tell us about the evolution of modern cellular organisms, contemporary biological molecules provide clues regarding the evolution of the earliest forms of life. In order to discuss precellular evolution in terms that are most accessible to a wide variety of disciplines, we use the phrase "molecular fossil" to describe the molecules that are central to this analysis. *A molecular fossil is any molecule whose contemporary structure or function provides a clue to its evolutionary history.* Molecular fossils are not to be confused with fossil molecules (DNA preserved in amber), or "living fossils" (slowly evolving species such as the coelacanth), or the physical fossils of cells and multicellular organisms that constitute the raw data for tracing more recent evolution. A molecular fossil is, of necessity, an abstraction rather than a tangible object: it records, embodies, and reflects ancient evolution but is not itself ancient.

Many Interesting Macromolecules Are Social and Are Therefore Constrained to Coevolve. There would be no such thing as a molecular fossil if evolution inevitably erased its own footsteps. But many biological macromolecules are social—they interact with other macromolecules. These interactions constrain the evolution of any individual molecule as well as the ensemble of other molecules with which it interacts. As diagrammed in Figure 2, a social macromolecule can change only in ways that preserve its ability to interact with its important partners. A change in any partner that alters one of these interactions will be tolerated only if all other partners change accordingly. Coevolution of this kind is possible for a molecule that interacts with one or a few different partners but becomes more difficult as the size of the ensemble

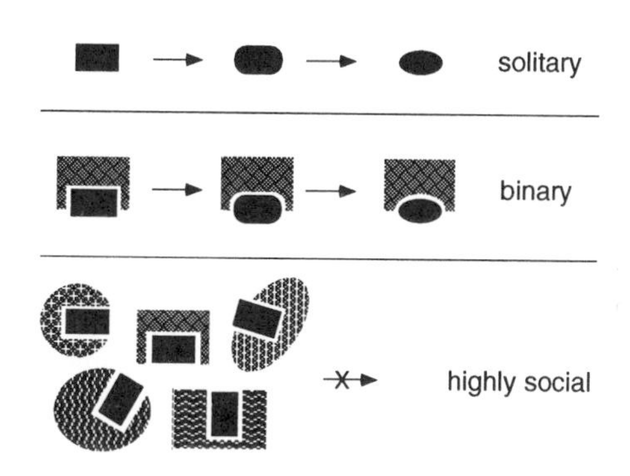

FIGURE 2 Social macromolecules. Evolution of a solitary macromolecule is almost unconstrained as long as it remains functional. The larger the interacting ensemble of macromolecules, the more difficult it is for any one of them to change.

of partners increases; the larger the ensemble, the deeper the valleys in the adaptive landscape. The most interactive structures and functions are thus the most likely to be preserved in their original forms—they are effectively frozen in time.

RESULTS AND DISCUSSION

The Earliest Genomic Tags. We now consider replication in the RNA world (Gilbert, 1986), an era that predates the evolution of either DNA or templated protein synthesis. In these simpler times, enzymes made of RNA replicated genomes made of RNA. Nonetheless, early replicases and genomes had to confront many of the same problems faced by contemporary RNA genomes. Two of these problems are immediately evident:

- Specificity of replication. How did genomes that were destined for replication distinguish themselves from junk RNA or catalytic RNAs that should not be replicated?
- The "telomere problem" (Watson, 1972). How did ancient genomes and replicases prevent loss of 3′ terminal sequences during successive rounds of replication?

The replicative strategy of the contemporary bacteriophage Qβ suggests a single, powerful solution to both problems. Qβ is a (+)-strand RNA phage. As shown in Figure 3, at the very 3′ terminus of Qβ is a tRNA-like structure, which ends in the sequence CCA. The 3′-terminal tRNA-like structure in the Qβ (+)-strand genome serves as a recognition element for the replicase, which initiates synthesis of the (−)-strand at the penultimate C of the CCA terminus. The tRNA-like structure thus ensures specificity of replication. Furthermore, the 3′-terminal CCA of Qβ can also function, at least in principle, like a modern telomere: loss of part or all of the CCA sequence could be restored by the CCA-adding enzyme, tRNA nucleotidyltransferase. Regeneration of a CCA terminus

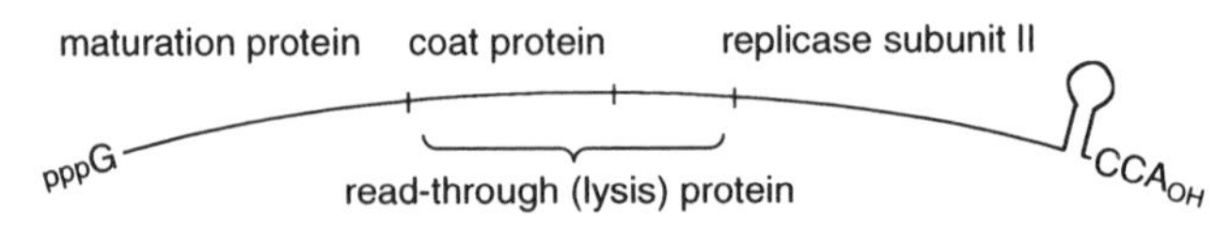

FIGURE 3 Qβ bacteriophage genome. The genome of Qβ bacteriophage is a single-stranded RNA with a 3′-terminal tRNA-like genomic tag. The genome also serves as a messenger for four phage proteins, including the catalytic subunit II of Qβ RNA replicase.

by this enzyme has in fact been demonstrated for brome mosaic virus (BMV), a plant virus with a 3′-terminal tRNA-like structure very similar to that of Qβ (Rao *et al.*, 1989).

These observations about the Qβ genome led us to propose that the first tRNA-like structures arose as "genomic tags" that marked the 3′ ends of ancient RNA genomes for replication by RNA enzymes in the RNA world (Weiner and Maizels, 1987; Maizels and Weiner, 1993). The simplest such tags would have been the predecessors of the "top half" of modern tRNA, consisting of a coaxial stack of the TΨC arm on the acceptor stem (see Figure 5). The presence of such 3′ terminal tRNA-like structures in two different kingdoms—contemporary bacterial viruses like Qβ and plant viruses such as turnip yellow mosaic virus (TYMV) and BMV (reviewed by Hall, 1979; Haenni *et al.*, 1982; Guerrier-Takada *et al.*, 1988)—suggests that these structures date back at least as far as the progenote. The role of tRNA in replication appears to have arisen much earlier, however. As we discuss in greater detail below, molecular fossil evidence suggests that tRNA-like structures were first used for replication of RNA genomes by RNA enzymes in the RNA world and then persisted through the transition from RNA to DNA genomes.

An Early Origin for tRNA Rationalizes the 5′ and 3′ Processing of tRNAs in Modern Cells. The genomic tag model immediately explains the existence of two enzymes with otherwise puzzling activities in contemporary tRNA processing, RNase P and tRNA nucleotidyltransferase. Transcription of tRNA typically initiates not at the 5′ end of the functional molecule, but at a site upstream. The 5′ leader sequences are then removed from the "pre-tRNA" by an endonucleolytic cleavage catalyzed by RNase P. RNase P is a ribonucleoprotein, but the RNA component alone is capable of catalysis (Guerrier-Takada *et al.*, 1983). The presence of a catalytic RNA component suggests that RNase P activity arose in the RNA world (Alberts, 1986); use of an RNA component cannot be explained by the need to recognize so many different species of tRNA, since elongation factor Tu accomplishes the same task without the help of RNA. Further attesting to the ancient origin of RNase P, the structure it recognizes is highly conserved: the *Escherichia coli* enzyme can cleave the 3′ tRNA-like structure of TYMV (Green *et al.*, 1988; Guerrier-Takada *et al.*, 1988; Mans *et al.*, 1990). But why would a tRNA processing enzyme be present in the RNA world? We suggest that the first function of RNase P may have been to free catalytic RNAs from the 3′-terminal genomic tags required for replication; cleavage may have been necessary to activate catalytic function or to prevent replication from interfering with catalysis. If this is the case, it would explain the puzzling fact that tRNAs undergo 5′ processing at all: there is no reason *a priori* why the mature 5′ ends of modern tRNAs could not be generated directly by transcription, yet there is only one instance known in which this is so (Lee *et al.*, 1989).

If early tRNAs functioned as telomeres, it would also explain the surprising fact that all cells contain tRNA nucleotidyltransferase, an activity that can regenerate the 3′-terminal CCA sequence of tRNAs. This activity can be viewed as a telomerase, responsible for maintaining the integrity of the terminal CCA. Interestingly, although eubacteria and archaea encode the 3′ terminal CCA of tRNA, eukaryotes do not, and they must rely instead on tRNA nucleotidyltransferase to produce mature tRNAs (Palmer *et al.*, 1992). The primitive state is not yet known, but as tRNA nucleotidyltransferase is present in all cells, it seems most plausible that genomically encoded CCA was devised later to circumvent a slow or inefficient step in tRNA processing. This may also explain why the telomerase function of tRNA nucleotidyltransferase can in certain cases be augmented by the replicase itself. A number of modern RNA and DNA polymerases, including Qβ RNA replicase (Blumenthal and Carmichael, 1979) and *Taq* DNA polymerase will add an untemplated A to the 3′ end of a newly synthesized molecule (for example, see Tse and Forget, 1990). Polyadenylation of polymerase II transcripts, which often occurs following a CA dinucleotide (Wigley *et al.*, 1990; Raabe *et al.*, 1993), may be viewed as an exaggerated example of this.

Viruses May Be Clues to Early Evolution. The Qβ RNA genome is replicated by an enzyme composed of four subunits. Only subunit II is encoded by the phage genome; the other three subunits—ribosomal protein S1 and elongation factors Tu and Ts (Blumenthal and Carmichael, 1979)—are components of the translation apparatus. If Qβ is viewed as no more than a cellular parasite, then this simple bacteriophage appears to have stolen elements of the protein synthesis machinery for its replication. However, if tRNAs arose early to function in replication, it would be natural for factors that recognized tRNA to accompany this molecule as it took on a new role in translation. The presence of translation factors in an RNA replicase may therefore be viewed as evidence that these two processes have long been intimately connected.

In addition, the notion that some modern viruses still reveal their ancient origins allows us to see viruses in a new way. Viruses usually command our attention as vectors of disease, and the extraordinary genomic diversity of modern viruses is rarely appreciated by nonvirologists. There are double-stranded viruses, single-stranded viruses, circular viruses and linear viruses, RNA viruses, DNA viruses, viruses made of RNA that replicate through DNA intermediates, and viruses made of both RNA and DNA. Why such diversity? We suggest that viruses may have evolved early and that their genomic diversity reflects the variety of replication strategies available before large DNA genomes became the cellular norm. This leads us to predict that the study of modern viruses will provide further insight into early evolution.

Transitional Genomes as Clues to Early Replication Strategies. If the original role of tRNA-like structures was in replication, as suggested by the single-stranded bacteriophage and plant virus genomes, one might expect to find additional examples of contemporary genomes in which tRNA plays that same role. When we first proposed the genomic tag hypothesis (Weiner and Maizels, 1987), only one other example of tRNA involvement in replication was known: In modern retroviruses, tRNAs function as primers for initiation of cDNA synthesis by the retroviral reverse transcriptase. Over the past few years, additional novel replication strategies have been described that employ tRNA-like structures. These appear to link replication of single-stranded RNA viruses with retroviral replication and with the synthesis of modern chromosomal telomeres. In each of these instances, a genomic RNA replicates *via* a DNA intermediate. We call these "transitional genomes," because they can be viewed as reenacting the transition from an RNA world to the contemporary DNA world.

Three different sorts of transitional genomes appear to link the function of tRNA in replication of RNA genomes with its role in replication of contemporary DNA genomes, as shown in Figure 4. The example most similar to the (+)-strand RNA viruses Qβ and TYMV is the Mauriceville plasmid of *Neurospora* mitochondria (Maizels and Weiner, 1987; Kuiper and Lambowitz, 1988; Akins *et al.*, 1989). This double-stranded DNA plasmid replicates in a most unusual way. First, rolling-circle transcription of the plasmid generates a multimeric RNA (+)-strand. The multimer is cleaved to produce full-length monomeric RNA transcripts, each with a 3′-terminal tRNA-like structure ending in CCACCA, a genomic tag with a reiterated CCA terminus (see Figure 4). The 3′-terminal genomic tag of the monomeric (+)-strand RNA then serves as the initiation site for replication, but in this case the template is copied not into (−)-strand RNA by an RNA replicase, as for Qβ and TYMV, but into cDNA by a reverse transcriptase. Moreover, the reverse transcriptase is encoded by the monomeric (+)-strand RNA, which doubles as an mRNA. The similarities between the replicative strategies of bacteriophage Qβ and the Mauriceville plasmid are remarkable: a full-length (+)-strand RNA with a 3′-terminal genomic tag encodes the enzyme which copies the genome starting at the penultimate C of the

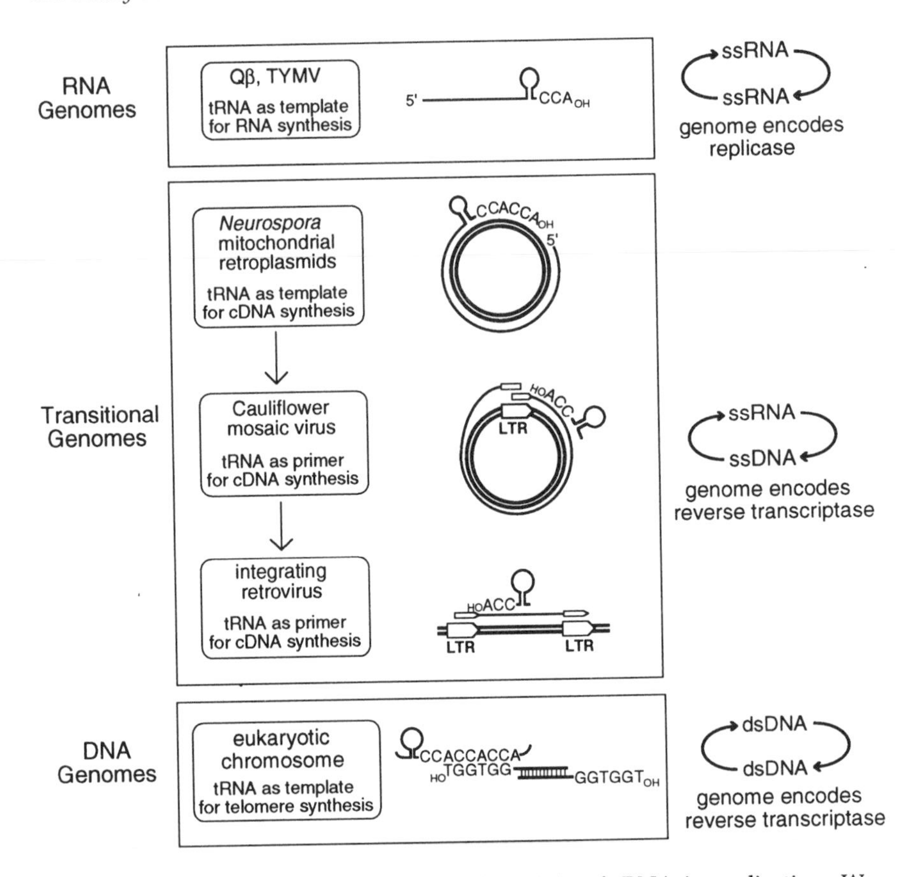

FIGURE 4 A functional phylogeny for the origin of tRNA in replication. We propose that tRNA began as a 3′-terminal genomic tag on single-stranded RNA genomes where it serves as template for initiation of RNA synthesis. The genomic tag was preserved in transitional replication strategies where it serves as template for initiation of DNA synthesis on the single-stranded RNA genome. Still later, the genomic tag was transformed into the primer for initiation of DNA synthesis in retroviral-like replication strategies. ss, Single-stranded; ds, double-stranded.

genomic tag. Thus the Mauriceville retroplasmid is best described as an RNA genome replicating through a DNA intermediate, with the plasmid DNA serving the same role as the DNA provirus does in the life cycle of a modern retrovirus.

The enzymology of replication of the Mauriceville retroplasmid provides further support for the transitional status of this plasmid genome (Wang and Lambowitz, 1993). The Mauriceville reverse transcriptase initiates cDNA synthesis without a primer—a feat previously thought to be impossible for a DNA polymerase. This suggests that the transition from RNA to DNA genomes did not require invention of a novel priming activity, but simply the transformation of an RNA polymerase into a DNA polymerase. The ability of Qβ replicase to misincorporate DNA bases in the presence of Mn^{2+} (Blumenthal and Carmichael, 1979) suggests that such a transformation is not implausible.

From Template to Primer. In Qβ, TYMV, and the Mauriceville retroplasmid, tRNA is the template for replication. In contrast, tRNA functions not as template but as primer in other transitional genomes which probably arose later. Examples of such genomes are cauliflower mosaic virus (CaMV) and vertebrate retroviruses. The CaMV genome is a circular duplex DNA that replicates as an extrachromosomal element without ever integrating into chromosomal DNA (see Figure 4). Transcription of the viral genome initiates at a unique regulatory region which, like a retroviral long terminal repeat (LTR), consists of an RNA polymerase II promoter just upstream from a polyadenylation signal. The polyadenylation signal is too close to the promoter to function efficiently, so transcription bypasses the signal the first time around and generates a terminally redundant transcript that is slightly longer than full length. This terminally redundant genomic (+)-strand RNA is then converted to a cDNA by the CaMV-encoded reverse transcriptase using tRNA as the primer (Hohn *et al.*, 1985; Covey and Turner, 1986; Sanfacon and Hohn, 1990). The CaMV replication scheme is very similar to that used by retroviruses (also shown in Figure 4), except that subsequent integration of retroviral DNA into the host chromosome generates a DNA provirus with a copy of the regulatory region at either end (hence, LTR). The terminally redundant DNA provirus is topologically equivalent to the circular CaMV genome, and transcription from the upstream promoter of the provirus to the distal polyadenylation signal generates the terminally redundant genomic RNA. Transcription of diverse, but related (Xiong and Eickbush, 1990), retroviral elements is also primed by tRNA (Kikuchi *et al.*, 1986; Chapman *et al.*, 1992).

Specificity and Catalysis Reside in Separate Polymerase Domains. The suggestion that tRNA functioned first as template for the initiation of replication and later as primer is consistent with the domain structure of contemporary polymerases. Many polymerases have two structural domains, one that is catalytic and another that determines template specificity. The classical example of this separation of functions is the σ factor of *E. coli* RNA polymerase (reviewed by Jaehning, 1991). Similarly, in Qβ replicase it is elongation factor Tu that recognizes that

3′-terminal tRNA-like structure, not the bacteriophage-encoded catalytic subunit (Blumenthal and Carmichael, 1979). For tRNA to evolve from template to primer, the replication enzyme would consist—like Qβ replicase—of two domains, one catalytic and the other recognizing the 3′-terminal tRNA-like tag. The relative orientation of these two domains would then determine whether the tRNA was used as template or as primer. The remarkable ability of the Mauriceville reverse transcriptase to use RNA either as primer or as template, at least *in vitro*, suggests that the transition from template to primer may have been straightforward (Wang and Lambowitz, 1993).

A Functional Phylogeny for the Evolution of tRNA in Replication: RNA Genomes to Modern DNA Telomeres. The function of tRNA in the replication of RNA genomes and transitional genomes leads us to propose a phylogeny for the evolution of tRNA in replication. As shown in Figure 4, tRNA-like structures first arose in ancient RNA genomes, where they served as *templates* for the initiation of replication and also functioned as primitive telomeres. These tRNA-like structures persisted during the evolution of DNA, and they are immortalized today in transitional genomes, RNA genomes that replicate *via* a DNA intermediate. In the transitional genomes, tRNAs functioned first as template and later as primer for synthesis of a cDNA copy by a reverse transcriptase encoded by the genomic RNA itself.

The role of tRNA in replication is not restricted to viruses and extrachromosomal elements but extends to cellular chromosomes as well. The termini of modern chromosomes are replenished by an enzyme called telomerase, which adds species-specific T_nG_m repeats, one nucleotide at a time, to an appropriate T_nG_m primer (reviewed by Blackburn, 1991). Telomerase is a ribonucleoprotein, and its RNA component serves as a built-in template for sequence addition by the reverse transcriptase-like protein subunit. In the *Tetrahymena* telomerase, for example, a built-in template containing the sequence 5′-$CA_2C_4A_2$ specifies addition of 5′-T_2G_4 telomeric repeats. There is an uncanny resemblance between the action of telomerase and the copying of the reiterated CCACCA terminus of a tRNA-like genomic tag by the Mauriceville reverse transcriptase. Moreover, the Mauriceville reverse transcriptase can, like telomerase, initiate at an *internal* CCA sequence by using a DNA primer (Wang and Lambowitz, 1993). The genomic tag hypothesis suggests that telomere addition can be viewed, from an evolutionary perspective, as abortive replication. Furthermore, if ancient tRNA-like structures were the predecessors of the built-in template in contemporary telomerases, it would explain why modern telomere sequences are variations on a C_nA_m repeat motif.

How Did tRNA Come to Play a Role in Translation? Covalent linkage of a basic amino acid to a 3′-terminal tRNA-like genomic tag might have improved the efficiency or specificity of replication in an RNA world, perhaps by permitting the negatively charged RNA replicase to bind more tightly to the negatively charged RNA genome. Alternatively, if aminoacylation *interfered* with replication, charging could have limited the number of genomes in the replicative pool or prevented free genomic tags from competing for the replicase. The aminoacylation activity could have arisen very early, even as a variant of the replicase itself. Aminoacylation chemically resembles RNA polymerization, and a variant replicase could have evolved to catalyze aminoacylation, just as a group I ribozyme, which naturally catalyzes phosphoester bond transfer, can be redesigned to catalyze reactions at a carbon center (Piccirilli *et al.*, 1992). The specificity with which modern group I ribozymes bind L-arginine leaves little doubt that an aminoacyl-tRNA synthetase made of RNA could charge tRNA with considerable specificity (Connell *et al.*, 1993). In any case, as we discuss in greater detail elsewhere (Maizels and Weiner, 1987), replication would have provided the driving force for the first two steps in the evolution of protein synthesis.

The apparent diversity in size and quaternary structure of modern aminoacyl-tRNA synthetases has long been puzzling. All these enzymes perform the same two-step reaction using an enzyme-bound aminoacyl-adenylate intermediate, and one might therefore have expected that all would be descended from a single ancestral protein. This mystery was refined but not clarified by the realization that modern aminoacyl-tRNA synthetases can be divided into two structurally and functionally distinct classes: synthetases with the classical Rossman nucleotide-binding fold charge the 2′ hydroxyl of tRNA, and synthetases with a seven-stranded antiparallel β-sheet generally charge the 3′ hydroxyl (Cusack *et al.*, 1990; Eriani *et al.*, 1990; Ruff *et al.*, 1991; Cavarelli *et al.*, 1993). However, if aminoacyl-tRNA synthetases first arose in an RNA world, as we suggested (Weiner and Maizels, 1987), and were then transformed by stepwise replacement of RNA with protein as envisioned by White (White, 1982), even a single RNA enzyme could give rise to multiple protein enzymes because there is unlikely to be a unique path for replacement of RNA by protein (Weiner and Maizels, 1987; Benner *et al.*, 1989).

The Top Half of tRNA Is Ancient. The tRNA-like structures in early genomes may have consisted simply of a coaxial stack of the TΨC arm on the CCA acceptor stem. We base this suggestion on two different kinds of evidence (Figure 5 and Table 1). First, the top half of modern

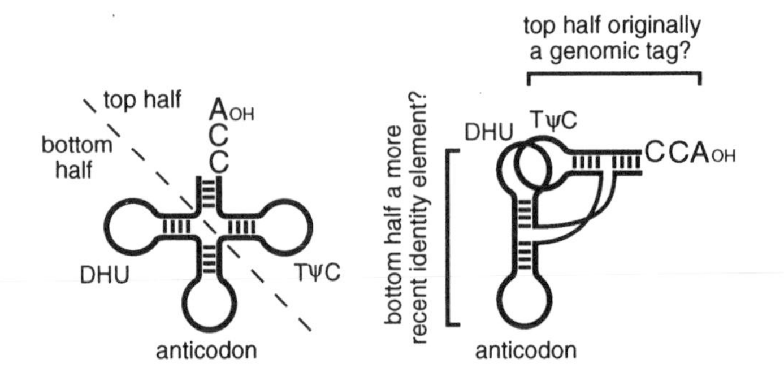

FIGURE 5 The two halves of contemporary tRNA. The "top half" of tRNA is structurally and functionally independent and may be more ancient than the "bottom half" of the molecule.

tRNA is an independent structural domain that is recognized by RNase P (McClain *et al.*, 1987; Yuan and Altman, 1994), Tu (Rasmussen *et al.*, 1990), tRNA synthetases (McClain, 1993; Schimmel *et al.*, 1993; Saks *et al.*, 1994), and perhaps even ribosomal RNA (Noller *et al.*, 1992). The importance of this domain in almost all macromolecular interactions involving tRNA suggests that it is ancient, as does its structural *independence* from the bottom half of the molecule. Second, the ability of the cell to distinguish each tRNA from all the others—solving what is usually referred to as the tRNA identity problem—depends to a surprising extent on the identity of specific nucleotides in the top half of the molecule (McClain, 1993; Schimmel *et al.*, 1993; Saks *et al.*, 1994), including the "discriminator base" just inboard from the CCA terminus (Crothers *et al.*, 1972). This suggests that the identity of some tRNAs, and perhaps the specificity of the cognate aminoacyl-tRNA synthetases, was established before the bottom half of tRNA was incorporated into the molecule. Whether the bottom half of tRNA arose as an expansion loop within the top half or as an independent structural and functional domain that was subsequently incorporated into the top half is a question that future work may be able to resolve.

TABLE 1 Enzyme activities for which the top half of tRNA is the primary determinant of recognition

RNase P	McClain *et al.*, 1987; Yuan and Altman, 1994
elongation factor Tu	Rasmussen *et al.*, 1990
tRNA synthetases	Schimmel *et al.*, 1993; Saks *et al.*, 1994
CCA-adding enzyme	Li and Thurlow
ribosomal RNA?	Noller *et al.*, 1992

CONCLUSIONS AND FUTURE PROSPECTS

The ubiquity and conservation of tRNA in the replication strategies of a variety of contemporary genomes suggest a functional phylogeny for tRNA. This phylogeny is unique in placing the origin of tRNA in replication, prior to the advent of templated protein synthesis. In this scenario, aminoacyl-tRNA synthetase activities would have arisen next, to facilitate or regulate replication, and both tRNA and the aminoacyl-tRNA synthetase activities would have predated the anticodon and mRNA. A corollary is that the top half of modern tRNA may have had a more ancient origin than the bottom half bearing the anticodon.

The genomic tag hypothesis has "explanatory power" (Popper, 1963). It makes sense of—and establishes possible relationships between—otherwise puzzling structures and functions including RNase P, the CCA-adding enzyme, telomerase, contemporary synthetases, and the terminal tRNA-like structures themselves. It is also robust. A number of key experiments alluded to above in support of the genomic tag model were carried out after our original proposal (Weiner and Maizels, 1987), including studies of the Mauriceville retroplasmid (Wang and Lambowitz, 1993), the internal RNA template of telomerase (Blackburn, 1991), cleavage of plant virus tRNA-like structures by RNase P (Green *et al.*, 1988; Guerrier-Takada *et al.*, 1988; Mans *et al.*, 1990), stereospecific binding of an amino acid by RNA (Connell *et al.*, 1993), the ability of a ribozyme to work on a carbon center (Piccirilli *et al.*, 1992), and the division of contemporary synthetases into two classes (Eriani *et al.*, 1990).

What do we expect to learn from this model in the future? One prediction is that there are likely to be other transitional genomes that employ tRNA or tRNA-like structures in their replication. Another is that detailed functional and structural studies of contemporary tRNAs (White, 1982; Pan *et al.*, 1991) will further support the independence of the top and bottom halves of the molecule, explain why contemporary tRNA is a cloverleaf rather than the pseudoknotted structure found in plant viruses (Mans *et al.*, 1990), and unlock the evolutionary history that must lie in the location and function of the (almost) universally modified bases in tRNA. But especially exciting is the possibility that plausible phylogenies will emerge for other key biochemical pathways, grounded in the structure and function of contemporary molecules.

SUMMARY

We propose a phylogeny for the evolution of tRNA that is based on the ubiquity and conservation of tRNA-like structures in the *replication* of contemporary genomes. This phylogeny is unique in suggesting that the function of tRNA in replication dates back to the very beginnings of life on earth, before the advent of templated protein synthesis. The origin we propose for tRNA has distinct implications for the order in which other components of the modern translational apparatus evolved. We further suggest that the "top half" of modern tRNA—a coaxial stack of the acceptor stem on the TΨC arm—is the ancient structural and functional domain and that the "bottom half" of tRNA—a coaxial stack of the dihydrouracil arm on the anticodon arm—arose later to provide additional specificity.

REFERENCES

Akins, R. A., Kelley, R. L. & Lambowitz, A. M. 1989.Characterization of mutant mitochondrial plasmids of *Neurospora spp.* that have incorporated tRNAs by reverse transcription. *Mol. Cell. Biol.* **9,** 678–691.

Alberts, B. M. 1986. The function of the hereditary materials: Biological catalyses reflect the cell's evolutionary history. *Am. Zool.* **26,** 781–796.

Benner, S. A., Ellington, A. D. & Tauer, A. 1989. Modern metabolism as a palimpsest of the RNA world. *Proc. Natl. Acad. Sci. USA* **86,** 7054–7058.

Blackburn, E. H. 1991. Structure and function of telomeres. *Nature (London)* **350,** 569–573.

Blumenthal, T. & Carmichael, G. C. 1979. RNA replication: Function and structure of Qβ replicase. *Annu. Rev. Biochem.* **48,** 525–548.

Cavarelli, J., Rees, B., Ruff, M., Thierry, J. C. & Moras, D. 1993.Yeast tRNA(Asp) recognition by its cognate class II aminoacyl-tRNA synthetase. *Nature (London)* **362,** 181–184.

Cech, T. R., Zaug, A. J. & Grabowski, P. J. 1981. In vitro splicing of the ribosomal RNA precursor of Tetrahymena: Involvement of a guanosine nucleotide in the excision of the intervening sequence. *Cell* **27,** 487–496.

Chapman, K. B., Bystrom, A. S. & Boeke, J. D. 1992. Initiator methionine tRNA is essential for Ty1 transposition in yeast. *Proc. Natl. Acad. Sci. USA* **89,** 3236–3240.

Connell, G. J., Illangesekare, M. & Yarus, M. (1993) Three small ribooligonucleotides with specific arginine sites. *Biochemistry* **32,** 5497–5502.

Covey, S. N. & Turner, D. S. 1986. Hairpin DNAs of cauliflower mosaic virus generated by reverse transcription *in vivo*. *EMBO J.* **5,** 2763–2768.

Crothers, D. M., Seno, T. & Söll, D. G. 1972. Is there a discriminator base in transfer RNA? *Proc. Natl. Acad. Sci. USA* **69,** 3063–3067.

Cusack, S., Berthet-Colominas, C., Hartlein, M., Nassar, N. & Leberman, R. 1990. A second class of synthetase structure revealed by X-ray analysis of *Escherichia coli* seryl-tRNA synthetase at 2.5 A. *Nature (London)* **347,** 249–255.

Eriani, G., Delarue, M., Poch, O., Gagloff, J. & Moras, D. 1990. Partition of tRNA synthetases into two classes based on mutually exclusive sets of sequence motifs. *Nature (London)* **347,** 203–206.

Gilbert, W. 1986. The RNA world. *Nature (London)* **319,**618.

Green, C. J., Vold, B. S., Morch, M. D., Joshi, R. L. & Haenni, A. L. 1988. Ionic conditions for the cleavage of the tRNA-like structure of turnip yellow mosaic virus by the catalytic RNA of RNase P. *J. Biol. Chem.* **263,** 11617–11620.

Guerrier-Takada, C., Gardiner, K., Marsh, T., Pace, N. & Altman, S. 1983. The RNA moiety of RNAase P is the catalytic subunit of the enzyme. *Cell* **53,** 267–272.

Guerrier-Takada, C., van Belkum, A., Pleij, C. W. A. & Altman, S. 1988. Novel reactions of RNAase P with a tRNA-like structure in turnip yellow mosaic virus. *Cell* **53,** 267–272.

Haenni, A.-L., Joshi, S. & Chapeville, F. 1982. tRNA-like structures in the genomes of RNA viruses. *Prog. Nucleic Acid Res. Mol. Biol.* **27,** 85–104.

Hall, T. C. 1979. Transfer RNA-like structures in viral genomes. *Int. Rev. Cytol.* **60,** 1–26.

Hohn, T., Hohn, B. & Pfeiffer, P. 1985. Reverse transcription in CaMV. *Trends Biochem. Sci.* **10,** 205–209.

Jaehning, J. A. 1991. Sigma factor relatives in eukaryotes. *Science* **253,** 859.

Kikuchi, Y., Ando, Y. & Shiba, T. 1986. Unusual priming mechanism of RNA-directed DNA synthesis in *copia* retrovirus-like particles. *Nature (London)* **323,** 824–826.

Kuiper, M. T. R. & Lambowitz, A. M. 1988. A novel reverse transcriptase activity associated with mitochondrial plasmids of *Neurospora. Cell* **55,** 693–704.

Li, Z. and Thurlow, D. L. Effects of base sequence in the loops of 20 RNA minihelices on recognition by ATP/CTP:tRNA nucleotidyltransferase.

Maizels, N. & Weiner, A. M. 1987. Peptide-specific ribosomes, genomic tags and the origin of the genetic code. *Cold Spring Harbor Symp. Quant. Biol.* **52,** 743–749.

Maizels, N. & Weiner, A. M. 1993. in *The RNA World,* eds.Gesteland, R. F. & Atkins, J. F. (Cold Spring Harbor Lab. Press, Plainview, NY), pp. 577–602.

Mans, R. M., Guerrier-Takada, C., Altman, S. & Pleij, C. W. 1990. Interaction of RNase P from *Escherichia coli* with pseudo knotted structures in viral RNAs. *Nucleic Acids Res.* **18,** 3479–3487.

McClain, W. H. 1993. Rules that govern tRNA identity in protein synthesis. *J. Mol. Biol.* **234,** 257–280.

McClain, W. H., Guerrier-Takada, C. & Altman, S. 1987. Model substrates for an RNA enzyme. *Science* **238,** 527–530.

Noller, H. F., Hoffarth, V. & Zimniak, L. 1992. Unusual resistance of peptidyl transferase to protein extraction procedures. *Science* **256,** 1416–1419.

Palmer, J. R., Baltrus, T., Reeve, J. N. & Daniels, C. J. 1992. Transfer RNA genes from the hyperthermophilic Archaeon, *Methanopyrus kandleri. Biochim. Biophys. Acta* **1132,** 315–318.

Pan, T., Gutell, R. R. and Uhlenbeck, O. C. 1991. Folding of circularly permuted transfer RNAs. *Science* **254,** 1361–1364.

Piccirilli, J. A., McConnell, T. S., Zaug, A. J., Noller, H. F. & Cech, T. R. 1992. Aminoacyl esterase activity of the *Tetrahymena* ribozyme. *Science* **256,** 1420–1424.

Popper, K. R. (1963). *Conjectures and Refutations* (Rutledge and Kegan Paul, London).

Raabe, T., Bollum, F. J. & Manley, J. L. 1993. Primary structure and expression of bovine poly(A) polymerase. *Nature (London)* **353,** 229–234.

Rao, A. L. N., Dreher, T. W., Marsh, L. E. & Hall, T. C. 1989. Telomeric function of the tRNA-like structure of brome mosaic virus RNA. *Proc. Natl. Acad. Sci. USA* **86,** 5335–5339.

Rasmussen, N. J., Wikman, F. P. & Clark, B. F. 1990. Crosslinking of tRNA containing a long extra arm to elongation factor Tu by *trans*-diamminedichloroplatinum(II). *Nucleic Acids Res.* **18,** 4883–4890.

Ruff, M., Krishanswamy, S., Boeglin, M., Poterszman, A., Mitschler, A., Podjarny, A., Rees, B., Thierry, J. C. & Moras, D. 1991. Class II aminoacyl transfer RNA synthetases: crystal structure of yeast aspartyl-tRNA synthetase complexed with tRNA(Asp). *Science* **252,** 1682–1689.

Saks, M. E., Sampson, J. R. & Abelson, J. N. 1994. The transfer RNA identity problem: a search for rules. *Science* **263,** 191–197.

Sanfacon, H. & Hohn, T. 1990. Proximity to the promoter inhibits recognition of cauliflower mosaic virus polyadenylation signal. *Nature (London)* **346,** 81–84.

Schimmel, P., Giege, R., Moras, D. & Yokoyama, S. 1993. An operational RNA code for amino acids and possible relationship to the genetic code. *Proc. Natl. Acad. Sci. USA* **90,** 8763–8768.

Tse, W. T. & Forget, B. G. 1990. Reverse transcription and direct amplification of cellular RNA transcripts by Taq polymerase. *Gene* **88,** 293–296.

Wang, H. & Lambowitz, A. M. 1993. The Mauriceville plasmid reverse transcriptase can initiate cDNA synthesis de novo and may be related to reverse transcriptase and DNA polymerase progenitor. *Cell* **75,** 1071–1081.

Watson, J. D. 1972. Origin of concatemeric T7 DNA. *Nature New Biol.* **239,** 197–201.

Weiner, A. M. & Maizels, N. 1987. tRNA-like structures tag the 3′ ends of genomic RNA molecules for replication: Implications for the origin of protein synthesis. *Proc. Natl. Acad. Sci. USA* **84,** 7383–7387.

White, H. G., III 1982. Evolution of coenzymes and the origin of pyridine nucleotides. In *The Pyridine Nucleotide Coenzymes*, eds. Everse, J., Anderson, B. & You, K. (Academic, New York), pp. 1–17.

Wigley, P. L., Sheets, M. D., Zarkower, D. A., Whitmer, M. E. & Wickens, M. 1990. Polyadenylation of mRNA: minimal substrates and a requirement for the 2′ hydroxyl of the U in AUAAA. *Mol. Cell. Biol.* **10,** 1705–1713.

Woese, C. (1990). Evolutionary questions: the "progenote". *Science* **247,** 789.

Woese, C. R., Kandler, O. & Wheelis, M. L. 1990. Towards a natural system of organisms: proposal for the domains Archaea, Bacteria, and Eucarya. *Proc. Natl. Acad. Sci. USA* **87,** 4576–4579.

Xiong, Y. & Eickbush, T. H. 1990. Origin and evolution of retroelements based upon their reverse transcriptase sequences. *EMBO J.* **9,** 3353–3362.

Yuan, Y. & Altman, S. 1994. Selection of guide sequences that direct efficient cleavage of mRNA by human ribonuclease P. *Science* **263,** 1269–1273.

3

Disparate Rates, Differing Fates: Tempo and Mode of Evolution Changed from the Precambrian to the Phanerozoic

J. WILLIAM SCHOPF

When G. G. Simpson wrote *Tempo and Mode* (Simpson, 1944), fossil evidence of the history of life consisted solely of that known from sediments of the Phanerozoic eon, the most recent 550 million years (Ma) of geologic time (Figure 1). Thus, Simpson's views of the evolutionary process were based necessarily on Phanerozoic life—the familiar progression from seaweeds to flowering plants, from trilobites to humans—a history of relatively rapidly evolving, sexually reproducing plants and animals successful because of their specialized organ systems (flowers, leaves, teeth, limbs) used to partition and exploit particular environments. In short, Simpson elucidated "normal evolution" played by the "normal rules" of the game—*speciation, specialization, extinction.*

Although certainly applicable to the megascopic eukaryotes of the Phanerozoic, there is reason to question whether these well-entrenched rules apply with equal force to the earlier and very much longer Precambrian phase of microbe-dominated evolutionary history (Figure 1). In place of sexual multicellular plants and animals, the biota throughout much of the Precambrian was dominated by simple nonsexual prokaryotes. Rather than evolving rapidly, many Precambrian microbes evidently evolved at an astonishingly slow pace. And instead

J. William Schopf is professor of paleobiology in the Department of Earth and Space Sciences and director of the Center for the Study of Evolution and the Origin of Life at the University of California, Los Angeles.

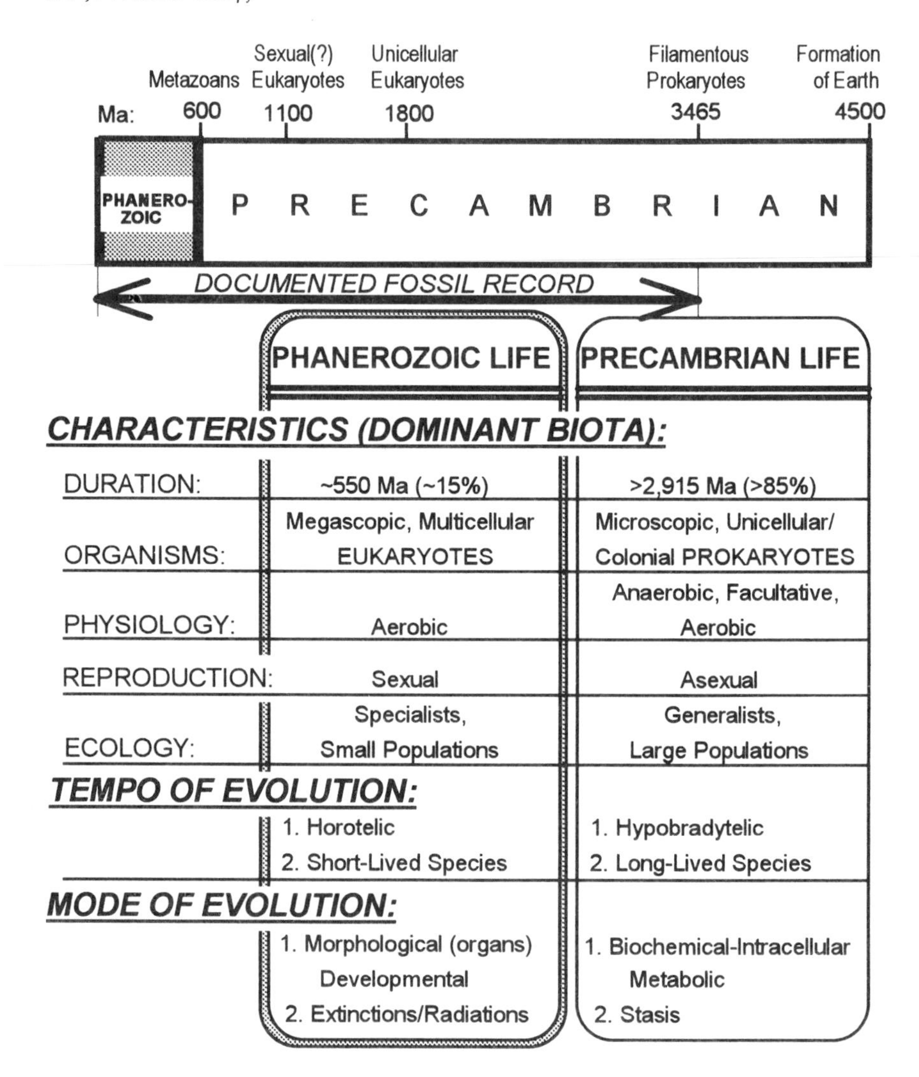

FIGURE 1 Comparison of the Phanerozoic and Precambrian histories of life.

of having specialized organ systems for exploitation of specific ecologic niches, members of the most successful group of these early-evolving microorganisms—photoautotrophic cyanobacteria—were ecologic generalists, able to withstand the rigors of a wide range of environments. In contrast with normal evolution, the "primitive rules" of prokaryotic evolution appear to have been *speciation*, *generalization*, and *exceptionally long-term survival*.

That there is a distinction in evolutionary tempo and mode between the Phanerozoic and Precambrian histories of life is not a new idea (Schopf, 1978), but it is one that has recently received additional impetus (Schopf, 1992a) and therefore deserves careful scrutiny. However, evaluation of this generalization hinges critically on the quality and quantity of the fossil evidence available, and because active studies of the Precambrian fossil record have been carried out for little more than a quarter century (Schopf, 1992b) a thorough comparison of the early history of life with that of later geologic time is not yet possible. Therefore, as a first approximation, the approach used here is to analyze the known fossil record of Precambrian cyanobacteria: well-studied, widespread, abundant, commonly distinctive, and evidently dominant members of the early prokaryotic biota. Microscopic fossils regarded as members of other prokaryotic groups are also known from the Precambrian (Schopf, 1992c, 1992d), but their documented record is minuscule. Hence, conclusions drawn here about the early fossil record apply strictly to free-living cyanobacteria, the evolutionary history of which may or may not be representative of prokaryotes in general. To evaluate the generalization, two central questions must be addressed. First, was the tempo of Precambrian cyanobacterial evolution markedly slower than that typical of Phanerozoic eukaryotes? Second, if so, how can this difference be explained?

Tempos of Evolution

In *Tempo and Mode*, Simpson coined terms for three decidedly different rate distributions in evolution, inferred from morphological comparisons of Phanerozoic and living taxa: tachytelic, for "fast"-evolving lineages; horotelic, the standard rate distribution, typical of most Phanerozoic animals; and bradytelic, for "slow" morphological evolution (Simpson, 1944). Included among the bradytelic lineages are so-called living fossils (such as linguloid brachiopods, horseshoe crabs, coelacanth fish, crocodilians, opossums), "groups that survive today and show relatively little change since the very remote time when they first appeared in the fossil record" (Simpson, 1944, p. 125). Simpson's bradytely closely approximates Ruedemann's earlier developed concept of "arrested evolution" (Ruedemann, 1918, 1922a, 1922b), both based on comparison of modern taxa with fossil forms that are virtually indistinguishable in morphology but are 100 Ma or more older.

Hypobradytely. Recently, a fourth term—hypobradytely—has been added to this list of rate distributions (Schopf, 1987) "to refer to the exceptionally low rate of evolutionary change exhibited by cyanobacterial taxa, morphospecies that show little or no evident morphological change over many hundreds of millions of years and commonly over more than one or even two thousand million years" (Schopf, 1992e, p. 596). Following Simpson's lead, hypobradytely is based strictly on morphological comparison of living and fossil taxa. Other data, such as chemical biomarkers (Schopf, 1992e), carbon isotopic compositions (Schidlowski *et al.*, 1983), and environmental distributions (Knoll and Golubic, 1992), can provide insight on the paleophysiology of fossil cyanobacteria, but the concept of hypobradytely does not necessarily imply genomic, biochemical, or physiological identity between modern and fossil taxa. The concept can be applied to cyanobacteria because the morphologic descriptors and patterns of cell division used to differentiate taxa at various levels of the taxonomic hierarchy are preservable in ancient sediments (Golubic, 1976a; Schopf, 1992d, 1992e); as emphasized by Knoll and Golubic (1992, p. 453), "Essentially all of the salient morphological features used in the taxonomic classification of living cyanobacteria can be observed in well-preserved microfossils."

Interpreted as evidencing exceptional "morphological evolutionary conservatism," detailed genus- and species-level similarities between Precambrian and extant cyanobacteria, both for filamentous (oscillatoriacean) and spheroidal (chroococcacean) taxa, were documented more than 25 years ago (Schopf, 1968). Since that time, such similarities have been recognized repeatedly. Indeed, it has become common practice for Precambrian paleobiologists to coin generic names intended to denote similarity or inferred identity between ancient cyanobacterium-like fossils and their modern morphological analogs by adding appropriate prefixes (palaeo-, eo-) or suffixes (-opsis, -ites) to the names of living cyanobacterial genera (Table 1). The validity of such comparisons is variable, and although more than 40 such fossil namesakes have been proposed (for genera referred to diverse cyanobacterial families; Table 1), their use does not in and of itself constitute compelling evidence of hypobradytely.

Caveats. Application of the concept of hypobradytely is not without potential pitfalls, three of which deserve particular mention. First, because of the enormous span of Precambrian time (Figure 1), and despite the notable paleontological progress of recent years (Schopf, 1992b), early biotic history is as yet very incompletely documented. In comparison with the vastly better documented record of Phanerozoic organisms—and even in geologic units of the Proterozoic (2500–550 Ma in age), by far the most studied portion of the Precambrian—the known cyanobacterial fossil record is scanty (for filamentous species amounting to ≈21 taxonomic occurrences per 50-Ma-long interval and, for spheroi-

TABLE 1 Precambrian generic namesakes, coined by various authors to suggest similarity to modern cyanobacterial genera (Mendelson and Schopf, 1992a)

Modern genus	Precambrian genus; author(s)	No. of species	Year published	Country of author(s)
	Family Oscillatoriaceae			
Lyngbya	*Palaeolyngbya*; Schopf	11	1968	USA
Lyngbya	cf. *Lyngbya*; Schopf, Xu, Xu, and Hsu	1	1984	USA, China
Microcoleus	*Eomicrocoleus*; Horodyski and Donaldson	1	1980	USA, Canada
Oscillatoria	*Archaeoscillatoriopsis*; Schopf	3	1993	USA
Oscillatoria	*Oscillatoriopsis*; Schopf	22	1968	USA
Oscillatoria	*Oscillatorites*; Schepeleva	1	1960	Russia
Oscillatoria	cf. *Oscillatoria*; Schopf and Sovietov	1	1976	USA, Russia
Phormidium	*Eophormidium*; Xu	3	1984	China
Schizothrix	*Schizothrix*; Edhorn	1	1973	Canada
Schizothrix	*Schizothropsis*; Xu	1	1984	China
Spirulina	*Palaeospirulina*; Edhorn	2	1973	Canada
Spirulina	*Spirillinema*; Shimron and Horowitz	1	1972	Israel
Spirulina	aff. *Spirulina*; Schopf and Blacic	1	1971	USA
	Family Chroococcaceae			
Anacystis	*Palaeoanacystis*; Schopf	8	1968	USA
Aphanocapsa	*Aphanocapsaopsis*; Maithy and Shukla	2	1977	India
Aphanocapsa	*Eoaphanocapsa*; Nyberg and Schopf	1	1984	USA
Aphanothece	*Eoaphanothece*; Xu	1	1984	China
Chroococcus	*Chroococcus*-like; Mendelson and Schopf	1	1982	USA
Eucapsis	*Eucapsamorpha*; Golovenoc and Belova	1	1985	Russia
Eucapsis	*Eucapsis?*; Licari, Cloud, and Smith	1	1969	USA
Gloeocapsa	*Eogloeocapasa*; Golovenoc and Belova	1	1984	Russia
Gloeocapsa	*Gloeocapsa*-like; Zhang	1	1985	China
Merismopedia	cf. *Merismopedia*; Schopf and Fairchild	1	1973	USA, Brazil
Microcystis	*Eomicrocystis*; Maithy	2	1975	India

TABLE 1 (Continued)

Modern genus	Precambrian genus; author(s)	No. of species	Year published	Country of author(s)
	Family Chroococcaceae (continued)			
Microcystis	*Microcystopsis*; Xu	1	1984	China
Microcystis	*Palaeomicrocystis*; Maithy	2	1975	India
Synechococcus	*Eosynechococcus*; Hofmann	13	1976	Canada
	Family Entophysalidaceae			
Entophysalis	*Eoentophysalis*; Hofmann	6	1976	Canada
	Family Pleurocapsaceae			
Hyella	*Eohyella*; Zhang and Golubic	4	1987	China, USA
Myxosarcina	cf. *Myxosarcina*; Schopf and Fairchild	1	1973	USA, Brazil
Pleurocapsa	*Eopleurocapsa*; Liu	1	1982	China
Pleurocapsa	*Palaeopleurocapsa*; Knoll, Barghoorn, and Golubic	6	1975	USA
	Family Nostocaceae			
Anabaena	*Anabaenidium*; Schopf	4	1968	USA
Aphanizomenon	*Palaeoaphanizomenon*; Mikhailova	1	1986	Russia
Isocystis	*Palaeoisocystis*; Xu	2	1984	China
Nostoc	*Nostocomorpha*; Sin and Liu	1	1978	China
Nostoc	*Palaeonostoc*; Nautiyal	1	1980	India
Nostoc	*Veteronostocale*; Schopf	2	1968	USA
	Family Rivulariaceae			
Calothrix	*Palaeocalothrix*; Xu	2	1984	China
Rivularia	*Primorivularia*; Edhorn	3	1973	Canada
	Family Scytonemataceae			
Plectonema	*Eoplectonema*; Liu	1	1984	China
Scytonema	*Palaeoscytonema*; Edhorn	5	1973	Canada

dal species, ≈46 occurrences; Schopf, 1992c; Mendelson and Schopf, 1992a). Second, assignment of some Precambrian microbial fossils (evidently <10%) to the cyanobacteria can be quite uncertain, a problem that applies especially to minute morphologically simple forms (atypically small-diameter oscillatoriaceans and chroococcaceans, for example), which in the fossil state are essentially indistinguishable from various noncyanobacterial prokaryotes (Schopf, 1992d). Third, lack of change in the external form of morphologically simple prokaryotes may not necessarily reflect evolutionary stasis of their internal physiological machinery (the so-called Volkswagen syndrome; Schopf *et al.*, 1983). This last problem is especially difficult to evaluate, but it may not be of overriding importance—from early in the Proterozoic to the present, the same cyanobacterial families and many of the same (morphologically defined) genera and even species appear to have inhabited the same or closely similar environments, patterns of distribution that "provide proxy information on physiological attributes" (Knoll and Golubic, 1992, p. 451).

Cyanobacterial Hypobradytely

Are cyanobacteria hypobradytelic? Data bearing on this question fall into two classes. First, a large amount of evidence indicates that the proposition is plausible, in fact likely to be correct, but, because of a lack of accompanying relevant (chiefly environmental) information, this evidence is not fully compelling. Second, a small number of in-depth studies firmly support the proposition, but because of their limited number and restricted taxonomic scope these studies do not establish the proposition generally. Considered together, however, the two classes of data present a strong case.

Evidence of Plausibility. Shown in Figure 2 are paired examples of morphologically comparable Proterozoic and living cyanobacteria including specimens illustrating the rather commonly cited (Schopf, 1968; Schopf and Blacic, 1971; Golubic and Campbell, 1979; Campbell, 1979) similarity between fossil *Palaeolyngbya* (Figure 2*B*) and *Lyngbya* (Figure 2*A*), its modern morphological counterpart. Numerous other genus- and species-level comparisons have been drawn [for example, between ≈850-Ma-old *Cephalophytarion grande* and the modern oscillatoriacean *Microcoleus vaginatus* (Schopf, 1968; Campbell, 1979); and between ≈2-billion year (Ga)-old *Eosynechococcus moorei* and the living chroococcacean *Gloeothece coerulea* (*Gloeobacter violaceus*) (Golubic and Campbell, 1979)]. Examples such as these—and the fact that over the past quarter century such similarities have been noted repeatedly and regarded as biologically and taxonomically significant by a large number of workers in many countries (Table 1)—provide a powerful argument for the plausibility of cyanobacterial hypobradytely.

Plausibility of the concept is similarly indicated by quantitative studies recently carried out on large assemblages of modern and Precambrian microbes. Morphometric data (for such attributes as cell size, shape, and range of variability; colony form; sheath thickness and structure) were compiled for 615 species and varieties of living cyanobacteria (Schopf, 1992d) as well as for an extensive worldwide sample of Proterozoic

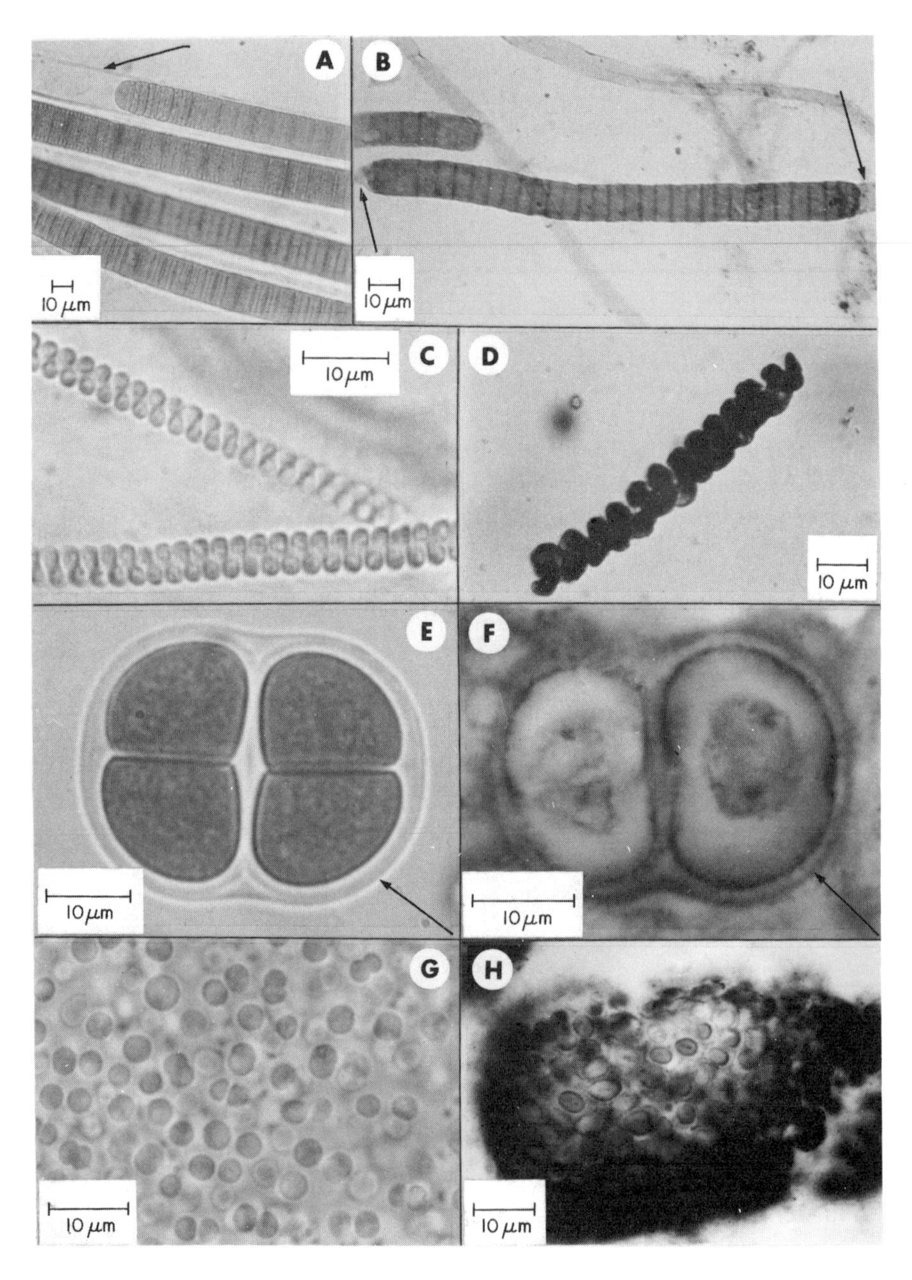

FIGURE 2 Comparison of living and Precambrian cyanobacteria. Living examples (*A, C, E,* and *G*) are from mat-building stromatolitic communities of northern Mexico. (*A*) *Lyngbya* (Oscillatoriaceae), encompassed by a cylindrical mucilaginous sheath (arrow). (*B*) *Palaeolyngbya,* similarly ensheathed (arrows),

cyanobacterium-like microfossils (Mendelson and Schopf, 1922a), both filamentous (650 taxonomic occurrences in 160 geologic formations) and spheroidal (1400 occurrences in 259 formations). To avoid confusion stemming from variations in taxonomic practice, fossils having the same or similar morphology (regardless of their binomial designations) were grouped together as informal species-level morphotypes designed to have ranges of morphologic variability comparable to those exhibited by living cyanobacterial species (Schopf, 1992c). Of the 143 informal species of filamentous microfossils thus recognized, 37% are essentially indistinguishable in morphology from established species of living (oscillatoriacean) cyanobacteria (Schopf, 1992c). Similarly, 25% of the 120 informal taxa of spheroidal fossil species have modern species-level (largely chroococcacean) morphological counterparts (Schopf, 1992c). Virtually all of the fossil morphotypes are referable to living genera of cyanobacteria, and the patterns and ranges of size distribution exhibited by taxa of cylindrical sheath-like Proterozoic fossils are essentially identical to those of the tubular sheaths that encompass trichomes of modern oscillatoriacean species (Figure 3).

About half of the ≈2000 fossil occurrences included in this morphometric study were reported from cherty carbonate stromatolites, with the remainder from clastic shales and siltstones. Data regarding the specific environmental settings represented by these lithologies are available for very few of the >300 fossiliferous geologic units considered. With varying degrees of uncertainty (but almost always without firm evidence), most of these strata have been assumed to represent relatively shallow water coastal marine facies, ranging from sabkhas and lagoons to mud flats and intertidal carbonate platforms. In such environments today, mat-building oscillatoriaceans (predominantly *Oscillatoria, Lyngbya, Phormidium, Spirulina, Microcoleus,* and *Schizothrix*) and subsidiary chroococcaceans (such as *Chroococcus, Aphanocapsa, Aphanothece,* and *Synechococcus*) are common (Golubic, 1976a, 1976b; Pierson *et al.*, 1992). As is shown in Figure 4, these same two families, as well as

FIGURE 2 (Continued) from the ≈950-Ma-old Lakhanda Formation of eastern Siberia. (*C*) *Spirulina* (Oscillatoriaceae). (*D*) *Heliconema*, a *Spirulina*-like cyanobacterium from the ≈850-Ma-old Miroedikha Formation of eastern Siberia. (*E*) *Gloeocapsa* (Chroococcaceae), a four-celled colony having a thick distinct encompassing sheath (arrow). (*F*) *Gloeodiniopsis*, a similarly sheath-enclosed (arrow) *Gloeocapsa*-like cyanobacterium, from the ≈1550-Ma-old Satka Formation of southern Bashkiria. (*G*) *Entophysalis* (Entophysalidaceae). (*H*) *Eoentophysalis*, an *Entophysalis*-like colonial cyanobacterium from the ≈2-Ga-old Belcher Group of Northwest Territories, Canada (Golubic and Hofmann, 1976).

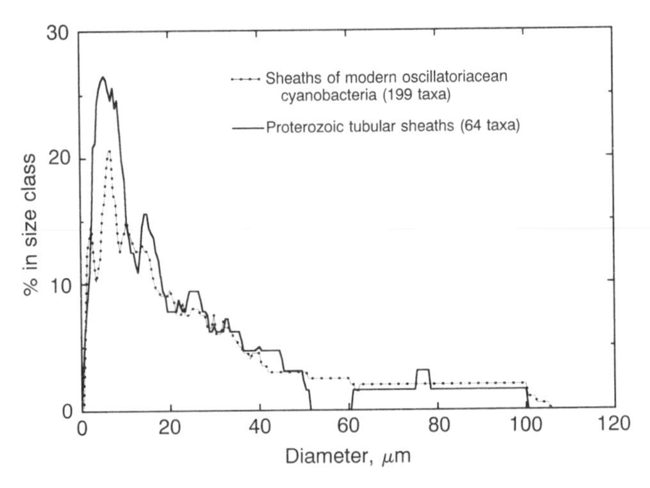

FIGURE 3 Comparison of patterns and ranges of size distribution of cylindrical sheath-like Precambrian fossils with those of the tubular sheaths of living oscillatoriacean cyanobacteria (Schopf, 1992d).

morphotypes referable to most of the same genera, are similarly common among the fossil taxa having living species-level counterparts.

In view of these data, it is difficult not to conclude that "the biological constitution of [Proterozoic] microbial mat communities was probably quite similar to that of modern communities in comparable environments" (Knoll, 1985, p. 411).

In-Depth Studies. In addition to studies of cellular morphology and likely (but not firmly established) broad-scope environmental comparisons, what is needed to move the hypobradytelic hypothesis from the plausible to the compelling are supporting data on the paleoenvironment and taphonomy of the fossils in question (Knoll and Golubic, 1992). A number of such in-depth studies have been carried out (Knoll *et al.*, 1975; Golubic and Hofmann, 1976; Knoll and Golubic, 1979; Green *et al.*, 1987; Green *et al.*, 1988), focusing on fossil representatives of two cyanobacterial families, the Entophysalidaceae and the Pleurocapsaceae, members of which are decidedly more distinctive morphologically than are the oscillatoriaceans and chroococcaceans discussed above. Golubic and Hofmann (1976) compared ≈2-Ga-old *Eoentophysalis belcherensis* (Figure 2*H*) with two modern entophysalidaceans (*Entophysalis major* and *Entophysalis granulosa*). They showed that not only are the fossil and modern species morphologically comparable (in cell shape and in form and arrangement of originally mucilaginous cellular envelopes) and that

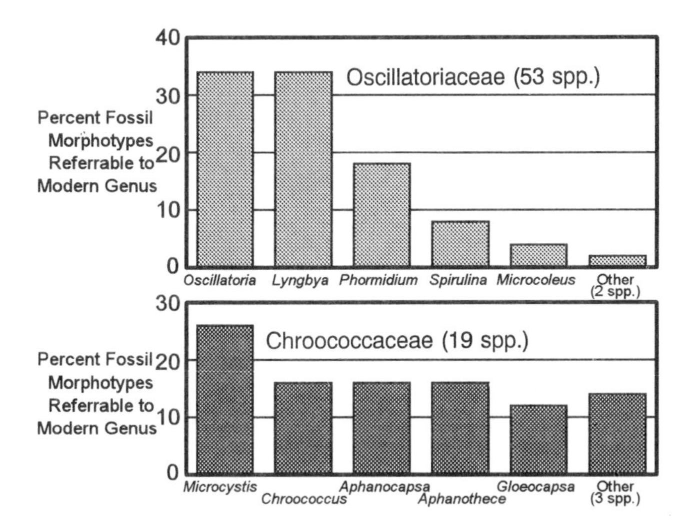

FIGURE 4 Distribution among modern oscillatoriacean and chroococcacean genera of Precambrian species-level morphotypes having living morphological counterparts (Schopf, 1992c, 1992d).

they exhibit similar frequency distributions of dividing cells and essentially identical patterns of cellular development (resulting from cell division in three perpendicular planes), but also that both taxa form microtexturally similar stromatolitic structures in comparable intertidal to shallow marine environmental settings, that they undergo similar postmortem degradation sequences, and that they occur in microbial communities that are comparable in both species composition and biological diversity. In a subsequent detailed study, Knoll and Golubic compared the morphology, cell division patterns, ecology, and postmortem degradation sequences of a second Precambrian entophysalidacean (≈850-Ma-old *Eoentophysalis cumulus*) with those of modern *E. granulosa* and concluded that the fossil "microorganism is identical in all its salient characteristics to members of the extant [cyanobacterial] genus" (Knoll and Golubic, 1979, p. 125).

Several species of fossil and living pleurocapsaceans have also been compared in detail. *Polybessurus bipartitus*, first reported from ≈770-Ma-old stromatolites of South Australia (Fairchild, 1975; Schopf, 1977), is a morphologically distinctive, gregarious, cylindrical fossil pleurocapsacean composed of stacked cup-shaped envelopes often extended into long tubes oriented predominantly perpendicular to the substrate. Specimens of this taxon occurring in rocks of about the same age in East Greenland were interpreted by Green *et al.* to be "a close morphological, reproductive, and behavioral counterpart" to populations of a species of the pleurocapsacean *Cyanostylon* present "in Bahamian environments similar to those in which the Proterozoic fossils occur" (Green *et al.*, 1987, p. 928). A second fossil pleurocapsacean described from the ≈770-Ma-old Australian deposit (*Palaeopleurocapsa wopfnerii*) has been compared by Knoll *et al.* with its living morphological and ecological analog (*Pleurocapsa fuliginosa*) and regarded as "further evidence of the evolutionary conservatism of [cyanobacteria]" (Knoll *et al.*, 1975, p. 2492). Two other species of morphologically distinctive fossil pleurocapsaceans (the endolithic taxa *Eohyella dichotoma* and *Eohyella rectroclada*), cited as "compelling examples of the close resemblance between Proterozoic prokaryotes and their modern counterparts" (Knoll *et al.*, 1986, p. 857), have been described by Green *et al.* from the East Greenland geologic sequence as being "morphologically, developmentally, and behaviorally indistinguishable" from living *Hyella* species of the Bahama Banks (Green *et al.*, 1988, pp. 837–838).

These in-depth studies of entophysalidaceans and pleurocapsaceans—involving analyses of environment, taphonomy, development, and behavior, in addition to cellular morphology—provide particularly convincing evidence of species-specific fossil-modern similarities.

Cyanobacteria are Hypobradytelic. Thus, numerous workers worldwide have noted and regarded as significant the detailed similarities in cellular morphology between Precambrian and extant cyanobacteria (Table 1; Figure 2). A substantial fraction of known Proterozoic oscillatoriacean and chroococcacean cyanobacteria have living species-level morphological counterparts (Figure 4), and almost all such fossils are referable to living cyanobacterial genera. And in-depth studies of several fossil-modern species pairs of morphologically distinctive entophysalidaceans and pleurocapsaceans permit detailed comparison of morphology, development, population structure, environment, and taphonomy, all of which show that for at least these taxa, ancient and modern cyanobacteria are essentially indistinguishable in salient characteristics.

Taken together, these observations support an obvious conclusion—the morphology (and evidently the physiology as well) of diverse taxa belonging to major cyanobacterial families evolved little or not at all over hundreds of millions, indeed thousands of millions of years. In comparison with the later history of life, this widespread hypobradytely is surprising. In Phanerozoic evolution, bradytelic stasis is notable principally because of its rarity (Simpson, 1944; Ruedemann, 1918, 1922a,

1922b), but in the Precambrian it seems to have been a general phenomenon characteristic of a group of prokaryotic microorganisms that dominated the Earth's biota, possibly even as early as 3.5 Ga ago (Schopf, 1993). Why have cyanobacteria evidently changed so little over their exceedingly long evolutionary history?

Survival of the Ecologically Unspecialized

To understand the underlying causes of cyanobacterial hypobradytely, it is instructive to review Simpson's thoughtful analysis in *Tempo and Mode,* for although he was unaware of the Precambrian prokaryotic fossil record, Simpson was much interested in slowly evolving (bradytelic) Phanerozoic lineages. In addition to noting (but dismissing) the possibility that "asexual reproduction (as inhibiting genetic variability)" might be conducive to slow evolution (Simpson, 1944, p. 137), he singled out two principal factors: large population size, and ecologic versatility, an exceptional degree of adaptation "to some ecological position or zone with broad . . . selective limits . . . a particular, continuously available environment" (Simpson, 1944, pp. 138, 140, 141). Because unusually slow evolution involves "not only exceptionally low rates of [evolutionary change] but also survival for extraordinarily long periods of time" (p. 138), and because "more specialized phyla tend to become extinct before less specialized," Simpson proposed "the rule of the survival of the relatively unspecialized" (Simpson, 1944, pp. 138, 143).

Although intended by Simpson to apply to Phanerozoic organisms, chiefly animals, these same considerations (with the addition of asexual reproduction) apply to Precambrian cyanobacteria. First, with regard to reproduction, cyanobacteria are strictly asexual, lacking even the parasexual processes known to occur in some other prokaryotes. Given the remarkable longevity of the cyanobacterial lineage and moderate or even low rates of mutation, however, the absence of sexually generated genetic variability cannot be the sole explanation for their hypobradytely. Second, like virtually all free-living microorganisms, cyanobacteria typically occur in local populations of large size. Coupled with their ease of dispersal (via water currents, wind, and hurricanes, for example) and for many species a resulting very wide (essentially cosmopolitan) geographic distribution, their large populations can also be presumed to have played a role in their evolutionary stasis (Figure 1). Third, and probably most important, however, is the ecologic versatility of the group.

Summarized in Table 2 are known ranges of survivability (and of growth under natural conditions) for modern oscillatoriaceans and

Table 2 Survival [and growth under natural conditions (boldface)] of oscillatoriacean and chroococcacean cyanobacteria*

Light intensity	1–5 $\mu E \cdot s^{-1} \cdot m^{-2}$	**50–60 $\mu E \cdot s^{-1} \cdot m^{-2}$**	**>2000 $\mu E \cdot s^{-1} \cdot m^{-2}$**	
Conditions	Cultures	**Optimum growth**	**Intertidal zone**	
Genera	1C; 2E, F	**1, 2**	**1, 2**	
Total salinity	**<0.001–0.1%**	**3.5%**	**27.5%**	100–200%
Conditions	**Freshwater**	**Marine**	**Great Salt Lake**	Salterns
Genera	**1, 2**	**1, 2**	**2D**	1B–D, F; 2F
Acidity/basicity	**pH 4**	**pH 7–10**	**pH 10–11**	
Conditions	**Hot springs**	**Optimum growth**	**Alkaline lakes**	
Genera	**2F**	**1, 2**	**1F; 2E**	
High temperature	**55–70°C**	**74°C**	111°C	112°C
Conditions	**Hot springs**	**Hot springs**	Dried	Dried
Genera	**1C, D, F; 2C, F**	**2F**	1E	1B
Low temperature	−269°C	−196°C	−55°C	**−2° to +4°C**
Conditions	Liquid He	Liquid H_2	Freeze-dried	**Antarctic lakes**
Genera	1C	1 D, E	1A	**1 C, D**
Desiccation	88	82 yr	**Absence of rainfall**	
Conditions	Dried	Dried	**Atacama Desert**	
Genera	1C	2E	**1E; 2B, D**	
Oxygen	**<0.01%**	**1%**	**20%**	**100%**
Conditions	**Anoxic lakes**	**Blooms, muds**	**Ambient O_2**	Cultures
Genera	**1, 2**	**1, 2**	**1, 2**	2B
Carbon dioxide	0.001%	**0.035%**	3.5%	40%
Conditions	Cultures	**Ambient CO_2**	Cultures	Cultures
Genera	2B	**1, 2**	1E; 2A, B, D, F	2B
Radiation	**Ultraviolet**	X-rays	γ-rays	Highly ionizing
Conditions	**290–400 nm[a]**	200 kr[b]	2560 kr[c]	Thermonuclear explosion
Genera	**1A, E; 2C**	1C	1B	1 B, E

Oscillatoriacean genera: 1A, *Lyngbya*; 1B, *Microcoleus*; 1C, *Oscillatoria*; 1D, *Phormidium*; 1E, *Schizothrix*; 1F, *Spirulina*. Chroococcacean genera: 2A, *Agmenellum*; 2B, *Anacystis*; 2C, *Aphanocapsa*; 2D, *Coccochloris*; 2E, *Microcystis*; 2F, *Synechococcus*.

[a]Absorbed by scytonemin pigment in encompassing sheaths.

[b]Twice as resistant as eukaryotic microalgae.

[c]Ten times as resistant as eukaryotic microalgae.

*Table from Abeliovich and Shilo, 1972; Brock, 1978; Cameron, 1963; Castenholz, 1969; Ciferri, 1983; Davis, 1972; Davison, 1991; Desikachary, 1959; Drouet, 1968; Drouet and Daily, 1973; Flowers and Evans, 1966; Fogg, 1973; Fogg *et al.*, 1973; Forest and Weston, 1966; Frémy, 1972; Fuhs, 1968; Garcia-Pichel and Castenholz, 1991; Godward, 1962; Grant and Tindall, 1986; Knoll and Bauld, 1989; Langworthy, 1978; Lloyd et al., 1977; Mizutani and Wada, 1982; Pardue *et al.*, 1976; Parker *et al.*, 1981; Schopf, 1974; Shields and Drouet, 1962; Vallentyne, 1963; Vincent *et al.*, 1993.

chroococcaceans, the most primitive (Giovannoni *et al.*, 1988) and commonly occurring (Schopf, 1992d) Precambrian cyanobacterial families. Similar tolerance is also exhibited by members of other cyanobacterial families. For example, a nostocacean was revived after more than a century of storage in a dried state (Davis, 1972) and a scytonematacean is reported to have maintained growth at pH 13 (Vallentyne, 1963). Thus, cyanobacteria exhibit notable ecologic flexibility, and even though no single oscillatoriacean or chroococcacean species is known to be capable of tolerating the total range of observed growth conditions (for example, thermophiles dominant in 70°C waters rarely grow below 50°C, and species adapted to highly alkaline lakes do not occur in acid hot springs), both groups include impressive ecologic generalists, able to thrive in virtually all present-day widespread environments (Table 2). Moreover, many of the oscillatoriacean and chroococcacean genera for which wide ecologic tolerance has been demonstrated (Table 2) are the same as those having species-level Precambrian-extant counterparts (Figure 4). Finally, numerous cyanobacteria, including both oscillatoriaceans and chroococcaceans (Stewart, 1980), are capable of fixing atmospheric nitrogen; provided with light, CO_2, a source of electrons (H_2, H_2S, H_2O), and a few trace elements, such cyanobacteria are highly effective colonizers, able to invade and flourish in a wide range of habitats.

The wide ecologic tolerance of cyanobacteria is almost certainly a product of their early evolutionary history. Fossil evidence suggests that oscillatoriaceans (Schopf, 1993) and chroococcaceans (Schopf and Packer, 1987) were extant as early as ≈3.5 Ga ago. If so, they must have originated and initially diversified in an oxygen-deficient environment, one lacking an effective UV-absorbing ozone layer. In such an environment, the ability to photosynthesize at low light intensities (Table 2) coupled with the presence of gas vesicles to control buoyancy (Jensen, 1993) would have permitted planktonic cyanobacteria to avoid deleterious UV by inhabiting the deep oceanic photic zone, just as *Synechococcus* does today. Similarly, numerous characteristics of living benthic mat-building cyanobacteria—effective DNA repair mechanisms, synthesis of UV-absorbing scytonemin, secretion of copious extracellular mucilage, phototactic motility, adherence to substrates, stromatolitic mat formation—initially may have been adaptations to cope with a high UV flux in near-shore shallow water settings. Adaptive radiation in an early oxygen-deficient environment is also suggested by the ability of cyanobacteria to live in either the presence or absence of oxygen (Table 2), their capability to switch between oxygenic and anoxygenic photosynthesis (Olson and Pierson, 1987), the occurrence of oxygen-sensitive nitrogenase in many taxa (Stewart, 1980), and the restriction of nitrogenase-protecting heterocysts to late-evolving members of the group (Giovannoni *et al.*, 1988). In addition, both the low affinity of cyanobacterial ribulose-bisphosphate carboxylase for CO_2 and the presence of intracellular CO_2-concentrating mechanisms (Badger, 1987) may reflect initial adaptation of the lineage to a CO_2-rich primordial environment (Kasting, 1992).

Finally, the remarkable hardiness of cyanobacteria—their ability to survive wide ranges of light intensity, salinity, temperature, and pH as well as prolonged desiccation and intense radiation (Table 2)—may be a product of their marked success in competing for photosynthetic space with other early-evolving microbes. Unlike the oxygen-producing photosynthesis based on chlorophyll *a* in cyanobacteria, that in all other photoautotrophic prokaryotes is anoxygenic and bacteriochlorophyll-based. Because biosynthesis of bacteriochlorophyll is inhibited by molecular oxygen (Olson and Pierson, 1987), oxygen-producing cyanobacteria would have rapidly supplanted oxygen-sensitive anoxygenic photoautotrophs throughout much of the global photic zone. As a result of the ease of their global dispersal and their success in competing for photosynthetic space, cyanobacteria presumably expanded into a broad range of habitable niches during an early, evidently rapid phase of adaptive radiation (Giovannoni *et al.*, 1988), evolving to become exceptional ecologic generalists. Thus, the ecologic versatility of cyanobacteria appears to hark back to an early stage of planetary history when they established themselves as the dominant primary producers of the Precambrian ecosystem.

In view of their evolutionary history, it is perhaps not surprising that Simpson's rule of survival of the (ecologically) relatively unspecialized is applicable to cyanobacteria, numerous taxa of which qualify as so-called living fossils. According to Stanley (1984, p. 280), such extraordinarily long-lived organisms "are simply champions at warding off extinction." If so, as has been previously suggested (Schopf, 1992e, p. 598), the "grand champions," over all of geologic time, must be the hypobradytelic cyanobacteria!

A Bipartite View of the History of Life

In broadbrush outline, biotic history thus seems divisible into two separate phases (Schopf, 1978, 1992a), each characterized by its own tempo and mode, each by its own set of evolutionary rules (Figure 1).

During the shorter more recent Phanerozoic eon, the history of life was typified by the horotelic evolution of dominantly megascopic, sexual, aerobic, multicellular eukaryotes based on alternating life cycle phases specialized either for reproduction or for nutrient assimilation.

Changes in the dominant (commonly diploid) phase resulted chiefly from structural modification of organ systems used to partition and exploit particular environments. In large part as a result of this ecologic specialization, the Phanerozoic was punctuated by recurrent episodes of extinction, each followed by the adaptive radiation of surviving lineages.

In contrast with Phanerozoic evolution, much of the earlier and decidedly longer Precambrian history of life was typified by the hypobradytelic evolution of dominantly microscopic, asexual, metabolically diverse, and commonly ecologically versatile prokaryotes, especially cyanobacteria. Evolutionary innovations were biochemical and intracellular. Once established, lineages exhibited long-term stasis. Extinction occurred rarely among prokaryotic ecologic generalists, evidently becoming a significant evolutionary force only late in the Precambrian and primarily affecting ecologically relatively specialized, large-celled eukaryotic phytoplankters (Mendelson and Schopf, 1992b; Vidal and Knoll, 1982; Schopf, 1992f).

Although as yet incompletely documented, this bipartite interpretation of evolutionary history seems consistent with the fossil record as now known. It remains to be established whether it, like Simpson's *Tempo and Mode*, will stand the test of time.

SUMMARY

Over the past quarter century, detailed genus- and species-level similarities in cellular morphology between described taxa of Precambrian microfossils and extant cyanobacteria have been noted and regarded as biologically and taxonomically significant by numerous workers worldwide. Such similarities are particularly well documented for members of the Oscillatoriaceae and Chroococcaceae, the two most abundant and widespread Precambrian cyanobacterial families. For species of two additional families, the Entophysalidaceae and Pleurocapsaceae, species-level morphologic similarities are supported by in-depth fossil-modern comparisons of environment, taphonomy, development, and behavior. Morphologically and probably physiologically as well, such cyanobacterial "living fossils" have exhibited an extraordinarily slow (hypobradytelic) rate of evolutionary change, evidently a result of the broad ecologic tolerance characteristic of many members of the group and a striking example of G. G. Simpson's [Simpson, G. G. (1944) *Tempo and Mode in Evolution* (Columbia Univ. Press, New York)] "rule of the survival of the relatively unspecialized." In both tempo and mode of evolution, much of the Precambrian history of life—that dominated by microscopic cyanobacteria and related prokaryotes—

appears to have differed markedly from the more recent Phanerozoic evolution of megascopic, horotelic, adaptationally specialized eukaryotes.

For helpful reviews of this manuscript, I thank W. Altermann, J. Bartley, J. Bragin, T. R. Fairchild, A. H. Knoll, D. M. Raup, B. N. Runnegar, E. Schultes, and J. Shen-Miller. This work was supported by National Aeronautics and Space Administration Grant NAGW-2147.

REFERENCES

Abeliovich, A. & Shilo, M. (1972) Photooxidative death in blue-green algae, *J. Bacteriol.* **111,** 682–689.

Badger, M. R. (1987) The CO_2-concentrating mechanism in aquatic phototrophs, in *The Biochemistry of Plants,* eds. Hatch, M. D. & Boardman, N. K. (Academic, New York), Vol. 10, pp. 219–274.

Brock, T. D. (1978) *Thermophilic Microorganisms and Life at High Temperatures* (Springer, New York).

Cameron, R. E. (1963) Morphology of representative blue-green algae, *Ann. N.Y. Acad. Sci.* **108** (2), 412–420.

Campbell, S. E. (1979) Soil stabilization by a prokaryotic desert crust: Implications for Precambrian land biota, *Origins Life* **9,** 335–348.

Castenholz, R. W. (1969) Thermophilic blue-green algae and the thermal environment, *Bacteriol. Rev.* **33,** 476–504.

Ciferri, O. (1983) *Spirulina,* the edible microorganism, *Microbiol. Rev.* **47,** 551–578.

Davis, J. S. (1972) Survival records in the algae, and the survival role of certain algal pigments, fat, and mucilaginous substances, *The Biologist* **54,** 52–93.

Davison, I. R. (1991) Environmental effects on algal photosynthesis: Temperature, *J. Phycol.* **27,** 2–8.

Desikachary, T. V. (1959) *Cyanophyta* (Indian Council Agricultural Res., New Delhi).

Drouet, F. (1968) *Revision of the Classification of the Oscillatoriaceae,* Academy of Natural Sciences, Philadelphia, Monograph 15 (Fulton, Lancaster, PA).

Drouet, F. & Daily, W. A. (1973) *Revision of the Coccoid Myxophyceae* (Hafner, New York).

Fairchild, T. R. (1975) *The Geologic Setting and Paleobiology of a Late Precambrian Stomatolitic Microflora from South Australia,* Ph.D. Dissertation (Univ. of California, Los Angeles).

Flowers, S. & Evans, F. R. (1966) The flora and fauna of the Great Salt Lake Region, Utah, in *Salinity and Aridity,* ed. Boyko, H. (Junk, The Hague, The Netherlands), pp. 367–393.

Fogg, G. E. (1973) Physiology and ecology of marine blue-green algae, in *The Biology of Blue-Green Algae,* eds. Carr, N. G. & Whitton, B. A. (Univ. of California Press, Berkeley, CA), pp. 268–378.

Fogg, G. E., Stewart, W. D. P., Fay, P. & Walsby, A. E. (1973) *The Blue-Green Algae* (Academic, New York).

Forest, H. S. & Weston, C. R. (1966) Blue-green algae from the Atacama Desert of northern Chile, *J. Phycol.* **2,** 163–164.

Frémy, P. (1972) *Cyanophycées des Côtes D'Europe* (Asher, Amsterdam).

Fuhs, G. W. (1968) Cytology of blue-green algae: Light microscopic aspects, in *Algae,*

Man, and the Environment, ed. Jackson, D. F. (Syracuse Univ. Press, Syracuse, NY), pp. 213–233.

Garcia-Pichel, F. & Castenholz, R. W. (1991) Characterization and biological implications of scytonemin, a cyanobacterial sheath pigment, *J. Phycol.* **27,** 395–409.

Giovannoni, S. J., Turner, S., Olsen, G. J., Barns, S., Lane, D. J. & Pace, N. R. (1988) Evolutionary relationships among cyanobacteria and green chloroplasts, *J. Bacteriol.* **170,** 3584–3592.

Godward, M. B. E. (1962) Invisible radiations, in *Physiology and Biochemistry of Algae*, ed. Lewin, R. A. (Academic, New York), pp. 551–566.

Golubic, S. (1976a) Organisms that build stromatolites, in *Stromatolites (Developments in Sedimentology 20)*, ed. Walter, M. R. (Elsevier, Amsterdam), pp. 113–126.

Golubic, S. (1976b) Taxonomy of extant stromatolite-building cyanophytes, in *Stromatolites (Developments in Sedimentology 20)*, ed. Walter, M. R. (Elsevier, Amsterdam), pp. 127–140.

Golubic, S. & Campbell, S. E. (1979) Analogous microbial forms in Recent subaerial habitats and in Precambrian cherts: *Gloeothece coerulea* Geitler and *Eosynechococcus moorei* Hofmann, *Precambrian Res.* **8,** 201–217.

Golubic, S. & Hofmann, H. J. (1976) Comparison of Holocene and mid-Precambrian Entophysalidaceae (Cyanophyta) in stromatolitic algal mats: Cell division and degradation, *J. Paleontol.* **50,** 1074–1082.

Grant, W. D. & Tindall, B. J. (1986) The alkaline saline environment, in *Microbes in Extreme Environments*, eds. Herbert, R. A. & Codd, G. A. (Academic, New York), pp. 25–54.

Green, J. W., Knoll, A. H., Golubic, S. & Swett, K. (1987) Paleobiology of distinctive benthic microfossils from the Upper Proterozoic Limestone-Dolomite "Series," central East Greenland, *Am. J. Bot.* **74,** 928–940.

Green, J. W., Knoll, A. H. & Swett, K. (1988) Microfossils from oolites and pisolites of the Upper Proterozoic Eleonore Bay Group, central East Greenland, *J. Paleontol.* **62,** 835–852.

Jensen, T. E. (1993) Cyanobacterial ultrastructure, in *Ultrastructure of Microalgae*, ed. Berner, T. (CRC, London), pp. 7–51.

Kasting, J. F. (1992) Proterozoic climates: The effect of changing atmospheric carbon dioxide concentrations, in *The Proterozoic Biosphere*, eds. Schopf, J. W. & Klein, C. (Princeton Univ. Press, Princeton, NJ), pp. 165–168.

Knoll, A. H. (1985) A paleobiological perspective on sabkhas, in *Ecological Studies, Vol. 53: Hypersaline Ecosystems*, eds. Friedman, G. M. & Krumbein, W. E. (Springer, New York), pp. 407–425.

Knoll, A. H., Barghoorn, E. S. & Golubic, S. (1975) *Paleopleurocapsa wopfnerii gen et sp. nov:* A Late precambrian alga and its modern counterpart, *Proc. Natl. Acad. Sci. USA* **72,** 2488–2492.

Knoll, A. H. & Bauld, J. (1989) The evolution of ecological tolerance in prokaryotes, *Trans. R. Soc. Edinburgh: Earth Sci.* **80,** 209–223.

Knoll, A. H. & Golubic, S. (1979) Anatomy and taphonomy of a Precambrian algal stromatolite, *Precambrian Res.* **10,** 115–151.

Knoll, A. H. & Golubic, S. (1992) Proterozoic and living cyanobacteria, in *Early Organic Evolution*, eds. Schidlowski, M., Golubic, S., Kimberley, M. M., McKirdy, D. M. & Trudinger, P. A. (Springer, New York), pp. 450–462.

Knoll, A. H., Golubic, S., Green, J. & Swett, K. (1986) Organically preserved microbial endoliths from the late Proterozoic of East Greenland, *Nature (London)* **321,** 856–857.

Langworthy, T. A. (1978) Microbial life in extreme pH values, in *Microbial Life in Extreme Environments*, ed. Kushner, D. J. (Academic, New York), pp. 279–315.

Lloyd, N. D. H., Cavin, D. T. & Culver, D. A. (1977) Photosynthesis and photorespiration in algae, *Plant Physiol.* **59,** 936–940.

Mendelson, C. V. & Schopf, J. W. (1992a) Proterozoic and selected Early Cambrian microfossils and microfossil-like objects, in *The Proterozoic Biosphere*, eds. Schopf, J. W. & Klein, C. (Cambridge Univ. Press, New York), pp. 865–951.

Mendelson, C. V. & Schopf, J. W. (1992b) Proterozoic and Early Cambrian acritarchs, in *The Proterozoic Biosphere*, eds. Schopf, J. W. & Klein, C. (Princeton Univ. Press, Princeton, NJ), pp. 219–232.

Mizutani, H. & Wada, E. (1982) Effect of high atmospheric CO_2 concentration on $\delta^{13}C$ of algae, *Origins Life* **12,** 377–390.

Olson, J. M. & Pierson, B. K. (1987) Evolution of reaction centers in photosynthetic prokaryotes, *Int. Rev. Cytol.* **108,** 209–248.

Pardue, J. W., Scalan, R. S., Van Baalen, C. & Parker, P. L. (1976) Maximum carbon isotope fractionation in photosynthesis by blue-green algae and a green alga, *Geochim. Cosmochim. Acta* **40,** 309–312.

Parker, B. C., Simmons, G. M., Jr., Love, G., Wharton, R. A., Jr., & Seaburg, K. G. (1981) Modern stromatolites in antarctic dry valley lakes, *BioScience* **31,** 656–661.

Pierson, B. K., Bauld, J., Castenholz, R. W., D'Amelio, E., Des Marais, D. J., Farmer, J. D., Grotzinger, J. P., Jørgensen, B. B., Nelson, D. C., Palmisano, A. C., Schopf, J. W., Summons, R. E., Walter, M. R. & Ward, D. M. (1992) Modern mat-building microbial communities: A key to the interpretation of Proterozoic stromatolitic communities, in *The Proterozoic Biosphere*, eds. Schopf, J. W. & Klein, C. (Cambridge Univ. Press, New York), pp. 245–342.

Ruedemann, R. (1918) The paleontology of arrested evolution, *N.Y. State Mus. Bull.* **196,** 107–134.

Ruedemann, R. (1922a) Additional studies of arrested evolution, *Proc. Natl. Acad. Sci. USA* **8,** 54–55.

Ruedemann, R. (1922b) Further notes on the paleontology of arrested evolution, *Am. Nat.* **56,** 256–272.

Schidlowski, M., Hayes, J. M. & Kaplan, I. R. (1983) Isotopic inferences of ancient biochemistries: Carbon, sulfur, hydrogen, and nitrogen, in *Earth's Earliest Biosphere*, ed. Schopf, J. W. (Princeton Univ. Press, Princeton, NJ), pp. 149–186.

Schopf, J. W. (1968) Microflora of the Bitter Springs Formation, Late Precambrian, central Australia, *J. Paleontol.* **42,** 651–688.

Schopf, J. W. (1974) The development and diversificaiton of Precambrian life, *Origins Life* **5,** 119–135.

Schopf, J. W. (1977) Biostratigraphic usefulness of stromatolitic Precambrian microbiotas: A preliminary analysis, *Precambrian Res.* **5,** 143–173.

Schopf, J. W. (1978) The evolution of the earliest cells, *Sci. Am.* **239,** 110–134.

Schopf, J. W. (1987) "Hypobradytely": Comparison of rates of Precambrian and Phanerozoic evolution, *J. Vertebr. Paleontol.* **7,** Suppl. 3, 25 (abstr.).

Schopf, J. W. (1992a) A synoptic comparison of Phanerozoic and Proterozoic evolution, in *The Proterozoic Biosphere*, eds. Schopf, J. W. & Klein, C. (Cambridge Univ. Press, New York), pp. 599–600.

Schopf, J. W. (1992b) Historical development of Proterozoic micropaleontology, in *The Proterozoic Biosphere*, eds. Schopf, J. W. & Klein, C. (Cambridge Univ. Press, New York), pp. 179–183.

Schopf, J. W. (1992c) Informal revised classification of Proterozoic microfossils, in *The*

Proterozoic Biosphere, eds. Schopf, J. W. & Klein, C. (Cambridge Univ. Press, New York), pp. 1119–1166.

Schopf, J. W. (1992d) Proterozoic prokaryotes: Affinities, geologic distribution, and evolutionary trends, in *The Proterozoic Biosphere*, eds. Schopf, J. W. & Klein, C. (Cambridge Univ. Press, New York), pp. 195–218.

Schopf, J. W. (1992e) Tempo and mode of Proterozoic evolution, in *The Proterozoic Biosphere*, eds. Schopf, J. W. & Klein, C. (Cambridge Univ. Press, New York), pp. 595–598.

Schopf, J. W. (1992f) Patterns of Proterozoic microfossil diversity: An initial, tentative, analysis, in *The Proterozoic Biosphere*, eds. Schopf, J. W. & Klein, C. (Princeton Univ. Press, Princeton, NJ), pp. 529–552.

Schopf, J. W. (1993) Microfossils of the Early Archean Apex chert: New evidence of the antiquity of life, *Science* **260,** 640–646.

Schopf, J. W. & Blacic, J. M. (1971) New microorganisms from the Bitter Springs Formation (Late Precambrian) of the north-central Amadeus Basin, Australia, *J. Paleontol.* **45,** 925–960.

Schopf, J. W., Hayes, J. M. & Walter, M. R. (1983) Evolution of Earth's earliest ecosystems: Recent progress and unsolved problems, in *Earth's Earliest Biosphere*, ed. Schopf, J. W. (Princeton Univ. Press, Princeton, NJ), pp. 361–384.

Schopf, J. W. & Packer, B. M. (1987) Early Archean (3.3-billion to 3.5-billion-year-old) microfossils from Warrawoona Group, Australia, *Science* **237,** 70–73.

Shields, L. M. & Drouet, F. (1962) Distribution of terrestrial algae within the Nevada Test Site, *Am. J. Bot.* **49,** 547–554.

Simpson, G. G. (1944) *Tempo and Mode in Evolution* (Columbia Univ. Press, New York).

Stanley, S. M. (1984) Does bradytely exist?, in *Living Fossils*, eds. Eldrige, N. & Stanley, S. M. (Springer, New York), pp. 278–281.

Stewart, W. D. P. (1980) Some aspects of structure and function in N_2-fixing cyanobacteria, *Annu. Rev. Microbiol.* **34,** 497–536.

Vallentyne, J. R. (1963) Environmental biophysics and microbial ubiquity, *Ann. N.Y. Acad. Sci.* **108** (2), 342–352.

Vidal, G. & Knoll, A. H. (1982) Radiations and extinctions of plankton in the Late Precambrian and Early Cambrian, *Nature (London)* **297,** 57–60.

Vincent, W. F., Castenholz, R. W., Downes, M. T. & Howard-Williams, C. (1993) Antarctic cyanobacteria: Light, nutrients, and photosynthesis in the microbial mat environment, *J. Phycol.* **29,** 745–755.

4

Proterozoic and Early Cambrian Protists: Evidence for Accelerating Evolutionary Tempo

ANDREW H. KNOLL

In the 50 years since G. G. Simpson published *Tempo and Mode in Evolution*, paleontological documentation of evolutionary history has improved substantially. Not only has the quality of stratigraphic and systematic data increased for animal, plant, and protistan taxa found in Phanerozoic* rocks; recent decades have witnessed a tremendous increase in the documented length of the fossil record. Speculation about a long pre-Cambrian history of life has been replaced by a palpable record of evolution that begins some 3000 Ma before the Cambrian explosion. In this paper, I examine the early fossil record of eukaryotic organisms, asking whether or not this longer record is amenable to the types of investigation used to estimate tempo in Phanerozoic evolution. Even though analysis is limited by incomplete sampling, patchy radiometric calibration, and taxonomic uncertainty, a robust pattern of increasing diversity and accelerating evolutionary tempo is evident.

Andrew H. Knoll is professor and chairman of the Department of Organismic and Evolutionary Biology at Harvard University, Cambridge, Massachusetts.

*The Phanerozoic Eon is one of the three major divisions of the geological time scale. Literally the age of visible animal life, the Phanerozoic Eon encompasses the past 545 million years (Ma), beginning at the start of the Cambrian Period. Earlier Earth history is divided between the Proterozoic (2500–545 Ma) and Archean (> 2500 Ma) eons.

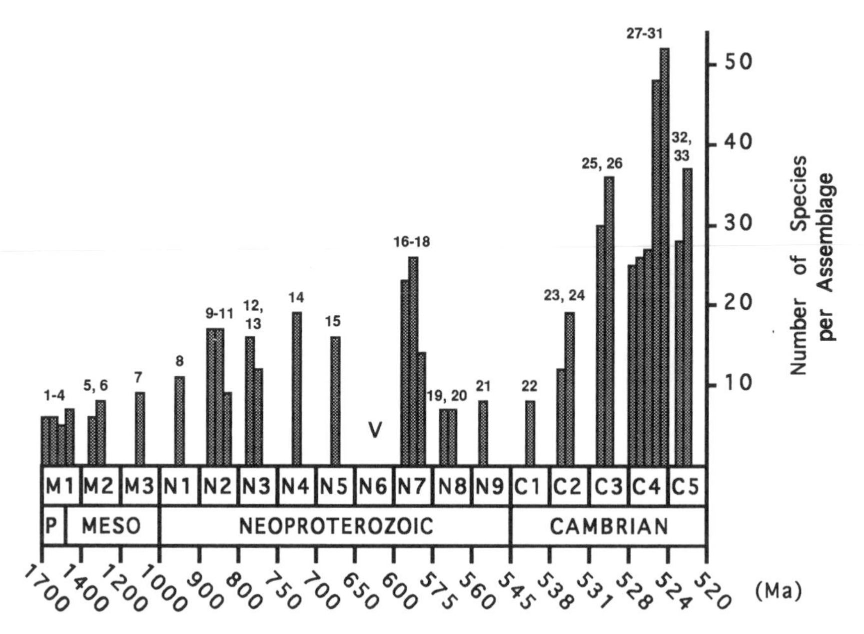

FIGURE 1 Species richness of selected protistan microfossil assemblages in 17 stratigraphic intervals running from the latest Paleoproterozoic era (P), through the Mesoproterozoic (Meso), Neoproterozoic, and Early Cambrian. Numbers 1–33 identifying assemblages refer to Table 1. V marks the Varanger ice age. Species richnesses are based on my taxonomic evaluation and do not necessarily reflect published tabulations. With a single exception (assemblage 31), all assemblages have been examined first-hand, resulting in uniform systematic treatment. Stratigraphic and systematic data for all figures and tables are available from the author.

The Nature and Limitations of the Record

Stratigraphic and Geochronometric Framework. The time interval considered here is 1700–520 Ma; that is, the latest Paleoproterozoic Eon to the end of the Early Cambrian Period (Figure 1 and Table 1). U-Pb dates on accessory minerals in volcanic rocks of known relationship to fossiliferous strata are limited for this interval—but then, such data are also limited for younger Paleozoic fossils on which much greater paleobiological demands are placed. Quantitative analysis of the Paleozoic fossil record is possible because a well-defined chronostratigraphic, or relative, time-scale has been calibrated by radiometric data in a few key sections.

The Proterozoic–Cambrian time scale is developing along the same

TABLE 1 Stratigraphic intervals used in analyses of tempo and representative acritarch assemblages

Interval (age in millions of years [Ma])			
Formation		Location	Reference
Late Paleoproterozoic- and Mesoproterozoic			
M1	(1700–1400 Ma)		
Satka [1]		Urals, Russia	Jank. 82
Bakal [2]		Urals, Russia	Jank. 82
Ust'-Il'ya [3]		Siberia	Veis. 92
McMinn [4]		Australia	Peat. 78
M2	(1400–1200 Ma)		
Omachtin [5]		Siberia	Vesi. 89
Zigazino-Kamarovsk [6]		Urals, Russia	Jank. 82
M3	(1200–1000 Ma)		
Baicaoping [7]		China	Yan. 92
Neoproterozoic			
N1	(1000–900 Ma)		
Lakhanda [8]		Siberia	Germ. 90
N2	(900–800 Ma)		
Miroyedikha [9]		Siberia	Germ. 90
Kwagunt [10]		Arizona, USA	Vida. 85
Dakkovarre [11]		Norway	Vida. 81
N3	(800–750 Ma)		
Andersby [12]		Norway	Vida. 81
Middle Visingsö [13]		Sweden	Vida. 76
N4	(750–700 Ma)		
Svanbergfjellet [14]		Svalbard	Butt. 94
N5	(700–650 Ma)		
Upper Visingsö [15]		Sweden	Vida. 76
N6	(650–600 Ma)*		
N7	(600–575 Ma; Volhyn)		
Pertatataka [16]		Australia	Zang. 92
Doushantuo [17]		China	Yin. 87
Kursovsky [18]		Siberia	Mocz. 93
N8	(575–560 Ma; Redkino)		
Redkino [19]		Baltic	Volk. 90
Mogilev/Nagoryany [20]		Ukraine	Asee. 83
N9	(560–545 Ma; Kotlin)		
Kotlin [21]		Baltic	Volk. 90
Early Cambrian			
C1	(545–538 Ma; Rovno)		
Rovno [22]		Baltic	Volk. 83
C2	(538–531 Ma; Lontova)		
Lontova [23]		Baltic	Volk. 83
Mazowsze [24]		Poland	Mocz. 91
C3	(531–528 Ma; Talsy)		
Talsy [25]		Baltic	Volk. 83
Lower Radzyń/Kaplanosy [26]		Poland	Mocz. 91

TABLE 1 (Continued)

Interval (age in Ma)			
Formation		Location	Reference
Early Cambrian (continued)			
C4	(528–524 Ma; Vergale)		
Middle Radzyń/Kaplanosy [27]		Poland	Mocz. 91
Qianzhisi [28]		China	Zang. 90
Tokammane [29]		Svalbard	Knol. 87
Vergale [30]		Baltic	Volk. 83
Buen [31]		Greenland	Vida. 93
C5	(524–520 Ma; Rausve)		
Upper Radzyń/Kaplanosy [32]		Poland	Mocz. 91
Rausve [33]		Estonia	Volk. 83

Assembly numbers in brackets refer to Figure 1.
*Interval includes Varanger ice age.

path (Harland *et al.*, 1990; Semikhatov, 1991; Knoll and Walter, 1992; Compston *et al.*, 1992; Bowring *et al.*, 1993). A biostratigraphic framework based on stromatolites, microfossils, and (in younger rocks) both the body and trace fossils of animals can be used to divide this nearly 1200-Ma expanse into recognizable intervals of various lengths. Complementing this is an increasingly well-supported chemostratigraphic framework based on the distinctive pattern of secular variation in the isotopic compositions of C and Sr in carbonate rocks (Kaufman and Knoll, in press). These data define the chronostratigraphic scale now being calibrated. Within the period under consideration, younger intervals are shorter than older ones, both because strong Neoproterozoic isotopic variation has no parallel in the Mesoproterozoic record and, more importantly, because of the finer biostratigraphic resolution in younger successions.

For the purposes of this analysis, I have divided the period from 1700 to 520 Ma into 17 intervals as shown in Table 1 and Figures 1–3. Table 1 and Figure 1 also show my placement of representative microfossil assemblages into these intervals. Others might estimate the ages of interval boundaries differently, and one or two assemblages might be moved to bins adjacent to those chosen here. However, no assemblage placement or estimate of interval duration is so egregiously uncertain as to affect the analysis in a substantial way. That is, relative to the strength and time scale of the pattern observed, uncertainties of time are acceptably small.

The Paleontological Data Base: Taxonomy. For the estimation of evolutionary tempo, I will restrict consideration to the organic-walled micro-

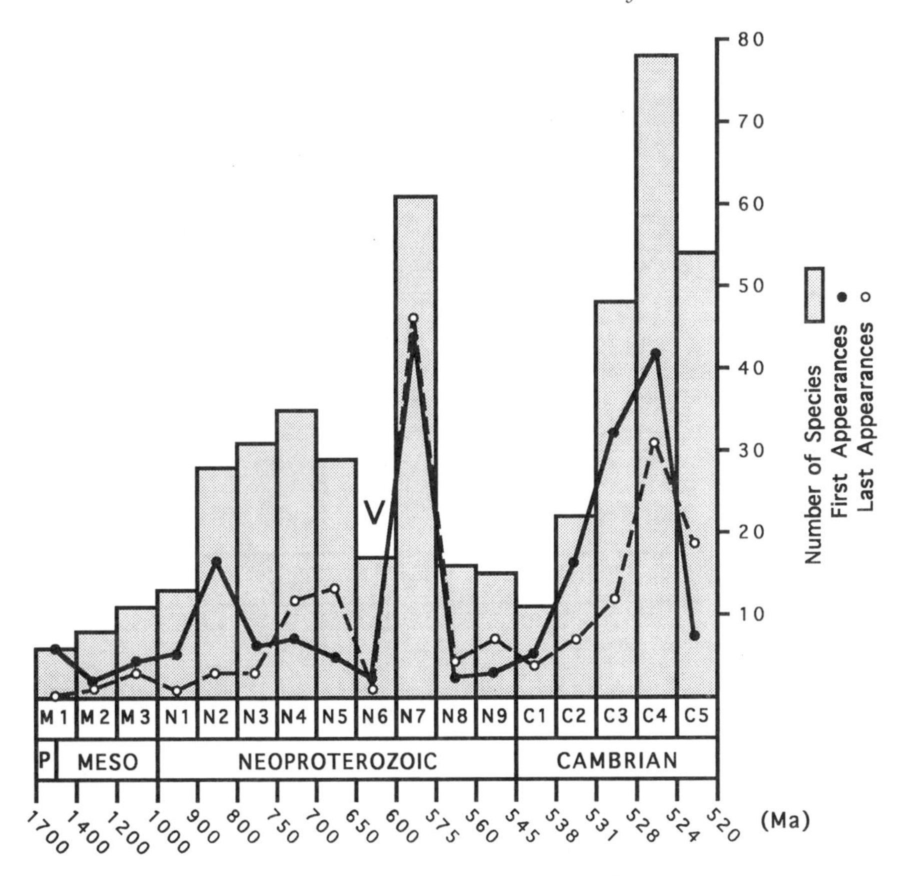

FIGURE 2 Total taxonomic richness (histogram), first appearances (solid dots), and last appearances (open dots) of protistan microfossil species for the 17 intervals recognized in this study (Tables 1 and 2). V marks the Varanger ice age. Abbreviations are as in Figure 1.

fossils known as acritarchs (Figure 4). Structural features leave little doubt that all or nearly all were eukaryotic. Most were the vegetative and reproductive walls of unicellular protists, although the reproductive cysts of multicellular algae and even egg cases of early animals may be included.

The total number of clades that contributed to the observed record is unknown, but probably small. Some of the Early Cambrian microfossils included here are clearly the phycomata of green algal flagellates (Tappan, 1980). (The phycoma is a nonmotile vegetative stage of the flagellates' life cycle characterized by a wall that contains the degrada-

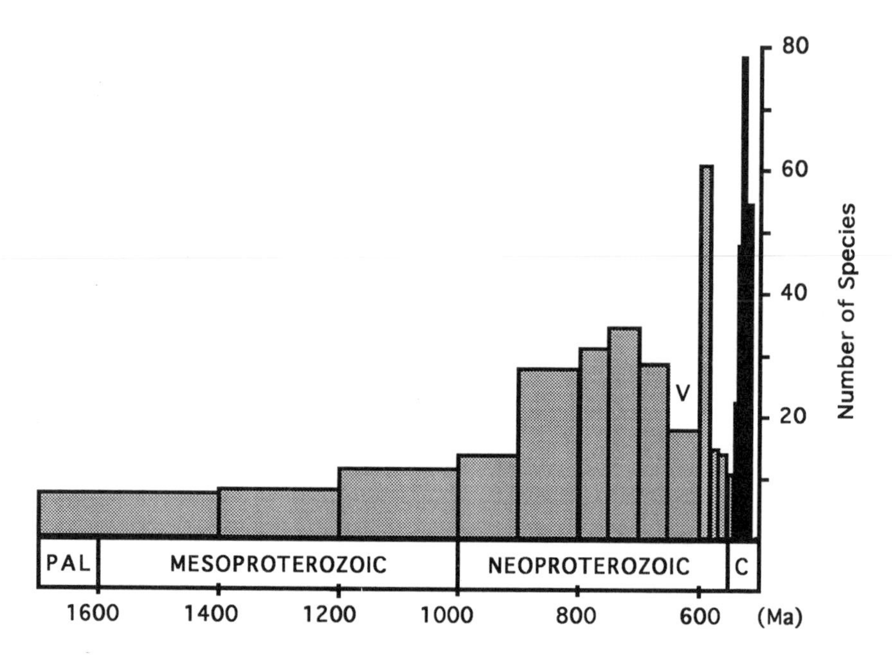

FIGURE 3 Histogram showing total species richness of protistan microfossils for the 17 intervals recognized in this study (Tables 1 and 2). The time scale along the abscissa is linear, underscoring the long initial interval of low diversity and the rapidity of later radiations. V marks the Varanger ice age. PAL, Paleoproterozoic; C, Cambrian.

tion-resistant polymer, sporopollenin.) Others, including most Neoproterozoic taxa, may also represent green algae (Tappan, 1980; Vidal and Knoll, 1983; Knoll *et al.*, 1991), but phylogenetic relationships have not been established unequivocally.

The pre-Ediacaran record of seaweeds is too patchy for meaningful evaluation of evolutionary tempo, but these fossils do provide a paleobiological context for the interpretation of microfossil assemblages. In particular, fossils of multicellular algae relate the latest Mesoproterozoic and early Neoproterozoic diversification of acritarchs to the biological differentiation of "higher" protists inferred from molecular phylogenies (Sogin *et al.*, 1989; Knoll, 1992a).

If the view of eukaryotic evolution provided by molecular phylogenies is reliable, many of the protistan phyla that differentiated during the Proterozoic are not represented in the fossil record. Therefore, care must be exercised in ascribing generality to the preserved record. The problem is well understood by invertebrate paleontologists who enjoy excellent preservation for only a few of the more than 30 phyla of invertebrate animals. Estimates of evolutionary tempo and the timing of diversification and extinction events are broadly similar across phyla for Phanerozoic invertebrates and protists with good fossil records. Therefore, the limited clade diversity of Proterozoic and Early Cambrian acritarchs may reflect a broader pattern of early eukaryotic evolution.

A second type of biological uncertainty concerns the interpretation of paleospecies. In studies of acritarchs, analyses are necessarily done at the species level, because biologically meaningful higher taxa have not been defined. The uncertain phylogenetic relationships of most forms exacerbate the common paleontological problem of relating paleospecies to biological species. While the paleontological use of the term species is convenient and accurate in the sense of "most inclusive diagnosable units," what we can really measure is diversity of morphology. Estimates of tempo are, therefore, to be viewed as rates of morphological diversification and turnover within a preservable subset of early eukaryotes.

The Paleontological Data Base: Sampling Quality. Several hundred Proterozoic and Lower Cambrian formations are known to contain protistan microfossils (Schopf and Klein, 1992); however, many assemblages are indifferently preserved, poorly described, and/or accompanied by inadequate stratigraphic and paleoenvironmental information. Assessment of sample quality is better based on those assemblages that are well preserved, meticulously monographed, and well buttressed by stratigraphic and sedimentological data (Figure 1).

Older assemblages are both less numerous and less diverse than those of younger intervals. One might, therefore, suppose that low observed diversity is a product of poor sampling or poor preservation; however, sampling adequacy is not simply a function of assemblages per interval. It is also dependent on quality of preservation, facies and/or paleogeographic heterogeneity, and rates of taxonomic turnover. If fossils are well preserved, cosmopolitan, and slowly evolving, a limited number of samples may be sufficient to characterize the paleobiology of an interval.

This appears to be the case for the Mesoproterozoic acritarch record. The quality of fossil preservation in Mesoproterozoic mudstones (e.g., Peat *et al.*, 1978) and silicified carbonates (e.g., Sergeev *et al.*, in press) matches the best seen in Neoproterozoic rocks, but the acanthomorphic (process or spine bearing) and other ornamented acritarchs seen in younger rocks of comparable environmental setting are not seen in these or any other rocks older than ca. 1100 Ma. In contrast, even metamorphosed Neoproterozoic rocks may contain ornamented acritarchs

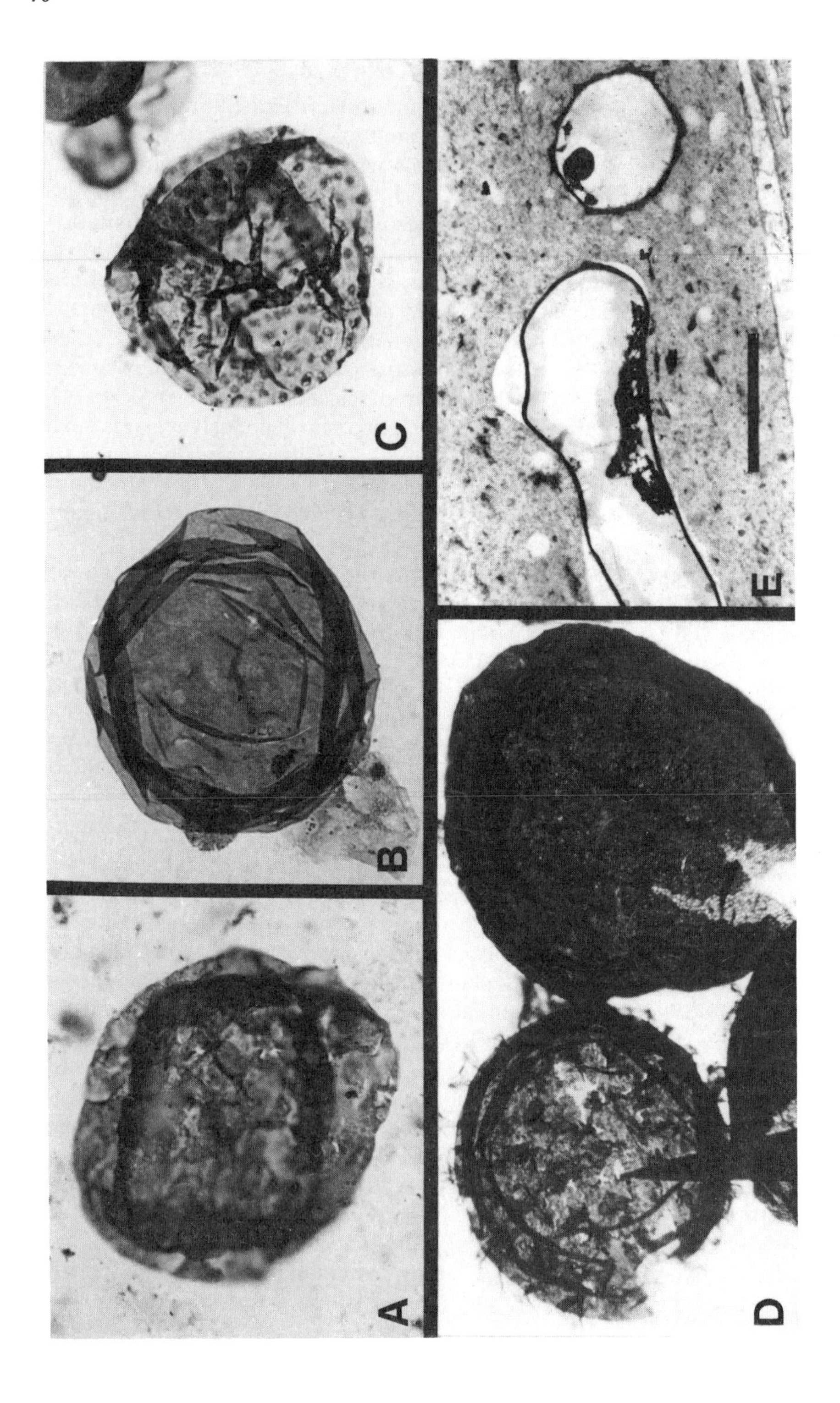
A
B
C
D
E

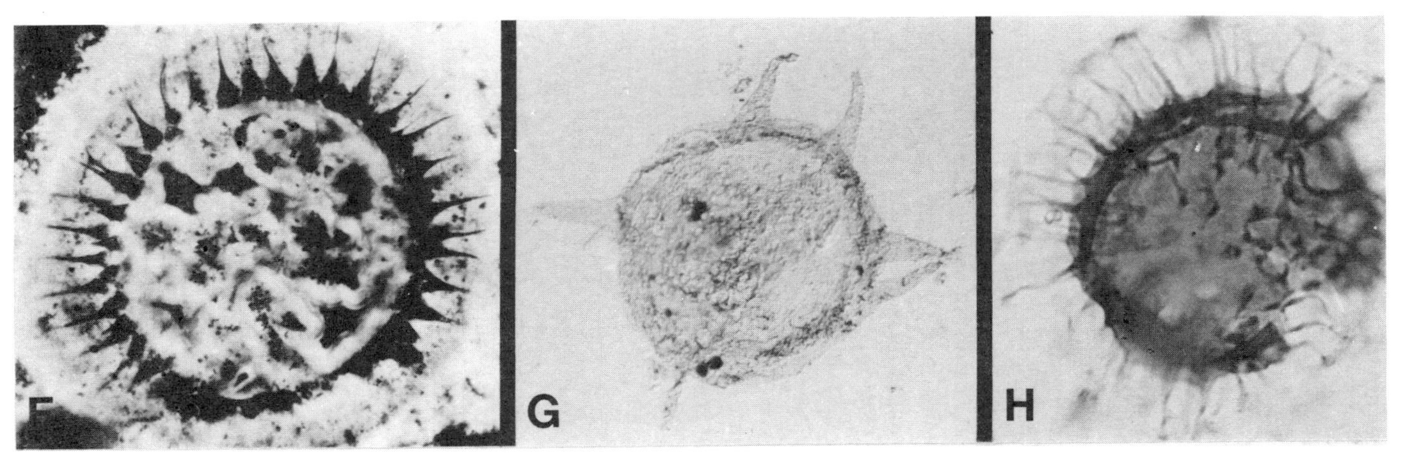

FIGURE 4 Representative Proterozoic and Early Cambrian acritarchs. *A–C* and the specimen on the left in *D* are spheromorphic; the specimen on the right in *D* and *E–H* are acanthomorphic. (*A*) *Leiosphaeridia* sp., Mesoproterozoic [M1] McMinn Formation, Australia. (*B*) *Leiosphaeridia crassa*, Neoproterozoic [N2] Miroyedikha Formation, Siberia. (*C*) *Kildinosphaera verrucata*, Miroyedikha Formation. (*D*) *Shuiyousphaeridium macroreticulata* (*Left*) and *Dictyosphaera incrassata* (*Right*), late Mesoproterozoic [M3] Baicaoping Formation, China. (*E*) *Trachyhystrichosphaera aimika*, Neoproterozoic [N4] Draken Formation, Svalbard. (*F*) *Tanarium densum*, Neoproterozoic [N7] Doushantuo Formation, China. (*G*) *Tanarium conoideum*, Neoproterozoic [N7] Kursovsky Formation, Siberia. (*H*) *Skiagia ciliosa*, Early Cambrian [C4] Tokammane Formation, Svalbard. (Bar in *E*: *A–C* and *H*, 25 μm; *D* and *G*, 60 μm; *E*, 300 μm; *G*, 100 μm.)

(Knoll, 1992b). Observations such as these suggest that differential preservation is not a principal determinant of observed diversity and turnover patterns.

In general, Proterozoic and Cambrian acritarch species have wide (and known) paleoenvironmental distributions and show little evidence of provincialism. This minimizes the likelihood that observed patterns are influenced strongly by differential sampling of facies among time intervals.

Perhaps the best indicators of sample quality are the degree of similarity among assemblages within an interval and the incremental taxonomic richness accompanying each new assemblage reported. The total number of assemblages known for the interval from 1700 to 1000 Ma is low, but the taxonomic similarity among samples is high. Insofar as knowledge of the age and environmental setting of an assemblage permits the prediction of taxonomic composition, the Proterozoic and Early Cambrian acritarch record appears to be sufficiently well sampled to permit the broad estimation of evolutionary tempo.

Despite my confidence that the existing record is governed more by evolution than by sampling, this paper should be read as a dispatch from the trenches and not as a definitive history. The events most likely to have escaped notice to date are short bursts of diversification and extinction of the type recorded in interval N7. The time intervals most likely to yield new assemblages that will modify the conclusions drawn here are those between 750 and 600 Ma ago, just prior to and including the Varanger ice age.

Early Eukaryotic Fossils: A Narrative Record

Acritarchs occur in rocks as old as 1900–1700 Ma (Zhang, 1986). The fossils are morphologically simple, but sedimentological distributions, size frequency distributions, and inferred excystment structures ally these remains to unequivocally eukaryotic microfossils that extend upward from this interval to the present. Independent evidence for the occurrence of late Paleoproterozoic to early Mesoproterozoic eukaryotes comes from significant sterane concentrations in bitumens (Summons and Walter, 1992) and problematic macrofossils (Walter *et al.*, 1990; Han and Runnegar, 1992). There is no reason to assume that these remains document the evolutionary first appearance of the Eucarya nor even any necessary reason to insist that they mark the emergence of clades capable of synthesizing preservable walls or cysts. What can be concluded is that eukaryotic organisms were significant parts of marine ecosystems in the late Paleoproterozoic Era and that the fossil record of earlier periods is poor.

Acritarchs are widespread and abundant in late Paleoproterozoic and Mesoproterozoic rocks, and in all known assemblages morphological diversity is limited to a few taxa of smooth-walled spheromorphs (leiosphaerids; Figure 4*A*) differentiated on the basis of size, spheroids bearing a single filament-like emergence, and/or somewhat lumpy or irregular vesicles (Keller and Jankauskas, 1982; Jankauskas, 1989).

Morphological diversification began in the late Mesoproterozoic Eon with the appearance of striated spheromorphic vesicles (*Valeria lophostriata*) and the first acanthomorphic acritarchs (Jankauskas, 1989; Knoll, in press). Chinese rocks poorly dated at ca. 1100 Ma contain the oldest known large (>100 μm) acanthomorphs (Yan and Zhu, 1992; Figure 4*D*)—a type of microfossil characteristically found in Neoproterozoic assemblages (Semikhatov, 1991; Knoll, in press). The 900- to 1000-Ma-old Lakhanda biota of Siberia (German, 1990) contains a moderate diversity of both acanthomorphs and lightly ornamented spheromorphs (Figure 4C). Latest Mesoproterozoic to early Neoproterozoic acritarch diversification is complemented by the first appearance of multicellular eukaryotes that can be placed in extant phyla. Red algae that display clear cellular differentiation are locally abundant in silicified peritidal carbonates of the Hunting Formation, arctic Canada (Butterfield *et al.*, 1990); probable chromophytic algae are beautifully preserved in Lakhanda mudstones (German, 1990); and several green algal taxa occur in the younger Svanbergfjellet Formation of Svalbard (Butterfield *et al.*, in press).

Acritarchs increase in both total and assemblage diversity in 900- to 800-Ma-old rocks (Figures 1–3), due largely to the differentiation of ornamented forms. Vase-shaped protistan tests also proliferate at this time. Most taxa that originated during the early-to-middle Neoproterozoic Era disappeared by the time of the great Varanger ice age (?650–590 Ma ago), but post-Varanger successions on three continents contain diverse assemblages of new and highly ornamented acritarchs (Yin, 1987; Knoll, 1992a; Zang and Walter, 1992; Moczydlowska *et al.*, 1993; Figure 4*F* and *G*). This postglacial diversification is all the more remarkable because it is so short-lived. Even exceptionally preserved latest Proterozoic acritarch assemblages are limited to a handful of leiosphaerids and small acanthomorphs. The extinction correlates stratigraphically with the appearance of diverse Ediacaran-type metazoans; where diverse acanthomorphs and Ediacaran remains occur in the same succession, the former lie stratigraphically beneath the latter. However, correlation to the independent chronostratigraphic record of C and Sr isotopic variation suggests that the two types of assemblage coexisted for a limited interval (Kaufman and Knoll, in press). Acritarchs again diversified rapidly during the Early Cambrian (Volkova *et al.*, 1983; Moczydlowska, 1991; Figure 4*H*).

Estimates of Evolutionary Tempo

Figures 1 and 2 depict assemblage and total diversity through the interval from 1700 to 520 Ma ago. (Figure 3 plots total diversity data on a linear time scale to show more clearly the length of the initial interval of low diversity and the rapidity of subsequent changes.) The similarity of the figures is not surprising, given the wide facies and geographic distributions of many taxa. Species richness began low and rose only slightly during the first 700–800 Ma of the acritarch record—an interval longer than the entire Phanerozoic Eon. A burst of first appearances 900–800 Ma nearly doubled both assemblage and total diversity, bringing them to a new level that would persist with limited change until the Varanger ice age. The figures show diversity peaking 750–700 Ma ago and then declining to a minimum during the Varanger interval. However, intervals N5 and N6 are the most poorly sampled of the entire period under consideration. Taxa whose currently known last appearance is in N4 or N5 may well be discovered in closer proximity to Varanger strata, while some of the many acritarch taxa whose first known appearance is in N7 may be found in earlier intervals. For example, the large acritarchs *Papillomembrana compta* and *Ericiasphaera spjeldnaessi*, both conspicuous components of N7 assemblages, occur in clasts of the Biskopås Comglomerate, Norway, that underlie Varanger tillites (Vidal, 1990). On the other hand, few pre-Varanger taxa occur in the beautifully preserved assemblages that characterize N7, and in places like northwestern Canada, assemblages deposited just before the ice age contain only characteristically pre-Varanger taxa (Allison and Awramik, 1989). Thus, the marked change in assemblage composition across the Varanger interval is probably a stable feature of the record, and the extinctions inferred from the figures may have been concentrated in a brief interval before or during the Varanger ice age.

The high diversity of immediately post-Varanger acritarch assemblages is apparent from the figures. A burst of first appearances lifted both assemblage and total species richness to their Proterozoic maxima, and an ensuing maximum in last appearances subsequently reduced diversity to levels resembling those of the Mesoproterozoic and earliest Neoproterozoic. Seventy-five percent of recorded species disappeared, including most if not all large morphologically distinctive forms.

The first four intervals of the Early Cambrian exhibit sharp increases in numbers of first appearances; species richness within assemblages eclipsed its Proterozoic maximum in C3 (ca. 531–528 Ma), and total diversity peaked one interval later (C4, ca. 528–524 Ma ago). Last

TABLE 2 Species richness, rates of origination, and rates of extinction for Proterozoic and early Cambrian acritarchs

Interval (Ma ago)	N	FA	FA/Ma	FA/sp/Ma	LA	LA/Ma	LA/sp/Ma
M1 (1700–1400)	6	6	0.02[a]	0.003[a]	0	0.00	0.000
M2 (1400–1200)	8	2	0.01	0.001	2	0.01	0.001
M3 (1200–1000)	11	5	0.03	0.004	3	0.02	0.002
N1 (1000–900)	13	5	0.05	0.005	1	0.01	0.001
N2 (900–800)	28	16	0.16	0.009	3	0.03	0.002
N3 (800–750)	31	6	0.12	0.004	3	0.06	0.002
N4 (750–700)	35	7	0.14	0.004	12	0.24	0.008
N5 (700–650)	28	5	0.10	0.004	13	0.26	0.010
N6 (650–600)	17	2	0.04	0.002	1	0.02	0.001
N7 (600–575)	60	44	1.76	0.060	46	1.84	0.060
N8 (575–560)	16	2	0.13	0.008	4	0.26	0.017
N9 (560–545)	15	3	0.20	0.015	6	0.40	0.030
C1 (545–538)	11	5	0.70	0.070	4	0.60	0.060
C2 (538–531)	22	16	2.30	0.185	7	1.00	0.080
C3 (531–528)	48	32	10.70	0.400	12	4.00	0.150
C4 (528–524)	78	42	10.50	0.198	31	7.75	0.150
C5 (524–520)	54	7	1.75	0.035	18	4.50	0.089

N, total species richness; FA, first appearance; LA, last appearance; FA (LA)/Ma, first (last) appearance per Ma; FA (LA)/Sp/Ma, first (last) appearance per species per Ma. For the calculation of per species rates of origination and extinction, standing diversity was taken to be the geometric mean of diversity at the beginning and end of each interval, making the simplifying assumption that all extinctions took place at the ends of intervals.
[a]All species present in M1 are counted as first appearances, but some may have originated earlier. Thus, calculated rates of first appearance for M1 may be too high.

appearances also increase throughout these intervals and exceed first appearances at the end of the Early Cambrian.

Table 2 shows calculated rates of cladogenetic evolution for each of the intervals under consideration. For intervals M1 though N1, both total and per taxon rates of first and last appearances are low, indicating not only that diversity was low but also that constituent species were long lasting. (The calculated rates of first appearance for M1 may be misleading, in that all species are recorded as first appearances. The presence of simple acritarchs in rocks that may be older than 1700 Ma indicates that at least some of these forms may have originated earlier.) By 900–800 Ma ago (N2), total rates of origination had increased by an order of magnitude to a level at which they remained for the duration of the pre-Varanger Neoproterozoic. Interestingly, after an increase during interval N2, *per taxon* rates of origination returned to levels comparable to earlier intervals; both total and per taxon extinction rates increased toward the Varanger ice age.

Another order of magnitude increase in origination and extinction

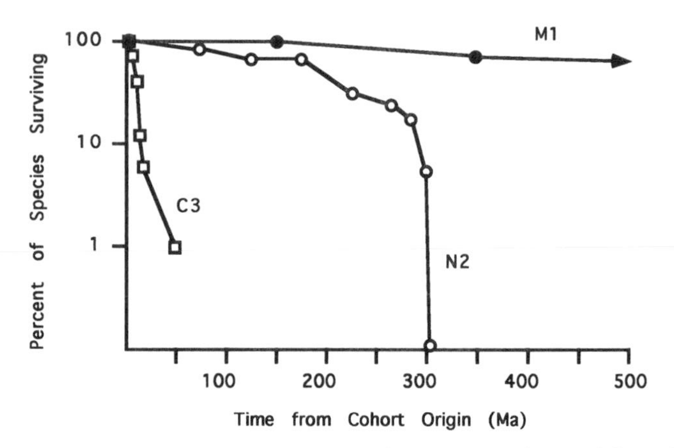

FIGURE 5 Cohort survivorship curves for species that originated during intervals M1 (1700–1400 Ma), N2 (900–800 Ma), and C3 (531–528 Ma). The abscissa denotes time since cohort origin.

rates attended the brief post-Varanger diversification event, after which terminal Proterozoic turnover returned to average Neoproterozoic levels. During the peak of the Cambrian acritarch radiation, origin and extinction rates both increased to levels an additional order of magnitude above the Neoproterozoic mean.

Cohort survivorship provides another means of evaluating evolutionary tempo (Van Valen, 1973; Raup, 1978, 1985). A comparison of the three cohort survivorship curves in Figure 5 shows that species originating in early Cambrian interval C3 turned over much more rapidly than those of Neoproterozoic cohort N2, which in turn decayed more quickly than Paleo- and Mesoproterozoic cohort M1. Very rough estimates of mean species duration and half-life (median species duration) confirm the order of magnitude increases in tempo between the Meso- and Neoproterozoic and again between the Neoproterozoic and Early Cambrian (Table 3). Thus, relative to earlier intervals, Cambrian acritarch assemblages contain more and more species that survive for shorter and shorter periods of time. Mean species duration and half-life for Cambrian acritarchs is similar to values computed for both younger protistan groups with good fossil records and Phanerozoic animal taxa (Table 3).

Discussion

How seriously should we take these figures? The general problems of sampling, data set size, and age estimation have already been noted. Imprecisions could easily alter estimates of tempo by a factor of two.

TABLE 3 Estimates of species durations

Taxonomic group	Half-life, Ma	Mean duration, Ma
Acritarch cohort M1	1390[a]	1960[a]
Acritarch cohort N2	75	102
Acritarch cohort C3	5.4	7.7
Planktonic foraminifera	5	7
Dinoflagellates	9	13
Diatoms	5.5	7.9
All invertebrates	7.7	11.1

Half-life (median species duration) and mean duration were calculated according to Raup (1978, 1985). Data for Phanerozoic protists and invertebrates are from Van Valen (1973) and Raup (1985).

[a]Of course, species that originated <1700 Ma ago cannot have true durations that exceed this age. Durations presented were calculated from the mean extinction rate (the slope of the cohort survivorship curve), which is very low.

However, it is unlikely that any combination of sampling, taxonomic, or geochronometric errors exerts a major control on the principal pattern revealed by this analysis—that of doubling in diversity and ten-fold increase in rates of origination and extinction near the Mesoproterozoic–Neoproterozoic boundary and again in the Early Cambrian. Indeed, this exercise quantifies what we have long known as biostratigraphers—that acritarch-based correlation is difficult among Mesoproterozoic successions, somewhat easier in the Neoproterozoic, and precise in the Lower Cambrian.

Comparisons with Previous Diversity Estimates. The diversity trends noted here are broadly similar to those outlined a decade ago by Vidal and Knoll (1983), indicating an overall stability of pattern despite substantial increases in the numbers of species and assemblages described. There is one difference between the two estimates, however, and it is a major one; Vidal and Knoll's compilation lacks any inkling of the short-lived diversity peak in N7. That peak first appears in the literature in 1988 in a figure by Zang (1988), who discovered highly diverse N7 acritarchs in the Pertatataka Formation, Australia. Since then, even more diverse assemblages have been recognized in rocks of this age (Figure 1). The N7 spike cautions us that despite the overall stability of Proterozoic and early Cambrian acritarch diversity trends, short-lived diversification and extinction episodes may be missed at current levels of temporal sampling density.

A different view of Proterozoic protistan diversity was presented by Schopf (1992), who showed a broad diversity peak 1000–850 Ma ago, followed by a strong and continuing decline until the end of the eon.

Schopf's compilations of mean assemblage diversity for plankton and eukaryotes emphasize the inferred early Neoproterozoic diversity peak even more strongly. This discrepancy arises for at least three reasons: (*i*) Schopf's estimates of species richness for early Neoproterozoic assemblages from Russia significantly exceed those accepted here, (*ii*) most of the fossils that determine the diversity levels of intervals N4 to N7 in the present paper do not appear in Schopf's data set, largely because of recent discovery, and (*iii*) Schopf's calculation of mean assemblage diversity is swamped by low diversity assemblages of limited paleobiological value. For these reasons, I believe that the diversity trends shown in Figures 1–3 of the present paper better reflect the known record of early protists.

Intimations of Mode? As noted above, the increase in acritarch diversity and tempo near the Mesoproterozoic–Neoproterozoic boundary coincides with the appearance of identifiable red, green, and probable chromophyte algae in the record. Branching patterns in molecular phylogenies of the eukaryotes suggest that these algal taxa, along with stramenopiles (ciliates, dinoflagellates, and plasmodia), fungi, and the ancestors of animals, diverged rapidly relatively late in the history of the domain (Sogin *et al.*, 1989). The paleontological data suggest that the radiation implied by molecular phylogenies occurred near the Mesoproterozoic–Neoproterozoic boundary; phylogenetic data, in turn, suggest possible explanations for the acceleration of evolutionary tempo documented by the fossils.

Nuclear introns, multicellular development that includes coordinated growth and cellular differentiation, and life cycles in which classical meiosis plays a prominent role are all characters displayed by higher eukaryotes but not earlier branching clades (Cleveland, 1947; Margulis *et al.*, 1989; Tibeyrence *et al.*, 1991; Palmer and Logsdon, 1991). The evolutionary relationships among these features are poorly understood, but possibly not coincidental. Either sexual life cycles or the exon shuffling made possible by introns could increase genetic variation and, thereby, accelerate evolutionary tempo (Schopf *et al.*, 1973; Knoll, 1992a). This would be true of nuclear introns whether they first evolved at the time of higher protistan differentiation (Palmer and Logsdon, 1991) or were simply retained more readily in lineages characterized by sexual life cycles (Hickey, 1982).

Given the population genetic possibilities of such changes, it is surprising that the greater increase in acritarch diversity and tempo is concentrated at the beginning of the Cambrian Period. At this time, there is no evidence of genetic reorganization. New faster evolving clades may enter the acritarch record, but groups such as the prasinophytes that appear to have been important on both sides of the Proterozoic–Cambrian boundary also document the acceleration of cladogenetic tempo. Of course, the sharp increase in acritarch diversity and turnover coincides with a comparable evolutionary burst in animals. The nearly simultaneous radiation in two such phylogenetically, developmentally, and trophically disparate groups suggests the importance of ecology in determining the tempo of Cambrian (and later) evolution. Evolving animals would have contributed in several ways to the complexity of environments perceived by acritarch-producing protists: for example, through predation, the disturbance of pre-existing physical environments, the creation of new physical environments, and the alteration of nutrient fluxes in marine platform and shelf waters. Diversifying protists would have had reciprocal effects on animals. Diversity levels reached by Early Cambrian animals and protists were later eclipsed by continuing diversification, but the increased rates of turnover established at this time have persisted for the past 500 Ma (Table 2; Van Valen, 1973; Raup, 1985).

This is interesting in light of evidence that turnover in Phanerozoic marine communities may be coordinated among species and concentrated at times of environmental disturbance represented sedimentologically by sequence boundaries (Brett *et al.*, 1990; Morris *et al.*, 1992; Miller, 1993). This suggests that the basal Cambrian increase in the biological complexity of environments may have lowered the response thresholds of populations to physical fluctuations, perhaps by decreasing population sizes and effective niche breadth.

The short-lived acritarch radiation in N7 stands out as anomalous. Is this when faster evolutionary tempo was established in protists, only to be cut off by mass extinction? Might it correspond to an epoch of cryptic animal diversification that presaged the Ediacaran faunas of the next interval? Is the acritarch diversification causally related to oceanographic changes that accompanied the end of the Varanger glaciation, and if so, why aren't comparable changes observed in the wake of earlier Neoproterozoic ice ages?

Conclusions

We still glimpse early biological history through a glass darkly, but broad patterns are beginning to come into focus. These patterns suggest that on the time scale of eukaryotic evolution as a whole, evolutionary tempo has increased episodically. Morphological diversity and turnover rates were low for the earliest recorded period of early protistan evolution, an interval that lasted longer than the entire Phanerozoic Eon. Near the Mesoproterozoic–Neoproterozoic boundary, the morphological diversity and turnover rates of acritarch-producing protists increased significantly, apparently as part of a larger increase in eukaryotic diversity that included heterotrophs as well as algae. Most notably, the Proterozoic and Early Cambrian record of acritarchs suggests that radiating animals had a profound effect on both diversity and turnover within clades already present in marine communities, implying an important role for ecology in fueling the Cambrian explosion and, perhaps, earlier protistan diversification.

SUMMARY

In rocks of late Paleoproterozoic and Mesoproterozoic age (ca. 1700–1000 million years ago), probable eukaryotic microfossils are widespread and well preserved, but assemblage and global diversities are low and turnover is slow. Near the Mesoproterozoic–Neoproterozoic boundary (1000 million years ago), red, green, and chromophytic algae diversified; molecular phylogenies suggest that this was part of a broader radiation of "higher" eukaryotic phyla. Observed diversity levels for protistan microfossils increased significantly at this time, as did turnover rates. Coincident with the Cambrian radiation of marine invertebrates, protistan microfossils again doubled in diversity and rates of turnover increased by an order of magnitude. Evidently, the Cambrian diversification of animals strongly influenced evolutionary rates within clades already present in marine communities, implying an important role for ecology in fueling a Cambrian explosion that extends across kingdoms.

I thank Zhang Yun, Yin Leiming, Kathleen Grey, Zang Wenlong, Malcolm Walter, Tadas Jankauskas, Nina Volkova, Tamara German, Alexei Veis, Vladimir Sergeev, Nicholas Butterfield, and—especially—Gonzalo Vidal and Malgorzata Moczydlowska for access to and discussions about Proterozoic and Cambrian microfossils. George Miklos, Kenneth Campbell, and J. William Schopf provided helpful criticisms of an earlier draft. This work was sponsored in part by National Aeronautics and Space Administration Grant NAGW-893 and National Science Foundation Grant BSR 90-17747.

REFERENCES

Allison, C. W. & Awramik, S. M. (1989) Organic-walled microfossils from earliest Cambrian or latest Proterozoic Tindir Group rocks, northwest Canada. *Precambrian Res.* **43,** 253–294.

Aseeva, E. A. (1983) Vendian microfossils of the Ukraine, in *Vend Ukrainy* [The Vendian of the Ukraine], eds. Velikanov, V. A., Aseeva, E. A. & Fedonkin, M. A. (Naukova Dumka, Kiev, pp. 102–127).

Bowring, S. A., Grotzinger, J. P., Isaachsen, C. E., Knoll, A. H., Pelechaty, S. M. & Kolosov, P. (1993) Calibrating rates of Early Cambrian evolution. *Science* **261,** 1293–1298.

Brett, C. E., Miller, K. B. & Baird, G. C. (1990) Paleocommunity temporal dynamics: the long-term development of multispecies assemblages. *Paleontol. Soc. Spec. Publ.* **5,** 178–209.

Butterfield, N. J., Knoll, A. H. & Swett, K. (1990) A bangiophyte red alga from the Proterozoic of arctic Canada. *Science* **250,** 104–107.

Butterfield, N. J., Knoll, A. H. & Swett, K. (1994) Paleobiology of the Neoproterozoic of arctic Canada. *Foss. Strata,* in press.

Cleveland, L. R. (1947) The origin and evolution of meiosis. *Science* **105,** 287–288.

Compston, W., Williams, I. S., Kirschvink, J. L., Zhang, Z. & Ma, G. (1992) Zircon U-Pb ages for the Early Cambrian time-scale. *J. Geol. Soc.* **149,** 171–184.

German, T. N. (1990) *Organic World Billion Year Ago* (Nauka, Leningrad).

Han, T. M. & Runnegar, B. (1992) Megascopic eukaryotic algae from the 2.1-billion-year-old Negaunee Iron-Formation, Michigan. *Science* **257,** 232–235.

Harland, W. B., Armstrong, R. L., Cox, A. V., Craig, L. E., Smith, A. G. & Smith, D. G. (1990) *A Geologic Time Scale 1989* (Cambridge Univ. Press, Cambridge, U.K.).

Hickey, D. (1982) Selfish DNA: a sexually-transmitted nuclear parasite. *Genetics* **101,** 519–531.

Jankauskas, T. V., ed. (1989) *Mikrofossilii dokembriya SSSR* [Precambrian Microfossils of the USSR] (Nauka, Moscow).

Jankauskas, T. V. (1982) Microfossils from the Riphean of the Southern Urals, in *Stratotip rifeya: Paleontologiya, Paleomagnetizm* [Stratotype of the Riphean: Paleontology, Paleomagnetism], ed. Keller, B. M. (Nauka, Moscow), pp. 84–120.

Kaufman, A. J. & Knoll, A. H. (1994) Neoproterozoic variations in the C-isotopic composition of seawater: implications for stratigraphy and biogeochemistry. *Precambrian Res.*, in press.

Keller, B. M. & Jankauskas, T. V. (1982) Microfossils in the Riphean stratotype section in the Southern Urals. *Int. Geol. Rev.* **24,** 925–933.

Knoll, A. H. (1992a) The early evolution of eukaryotes: a geological perspective. *Science* **256,** 622–627.

Knoll, A. H. (1992b) Vendian microfossils in metasedimentary cherts of the Scotia group, Prins Karls Forland, Svalbard. *Palaeontology* **25,** 751–774.

Knoll, A. H. (1994) Archean and Proterozoic paleontology, in *Palynology: Principles and Applications,* ed. Jansonius, J. & MacGregor, D. C. (Am. Assoc. Stratigraphic Palynologists Found., Tulsa, OK), in press.

Knoll, A. H. & Swett, K. (1987) Micropaleontology across the Precambrian-Cambrian boundary in Spitsbergen. *J. Paleontol.* **61,** 898–926.

Knoll, A. H. & Walter, M. R. (1992) Latest Proterozoic stratigraphy and Earth history. *Nature (London)* **356,** 673–678.

Knoll, A. H., Swett, K. & Mark, J. (1991) Paleobiology of a Neoproterozoic tidal flat/lagoonal complex: the Draken Conglomerate Formation, Spitsbergen. *J. Paleontol.* **65,** 531–570.

Margulis, L., Corliss, J. O., Melkonian, M. & Chapman, D. J., eds. (1989) *Handbook of Protoctista* (Jones & Bartlett, Boston).

Miller, W. (1993) Benthic community replacement and population responses. *N. Jb. Geol. Paläont. Abhandlangen* **188,** 133–146.

Moczydlowska, M. (1991) Acritarch biostratigraphy of the Lower Cambrian and the precambrian/Cambrian boundary in southeastern Poland. *Foss. Strata* **29,** 1–127.

Moczydlowska, M., Vidal, G. & Rudavskaya, V. A. (1993) Neoproterozoic (Vendian) phytoplankton from the Siberian Platform, Yakutia. *Palaeontology* **36,** 495–521.
Morris, P. J., Ivany, L. C. & Schopf, K. M. (1992) Paleoecological stasis in evolutionary theory. *Geol. Soc. Am. Abstr. Prog.* **24,** A313.
Palmer, J. D. & Logsdon, J. M. (1991) The recent origins of introns. *Curr. Opin. Genet. Dev.* **1,** 470–477.
Peat, C. R. Muir, M. D., Plumb, K. A., McKirdy, D. M. & Norvick, M. S. (1978) Proterozoic microfossils from the Roper Group, Northern territory, Australia. *BMR J. Aust. Geol. Geophys.* **3,** 1–17.
Raup, D. M. (1978) Cohort analysis of generic survivorship. *Paleobiology* **4,** 1–15.
Raup, D. M. (1985) Major features of the fossil record and their implications for evolutionary rate studies, in *Rates of Evolution*, eds. Campbell, K. S. W. & Day, M. F. (Allen & Unwin, London), pp. 1–14.
Schopf, J. W. (1992) Patterns of Proterozoic microfossil diversity; an initial, tentative, analysis, in *The Proterozoic Biosphere*, eds. Schopf, J. W. & Klein, C. (Cambridge Univ. Press, Cambridge, U.K.), pp. 529–552.
Schopf, J. W. & Klein, C., eds. (1992) *The Proterozoïc Biosphere* (Cambridge Univ. Press, Cambridge, U.K.).
Schopf, J. W., Haugh, B. N., Molnar, R. E. & Satterthwait, D. F. (1973) On the development of metaphytes and metazoans. *J. Paleontol.* **47,** 1–9.
Semikhatov, M. A. (1991) General problems of Proterozoic Stratigraphy in the USSR. *Sov. Sci. Rev. G* **1,** 1–192.
Sergeev, V. N., Knoll, A. H. & Grotzinger, J. P. (1994) Paleobiology of the Mesoproterozoic Billyakh Group, northern Siberia. *J. Paleontol.*, in press.
Simpson, G. G. (1944) *Tempo and Mode in Evolution* (Columbia Univ. Press, New York).
Sogin, M., Gunderson, J., Elwood, H., Alonso, R. & Peattie, D. (1989) Phylogenetic meaning of the kingdom concept: an unusual ribosomal RNA from *Giardia lamblia*. *Science* **243,** 75–77.
Summons, R. E. & Walter, M. R. (1992) Molecular fossils and microfossils of prokaryotes and protists from Proterozoic sediments. *Am. J. Sci.* **290-A,** 212–244.
Tappan, H. (1980) *The Paleobiology of Plant Protists* (Freeman, San Francisco).
Tibeyrence, M., Kjellberg, F., Arnaud, J., Oury, B., Darde, M. & Ayala, F. (1991) Are eukaryotic microorganisms clonal or sexual? A population genetics vantage. *Proc. Natl. Acad. Sci. USA* **88,** 5129–5133.
Van Valen, L. (1973) A new evolutionary law. *Evol. Theory* **1,** 1–30.
Veis, A. F. & Semikhatov, M. A. (1989) Lower Riphean Omakhtin association of microfossils in eastern Siberia, composition and conditions of formation. *Izv. Akad. Nauk SSSR Ser. Geol.* **5,** 36–54.
Veis, A. F. & Vorobyeva, N. G. (1992) Riphean and Vendian microfossils of the Anabar Massif. *Izv. Akad. Nauk SSSR Ser. Geol.* **8,** 114–130.
Vidal, G. (1976) Late Precambrian microfossils from the Visingsö Beds, southern Sweden. *Foss. Strata* **9,** 1–57.
Vidal, G. (1981) Micropaleontology and biostratigraphy of the Upper Precambrian and Lower Cambrian in East Finnmark, northern Norway. *Norges Geol. Unders. Bull.* **362,** 1–53.
Vidal, G. (1990) Giant acanthomorph acritarchs from the Upper Proterozoic in Southern Norway. *Palaeontology* **33,** 287–298.
Vidal, G. & Ford, T. D. (1985) Microbiotas from the late Proterozoic Chuar Group (northern Arizona) and Uinta Mountain Group (Utah) and their chronstratigraphic implications. *Precambrian Res.* **28,** 349–389.

Vidal, G. & Knoll, A. H. (1983) Proterozoic plankton. *Mem. Geol. Soc. Am.* **161,** 265–277.
Vidal, G. & Peel, J. S. (1993) Acritarchs from the Lower Cambrian Buen Formation in North Greenland. *Bull. Grønl. Geol. Unders.* **164,** 1–35.
Volkova, N. G. (1990) Acritarchs and other plant microfossils of the East European Platform, in *The Vendian System,* eds. Sokolov, B. S. & Iwanowski, A. B. (Springer, Berlin), Vol. 1, pp. 155–164.
Volkova, N. G. *et al.* (1983) Plant microfossils, in *Upper Precambrian and Cambrian Palaeontology of the East-European Platform,* eds. Urbanek, A. & Rozanov, A. Yu. (Wydawnictwa Geologiczne, Warsaw), pp. 7–46.
Walter, M. R., Du, R. & Horodyski, R. J. (1990) Coiled carbonaceous megafossils from the Middle Proterozoic of Jixian (Tianjin) and Montana. *Am. J. Sci.* **290-A,** 133–148.
Yan, Y. & Zhu, S. (1992) Discovery of acanthomorphic acritarchs in the Baicaoping Formation of Yongli, Shanxi, and its geological significance. *Acta Micropalaeontol. Sinica* **9,** 278–282.
Yin, L. (1987) Microbiotas of the latest Precambrian sequences. *Strat. Palaeontol. Systemic Bound. China Precamb.-Camb. Bound.* **1,** 415–494.
Zang, W. (1988) Dissertation (Australian National Univ., Canberra, Australia).
Zang, W. (1990) Sinian and Early Cambrian floras and biostratigraphy on the South China Platform. *Palaeontograph. Abteilung B* **224,** 75–119.
Zang, W. & Walter, M. R. (1992) Late Proterozoic and Cambrian microfossils and biostratigraphy, Amadeus basin, central Australia. *Mem. Assoc. Aust. Palaeontol.* **12,** 1–132.
Zhang, Z. (1986) Clastic facies microfossils from the Chuanlinggou Formation (1800 Ma) near Jixian, North China. *J. Micropalaeontol.* **5,** 9–16.

Part II

MACROEVOLUTION

The majority of skeletonized (readily preserved) animal phyla appear in the early Cambrian, in an exuberant burst of diversity long known as the Cambrian explosion. The evolution of so numerous and diverse body plans would seem to call for long time spans. At the time of the publication of *Tempo and Mode* and for the next two decades, the scarcity of the fossil record prior to the Cambrian was seen as the "explanation" of the Cambrian explosion. The explosion was only apparent. The evolution of the major body plans had come to be gradually, but the record was lacking. The eventual discovery of the Ediacaran faunas and of many Precambrian fossils failed to show a record of gradual emergence of the phyla. The evolution of the metazoan body plans and subplans happened rapidly, James W. Valentine tells us in Chapter 5. He combines fossil evidence from the Precambrian and early Cambrian with genetic and cell biology analysis of living forms, to reconstruct the evolutionary burst that created so much novelty, more than would ever appear at any other time.

Species extinction was long a neglected, if not totally ignored, subject of investigation for twentieth-century evolutionists. This is surprising, says David M. Raup in Chapter 6, because Darwin attached considerable significance to extinction, and because species extinctions have of necessity been just about as common as originations, living species representing the small surplus cumulated over millions of years. Raup concludes that mass extinctions have been of great consequence in restructuring the biosphere, because successful groups become eliminated, thereby empowering previously constrained groups to expand

and diversify. He also shows that species extinctions for the most part are not caused by natural selection.

Stephen Jay Gould, in Chapter 7, takes issue with Simpson's conclusion that paleontological processes can be accounted for by microevolutionary causes. Two major domains exist, he argues, where distinctive macroevolutionary theories are needed. One concerns nongradual transitions, such as punctuated equilibrium and mass extinction; the other calls for an expansion of the theory of natural selection to levels both below and above organisms.

Whence the topological configuration of vascular land plants? Physics, geometry, and computer simulations allow Karl J. Niklas to explore, in Chapter 8, the rules and significance of morphological variations. The more complex the functions that an organism must perform in order to grow, survive, and reproduce, the greater the diversity of morphological types that will satisfy the requirements. Unexpected is the additional conclusion that the number and accessibility of fitness optima also increase with the complexity of functions.

5

Late Precambrian Bilaterians: Grades and Clades

JAMES W. VALENTINE

The record of the first appearance of living phyla, classes, and orders can best be described in Wright's (1949) term as "from the top down" (Figure 1). Nearly all of the durably skeletonized (i.e., easily preservable) phyla appear in the Early Cambrian, body plans already in place so far as can be told, and then radiate into numbers of classes, and these into orders, so that the diversity peak of each lower taxonomic rank is shifted towards the present (Valentine, 1969, Erwin *et al.*, 1987). Far more evolution devoted to the rise of body plans and subplans is recorded during the Early Cambrian than during any subsequent geologic Epoch, producing a burst of novelty, termed the Cambrian explosion, that created the "tops." However, there must have been a buildup to those body plans at some time during the preceding Epochs, when the taxa were built "from the bottom up." Origination and extinction rates of families, genera, and species were highest early in metazoan history, but diversities then were low (Sepkoski, 1984, 1992, 1993; Valentine *et al.*, 1991). The relatively few body fossils known from the late Precambrian do not throw light on the sequence of evolutionary advances that led to the Cambrian taxa. The purpose of this paper is to characterize the evolution of metazoan body plans during the late Precambrian and Early Cambrian, with evidence drawn chiefly from Phanerozoic fossils and from living forms.

James W. Valentine is professor of integrative biology at the University of California, Berkeley.

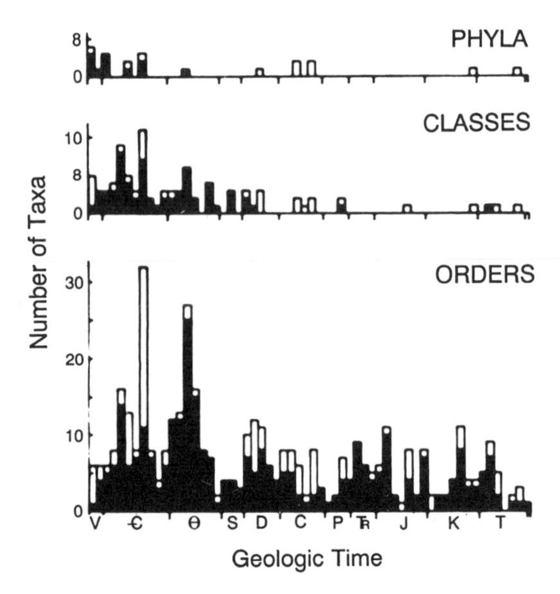

FIGURE 1 Histograms of first appearances of metazoan phyla, classes, and orders in the marine fossil record. Filled bars, durably skeletonized taxa; open bars, poorly or non-skeletonized taxa. Bar widths equal 10 million years (my). [Reprinted with permission, from Valentine, 1969 (copyright Society for the Study of Evolution).]

Appearance of Phyla in the Fossil Record

Ages of First Appearances. Figure 2 depicts the ages associated with the sequence of Stages and Series of the late Precambrian and Early Cambrian, with most terminology based on the Russian sequence (Bowring *et al.*, 1993). The past couple of decades have witnessed intense work on the early faunas, and during most of that time the base of the Tommotian has been taken as the base of the Cambrian. However, within the last few years new criteria have been developed and now the lowest Cambrian boundary is commonly based on the earliest appearance of the trace fossil *Phycodes pedum* (see Narbonne *et al.*, 1987). Choosing this boundary has lowered the base of the Cambrian, enlarging that Period by about one half (Figure 2). Despite this expansion of the Cambrian, new absolute age estimates have caused the length of time believed to be available for the Cambrian explosion to be shortened (Bowring *et al.*, 1993). The relationship of the dates given in Figure 2 to the boundaries of the Stages of the Lower Cambrian remains a difficult stratigraphic problem, but it is likely that the most critical Stages, the Tommotian and Atdabanian, are probably only 8–10 my in

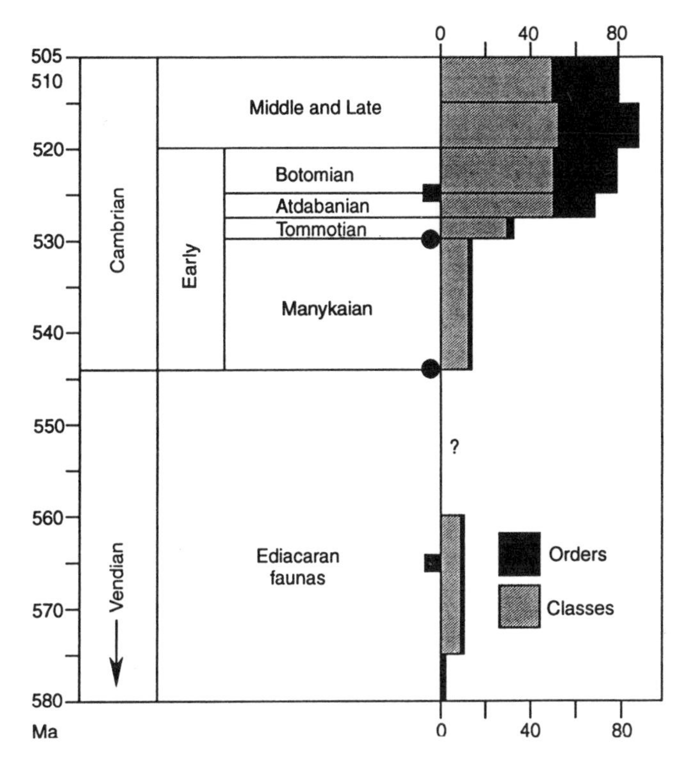

FIGURE 2 Geochronological time scale for the late Precambrian and Cambrian, with stratigraphic terms derived from the Russian geological column and with recorded metazoan diversity indicated by orders and classes (the earliest records are poorly constrained in time). Dates are from U-Pb zircon ages assigned to the stratigraphic levels indicated by filled circles or squares. [Reprinted with permission, from Bowring *et al.*, 1993 (copyright American Association for the Advancement of Science); diversity data from Sepkoski, 1992.]

duration; over 50 metazoan orders first appear in the record during that interval (Valentine *et al.*, 1991).

Vendian Faunas. The earliest fossils that may be metazoans are preserved in two modes. One is as body fossils, chiefly as impressions in this case. Most of these fossils somewhat resemble cnidarians but many cannot easily be assigned to living classes, and their affinities even as to kingdom are in dispute. Seilacher (1989) has suggested that most of them are not metazoans at all but represent a separate, extinct multicellular clade; they are suggested to have fed via symbionts as chemautotrophs (Seilacher, 1989), phototrophs, or osmotrophs (McMenamin, 1993). However, especially well-preserved specimens of "medusoid" forms show tentacles, which suggest prey capture, and there are other indications that feeding was by ingestion (Fedonkin, 1994). Other frond-like forms resemble sea pens; Conway Morris (1993) has studied similar forms from the Burgess Shale, wherein preservation is better, and has concluded that both Vendian and Burgess Shale forms are cnidarians. It seems likely on present evidence that most of the late Precambrian forms do have cnidarian affinities. Even so, the late Precambrian body fossils do not represent direct ancestors of any of the higher metazoans and do not help to resolve the puzzle of the origin of the remaining phyla. A few other body fossils may represent bilaterians (see below), but such an assignment is disputed also.

The other mode of preservation of late Precambrian forms is as trace fossils—markings made by animal activities (Glaessner, 1969; Fedonkin, 1985b; Crimes, 1989). Some traces are of winding, rather featureless, trails, but others display transverse rugae and contain pellets that can be interpreted as of fecal origin. The bilaterian nature of these traces is not in dispute. Furthermore, such traces must have been made by worms, some of which had lengths measured in centimeters, with through guts, which were capable of displacing sediment during some form of peristaltic locomotion, implying a system of body wall muscles antagonized by a hydrostatic skeleton. Such worms are more complex than flatworms, which cannot create such trails and do not leave fecal strings (Fedonkin and Runnegar, 1992; Fedonkin, 1994). It is among the trace makers that the ancestors of the Cambrian clades are most likely to be represented.

Early Cambrian Faunas. During the Manykaian Stage mineralized skeletons begin to appear, and during the Tommotian and Atdabanian they increase spectacularly in numbers and diversity (Bengtson and Conway Morris, 1992). Many of these fossils are dissociated sclerites that give few clues as to the nature of their scleritomes or of the animals that bore them. In a few cases, however, sclerites have been found in life associations or preserved with soft-bodied remains to provide indications of a body plan (Conway Morris and Peel, 1990; Hinz *et al.*, 1990; Ramsköld and Hou, 1991). Of living phyla, skeletons of brachiopods, mollusks, arthropods (trilobites), and echinoderms appear in the Tommotian and Atdabanian Stages, and nearly all durably skeletonized phyla are known by the end of the Early Cambrian. The exceptions are chordates (Middle Cambrian) and bryozoans (Early Ordovician); however, the body plans of neither of those phyla require a mineralized skeleton and both may have been present well before they appear as fossils. Body fossils of some phyla or subphyla lacking mineralized skeletons (priapulans, onychophoran relatives, etc.) are known from rocks of probable Atdabanian age from the Baltic Shield (Dzik and Lendzion, 1988 and refs. therein), China (Hou *et al.*, 1991), and Greenland (Conway Morris *et al.*, 1987), together with a variety of animals that cannot be assigned with any confidence to living phyla (*Dinomischus*, paleoscolicids, halkieriids, etc.).

Trace fossils in the Early Cambrian are greater in abundance and diversity than in the late Precambrian (Crimes, 1992). Penetrating vertical burrows, exceedingly rare and small earlier, become larger, longer, and more common, and bioturbation increases in depth and intensity in increasingly younger sediments (Crimes and Droser, 1992; Droser and Bottjer, 1993). The increased biological activity indicated by traces is consistent with that indicated by the explosion of body fossil types.

Middle Cambrian Faunas. The rate of appearance of novel body plans slows greatly during the Middle Cambrian with the exception of the fauna of the Burgess Shale and its correlatives, which create a diversity "spike" (Whittington, 1985; Conway Morris, 1992). The Burgess Shale fauna is exceptionally diverse and brings to light many of the less easily preserved members of the Cambrian fauna. The Chinese Atdabanian fauna is also exceptionally preserved and contains numbers of taxa in common with the Burgess Shale, suggesting that many of the Burgess Shale forms, or at least the higher taxa to which they belong, would be found to have originated in the Early Cambrian if that fauna were better known.

Summary. It is consistent with the fossil record that all of the body plans now ranked at the phylum level originated by the close of the Early Cambrian, although some that are not easily fossilized do not appear until later, and indeed some (platyhelminths, gnathostomulids, gastrotrichs, acanthocephalans, loriciferans, and kinorhynchs) are unknown as body fossils. As the very first bearers of any given body plan are unlikely to be found, any corrections for a smearing out of first appearances towards the Recent may add to the abruptness of the Cambrian explosion. The first appearances of fossil groups whose relationships are problematic but that have very distinctive skeletal or body plans and that may be phyla or subphyla are also concentrated during the Early Cambrian. The Cambrian explosion was geologically abrupt and taxonomically broad.

The Rise of Body-Plan Complexity

Metazoans evidently originated from unicellular (perhaps colonial) choanoflagellates or their allies (Wainright *et al.*, 1993). By the time of the Cambrian explosion, some metazoan bodies were as complex as primi-

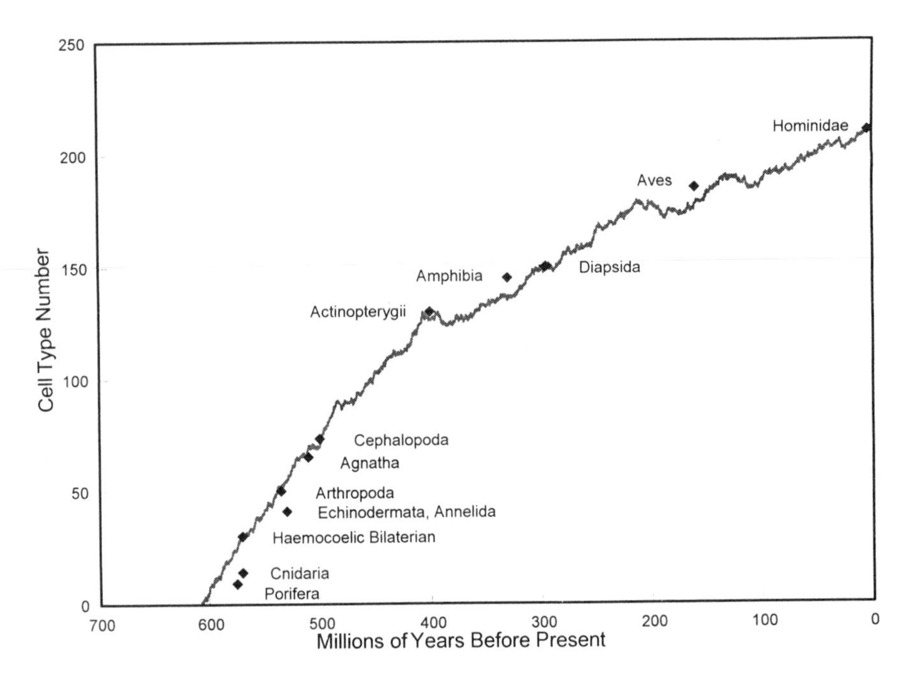

FIGURE 3 Estimated somatic cell-type number (except nerve cells) required by the body plans of various taxa, plotted against the time of origin of those taxa indicated by the fossil record. The irregular curved line represents the upper bound of cell-type number generated by a stochastic model based on passive increase or decrease of cell-type number within each lineage during diversification of lineages from 1 to 2000; the line is an average of five runs. [Reproduced with permission, from Valentine *et al.*, 1994 (copyright Paleobiology).]

Cambrian explosion, some metazoan bodies were as complex as primitive arthropods and other higher invertebrates. This rise in complexity is obscured in the fossil record. Perhaps the best practical index of body-plan complexity is cell-type number (Bonner, 1965; Sneath, 1964; Raff and Kaufman, 1983), which may be taken to have begun as two in metazoans and to have increased through time as more complex bodies evolved. Figure 3 depicts the estimated cell-phenotype numbers of the more complex body plans known at a given time during the history of the metazoans, plotted against the times of their first appearances as judged by the fossil record (Valentine *et al.*, 1994). A curve embracing the points should approximate the upper bound of body-plan complexity. Complexity increases may be forced, perhaps by natural selection, or may be passive, resulting from random opportunities to become either more or less complex (Fisher, 1986; Gould, 1988; McShea, 1991,

1993). In Figure 3, a computer model of random change in cell-type number in diversifying lineages over 4000 steps produced the upper bound shown by the shaded line (Valentine *et al.*, 1994). In that model the number of lineages began at one and increased logistically to 2000; the addition or subtraction of a cell type in each lineage in any step was treated as a Markov process. The model curve is scale to geological time by being pinned to landmarks at 30 and 210 cell types but is free elsewhere.

The chief caveats in interpreting this model relate to the estimates of cell-phenotype numbers indicated on Figure 3. The estimates represent a sampling of cell types and are made with a lumper's approach, and nerve cell types are not distinguished (Valentine *et al.*, 1994). For the simpler organisms the numbers may be fairly accurate, but for increasingly complex forms the figures increasingly underestimate true cell-type numbers. However, the intent is not to measure an increasing developmental complexity, or an increasing information content of metazoan genomes, but simply to reflect the gross morphological complexity of the body plans. Nevertheless, that portion of the curve represented by chordates should be treated with caution. As the early portion of the curve should be on the firmer footing, it is worth considering its implications for the early history of the metazoans and for the Cambrian explosion.

Although the model is not meant to replicate the history of metazoan complexity, the behavior of the upper bound suggests that no forcing mechanism may be necessary to account for the empirical complexity increase in early metazoan body plans. The relatively rapid initial increase in complexity, created partly by a "floor" of two cell types and partly by the diversification of clades, may have been a feature of real clades even in the absence of forcing agents. The origin of metazoans may thus be hypothesized to have been near 600 my ago, more recently than is usually supposed. The rise in complexity is parabolic in the model. At present there is no evidence of a major step in body-plan complexity during the Cambrian explosion. The implication is that when animals with, say, 45 cell types appear during the explosion, there were ancestors of that clade with 44 cell types (or in that general region) that we don't see, and complexities should dwindle, perhaps at an increasing rate as we go back farther in time. Again, the ancestors may be known today only by the late Precambrian traces.

Phylogenetic Models and Body Plans

The Molecular Phylogenetic Model. Expectations as to the body plans of late Precambrian bilaterians depend partly upon phylogenetic models.

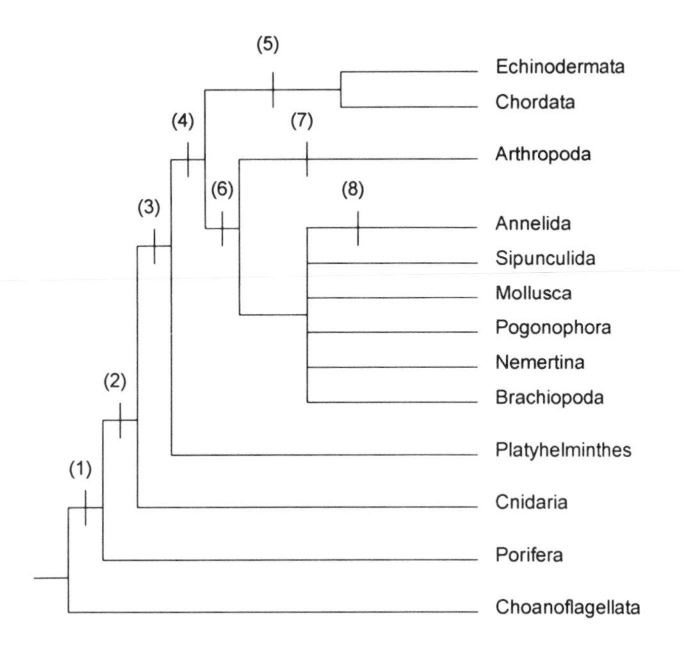

FIGURE 4 Branching sequence of lineages leading to some modern phyla as indicated by some of the more robust 16S and 18S rRNA molecular phylogenies, with a hypothesized sequence of introduction of some design elements in their body plans as indicated. Branch lengths or spacings are not scaled. Design elements are 1, collagen, tissues; 2, gastrulation, enteron; 3, mesoderm; 4, blood-vascular system; 5, oligomerous coelom; 6, hemocoel; 7, arthropodan segmentation; 8, annelidan segmentation. If the unresolved rRNA branching events leading to various protostomes occurred within proto-mollusks, it may not be possible in principle to characterize them by synapomorphies that are germane to the derived features of living phyla. (Molecular data from various sources but chiefly Wainright *et al.*, 1993; Field *et al.*, 1988; Lake, 1990; Turbeville *et al.*, 1991; Turbeville *et al.*, 1992; Raff, 1994; design element sequence chiefly from Valentine, 1992.)

There is hardly a scheme of relationships that hasn't been suggested by one worker or another; the models that have been most important have been reviewed by Willmer (1990). Comparative studies of 16S and 18S rRNA sequences from living phyla have provided new evidence as to ancestral branching patterns (e.g., Field *et al.*, 1988; Lake, 1990; Turbeville *et al.*, 1991, 1992; Wainright *et al.*, 1993; Wheeler *et al.*, 1993; Raff, 1994). While these findings are still provisional and involve some contradictions, most can be rationalized with morphologic and developmental evidence and accord well with the fossil evidence, so they will be used here as the basis of a phylogenetic scenario. Figure 4 summarizes some of the more robust findings by listing some phyla from which rRNAs were studied to the right and the inferred branching patterns of the rRNAs on the left, with the implied appearance of some design elements in metazoan body plans indicated. A few of the design elements require comment. For present purposes, the unqualified terms "hemocoel" and "coelom" refer to fluid-filled spaces usually used as hydrostatic skeletons—the former being developed from the topological position of the blastocoel, the latter being developed within mesoderm. Spaces devoted only to serving organs (commonly as ducts or as buffering against solid tissues) are termed "organ hemocoels" or "organ coeloms." "Seriation" refers to longitudinal repetition of organs, muscles, or other features but not necessarily in a correlated fashion, while "segmentation" refers to serial repetition with correlation among organ systems.

Many of the metazoan relationships indicated in Figure 4 are quite conventional, such as the sequence of choanoflagellates/poriferans/cnidarians/platyhelminths, the sister-group relationship of platyhelminths with higher metazoans, and the sister-group relationship of protostomes with deuterostomes. Less conventional, though hardly shocking, is the sister-group relationship of the arthropods with the unresolved cluster of protostomes, which includes mollusks and annelids. This is a particularly important branching, because it suggests that the last common ancestor of the protostomes was hemocoelic and that the coelom(s) of the deuterostomes arose independently of the coelom(s) of the protostomes. This branching also suggests that arthropod and annelid segmentation arose independently. The next deeper branching, between protostomes and deuterostomes, suggests that while their last common ancestor was not coelomic and need not have been hemocoelic, it did have a blood-vascular system. Other interpretations are possible but they are less parsimonious.

Late Precambrian Body Plans. The list of phyla on the right of Figure 4 indicates the body plans of the living organisms studied but certainly does not indicate the body plans of the common ancestors deeper within the branching pattern. On present evidence all of the branching between the choanoflagellate/poriferan and the arthropod/unresolved protostome branches occurred during the Precambian. The bilaterian traces of the Vendian must have been made by descendants of the platyhelminth/higher metazoan branch. The simple earliest trails require active worms with some hydrostatic system, possibly a tissue skeleton, or fluid in part; the relief indicated by these traces implies a non-flat body and that, in turn, a blood-vascular system. Such a worm, a vascularized "roundish flatworm," acoelomate (except possibly for organ spaces), probably appropriate ancestor for both protostomes and deuterostomes. Today the nearest approximations to a worm of this grade are found among the mollusks—aplacophorans and chitons—although these living groups have derived features that would have been absent in the Vendian. The more complex horizontal trails or shallow burrows appearing later imply a more efficient hydrostatic system and thus probably an ample hemocoel. Again, worms of this grade would probably be regarded as simplified mollusks, were they to be found alive and shoehorned into a living phylum.

If the sequence of introductions of design elements was truly parsimonious, the ancestral arthropods branched from this sort of protomollusk. A plausible scenario would begin with a seriated form, probably preferring hard grounds and supplementing a fundamentally peristaltic creeping locomotion by lateral body projections that served as accessory gripping mechanisms. Sclerotization may have begun as protection in such a habitat, since burrowing would not be possible, but flexibility of the body wall and peristaltic efficiency would have been sacrificed as it became heavier. Locomotory functions shifted to the lateral appendages, and a segmented anatomy evolved as series of muscles, nerves, and blood vessels developed to serve them. Flexibility of the trunk was maintained by jointing of the integument; this permitted vertical flexure when creeping over uneven substrates and perhaps lateral flexure to enhance locomotory power as needed through the use of the body wall musculature (Valentine, 1989). The jointing enhanced the segmented architecture. Body spaces continued to be developed on the site of the blastocoel: a hemocoel to provide hydrostatic functions and an enlarged organ hemocoel to bathe the heart. Jointing of the appendages occurred as they lengthened within a rigidifying exoskeleton. By the Tommotian, arthropod-type appendages were sufficiently well developed to permit their use in furrowing the substrate, presumably in search of food, which created a characteristic type of trace fossil (Crimes, 1989), probably reflecting an evolutionary radiation of jointed body types onto soft substrates.

The number of independent lineages that reached the arthropod condition has been in dispute. An analysis of the branching order of 12S rRNA within the arthropods (Ballard *et al.*, 1992) suggests that ancestors of the myriapods branched most deeply and were followed by branchings that produced the ancestors of onychophorans (a lobopodal group that does not have jointed appendages) and then of the chelicerates, crustaceans, and hexapods in that order. However, an 18S rRNA tree (Wheeler *et al.*, 1993) suggests that the onychophorans are sisters to a monophyletic arthropod clade, a view supported by a consensus tree that includes morphological and Ubiquitin sequence data. A number of fossil types that can be interpreted as lobopods are known from the Early (possibly Tommotian) and Middle Cambrian (Ramsköld and Hou, 1991; Bengtson *et al.*, 1986; Budd, 1993). Living lobopods have hemocoels but flexible body walls (and can therefore squeeze through narrow openings). These data leave open the question of the body plans associated with the branch points. Perhaps the most likely possibility is that the onychophoran and arthropodan ancestors branched from a hemocoelic, seriated form before jointed, sclerotized exoskeletons appeared.

Among the Vendian body fossils of uncertain afinity are a few that have sometimes been interpreted as segmented bilaterians, including sprigginids and vendomiids (Glaessner, 1984; Fedonkin, 1985a). These forms are serially constructed and, if bilaterians, have cephalic shields but lack appendages; they have also been reconstructed as frond-like forms (Seilacher, 1989; Bergström, 1989). In some of them at least, the putative jointing alternates across the midline, so that "segments" are laterally offset rather than continuous; however, offsetting of serial structures is known in many living bilaterians (e.g., muscle blocks in cephalochordates) (Jeffries, 1986). If these fossils are bilaterians, then they may belong to the grade of segmented, hemocoelic organisms postulated to have given rise to arthropods (Valentine, 1989, 1992).

Ancestral annelids may also have branched from proto-molluscan grade ancestors, but the formation of the annelid body plan involved the origin of the famous compartmented coelom, widely interpreted as a hydrostatic skeleton to aid in peristaltic burrowing (Clark, 1964) or horizontal plowing. Undoubted annelidan body fossils have not been found in rocks older than Middle Cambrian; the earliest penetrating burrows may have been made by so-called pseudocoelomates such as priapulids or paleoscolecids, both of which are known in the Lower Cambrian. Long tubes presumably constructed by vermiform organisms, the sabelliditids, occur in the Manykaian and are sometimes considered to be pogonophorans, but no body fossils of these forms are known. Like arthropods, annelids possess larval intramesodermal spaces. In various arthropod taxa the spaces are occluded, or become organ coeloms, or are incorporated into the blood-vascular system during development; in annelids they are lost during metamorphosis, although the adult coelom develops (by schizocoely) within tissues that are derived from larval mesoderm (Anderson, 1973). If the larval coeloms in these phyla are homologous they are presumably plesiomorphic. In the larvae of some phyla, an intramesodermal space serves as a nephridium (Ruppert and Balser, 1986), and possibly this was the original function in this case as well. The segmentation in annelids is not restricted to the locomotary system, although it does involve the parapodia and their musculature and its vascularization and innervation. Unlike the arthropods, annelidan gonads and nephridia are also seriated in concert with the coelomic compartmentalization. Annelidan and arthropodan segmentation, though associated with locomotion in each case, operate on different principles and involve some different organ systems including the fluid skeletons; there is little reason to regard them as homologous. Eernisse *et al.* (1992) have presented a cladistic analysis that arrives at similar conclusions (but see Backeljau *et al.*, 1993).

Brachiopods are commonly regarded as allied to phoronids and bryozoans, as all three phyla possess similarly regionated coeloms and lophophores; this relationship has not been corroborated by molecular techniques at this writing. According to the 18S rRNA tree, the brachiopods may have arisen from the last common arthropod–annelid ancestor and therefore from a proto-molluscan grade in the present scenario. It has been suggested (Valentine, 1992) that an unsegmented coelom was evolved within an ancestral worm clade for locomotion in soft sediments; direct peristalsis, which requires an unsegmented coelom, is a common locomotory technique in flocculent muds, for example (Elder, 1980). Radiation of this clade onto firmer substrates may then have produced sessile forms with regionated coeloms to serve both a trunk and a tentacular crown, the feeding lophophore. Perhaps annelids have also descended from a small-bodied ancestor with an unregionated coelom, with segmentation evolving for efficiency in locomotion.

Mollusks exhibit the body plan that is least derived from that of the postulated protostome ancestor, and most molluscan classes have only organ coeloms. However the cephalopods, which form the most advanced molluscan class and which appear only in the Upper Cambrian, have a well-developed coelomic space (a gonocoel?) that can be regarded as a novel evolutionary feature. The nemertine rhynchocoel, which has been demonstrated to be a coelom histologically (Turbeville, 1991), probably was derived independently.

Late Vendian and Manykaian bilaterian body plans are thus visualized as consisting of an array of vermiform types, including flatworms and "round flatworms" with blood-vascular systems, many with hemocoels or "pseudocoels," and some with seriation of one organ system or another (Valentine, 1989, 1990; Bergström, 1989). Organ coeloms were doubtless present in many lineages. Possibly some of these worms were incipiently segmented in the arthropod style. It seems likely that some worms had unregionated perivisceral coeloms, but it is doubtful that any had annelid-style coelomic segmentation. In this scenario, the body plans of these worms were based primarily upon adaptations to locomotory requirements on a variety of substrates, radiating from at least two proto-molluscan body plans, one with a simple blood vascular system and the other with a hemocoel. Once a successful new locomotory system was evolved on one substrate type, invasion and reinvasion of other substrates became possible and the diversity of types increased further. These worms have left us an array of trace fossils and evidently little else, except their descendants, among whom we number.

Tempo and Mode and Body Plans

The more complex of the late Precambrian worms must have been higher invertebrates in every sense, with, by invertebrate standards, sophisticated organ systems consisting of appropriately specialized tissues and these in turn composed of differentiated cell phenotypes that, judging by the body plans of living organisms, probably numbered in the 40s. Other worms were clearly less complex. For those worms that gave rise to Phanerozoic clades, it is expected that their primitive members were only minimally more complex than their ancestors, and it is plausible that the ancestral body plans were precursory to their Phanerozoic descendants. These considerations, and the evidence reviewed above, imply that many of the branching events required of metazoan diversifications occurred well before the onset of the Cambrian explosion. After the branchings, complexity continued to increase within many clades. For example, proto-mollusks gave rise to several distinctive hemocoelic and coelomic body plans, and early mollusks gave rise to the relatively complex cephalopods. The pattern of branching during the Vendian appears as a more-or-less orderly accumulation of a series of derived features. This has been the usual interpretation of the metazoan diversification pattern, more-or-less *faute de mieux*, though occasionally it has been challenged by proponents of a bush or grass-like pattern of many parallel lineages rising from a common ancestral clade (Nursall, 1962), which in the latest version is composed of flatworms (Willmer, 1990; Barnes *et al.*, 1993). It appears more likely that the body plans from which numbers of phyla evolved were proto-molluscan.

Internal Factors. The mode of evolution that might account for the observed explosive appearance of body plans in the fossil record must have involved changes in genetic regulation within many lineages, including repatternings of gene expression and respecifications of cell fates and movements, but perhaps with little increase in levels of cell differentiation. Evidence of the nature of the transcriptional regulation of these processes is beginning to accumulate. It is clearly of interest to study the similarities and differences among the regulators that mediate body-plan formation in metazoans, in the context of the molecular phylogenetic tree and of the sequence of morphological complexities outlined above, but such comparative data are still scarce. Many of the relevant data involve homeobox genes, best known for their roles in determining developmental patterns in *Drosophila* and mice (Akam, 1987; Shashikant *et al.*, 1991). It appears that these genes are widely (probably universally) distributed in metazoan phyla (McGinnis, 1985; Kenyon and Wang, 1991; Schierwater *et al.*, 1991; Webster and Mansour, 1992; Oliver *et al.*, 1992; Schummer *et al.*, 1992; Bartels *et al.*, 1993; Chalfie, 1993); they are highly conserved, and some are found in conserved clusters, such as the homeotic *Hox/HOM* genes. The anterior boundaries of the expression of these genes occur in a sequence along the anteroposterior body axis of bilaterians in the conserved order in which the genes occur in the cluster.

Most nonchordate phyla are thought to have only a single *Hox/HOM* cluster. Relative to higher metazoans, few *Hox/HOM* genes have been found in platyhelminths (Bartels *et al.*, 1993) and there may be only a small cluster in nematodes, although they have perhaps 60 homeobox genes (Chalfie, 1993; Bürglin and Ruvkun, 1993). So far as is known, most higher invertebrates have a single cluster; cephalochordates are inferred to have two clusters, agnathans to have three (or four), and mammals have four (Pendleton *et al.*, 1993). Most or all of the *Hox/HOM* genes within clusters evidently arose by duplications beginning with an ancestral homeobox gene, while multiple clusters such as found in chordates arose later by cluster duplication, homologous genes commonly being more similar among clusters than within them (Kappen *et al.*, 1989). Cartwright *et al.* (1993) have produced evidence from a survey of *Hox/HOM* genes in the chelicerate *Limulus polyphemus* that it may have four clusters. A form interpreted as an ancestral chelicerate (it lacks chelicerae) is described from the Middle Cambrian Burgess Shale (Briggs and Collins, 1988); whether any cluster duplication had occurred by then is unknown.

In addition to the homeotic genes, a cascading series of pattern-formation genes is expressed earlier in metazoan development and mediates the progressive regionation and differentiation of body form; in *Drosophila* these include maternal genes and gap, pairrule, and segment-polarity genes, expressed in that order (Akam, 1987; Ingham, 1988). Some of these genes, such as the segment-polarity gene *engrailed*, are known to be represented by homologs in, among other organisms, mice (see Patel *et al.*, 1989), and thus were presumably present in a common protostome–deuterostome ancestor. The segment-polarity gene *hedgehog* (*hh*) has been shown to have a homolog in chordates, *Sonic hedgehog* (*Shh*), that mediates polarity in the developing central nervous system, trunk, and limbs of various members of that phylum, activating Hox genes (Riddle *et al.*, 1993; Echelard *et al.*, 1993; Krauss *et al.*, 1993). It is an obvious hypothesis that a gene ancestral to *hh* and *Shh* was present and mediated polarizing activity in the last common protostome–deuterostome ancestor and is probably much more ancient.

It is possible to speculate on aspects of the general course of evolution of metazoan body plans, assuming that the preceding picture of early metazoan evolution is approximately correct. Homeobox genes presumably arose within protistans, but took on a role in the specification of cell fates, movements, and patterns as cell differentiation accompanied the rise of multicellular organisms. During the first few tens of millions of years of metazoan history a regulatory cascade was assembled, mediating the transcription of a growing morphological complexity. If the rise of complexity was as rapid as hypothesized here, it seems likely that many of these regulatory pathways were already present in protistans. The number of transcriptional regulators may have grown hand in hand with the upper bound of metazoan complexity. Bilaterians became increasingly differentiated along their anteroposterior body axes, and this trend may be reflected in the growth of the *Hox/HOM* cluster. The numbers of homeobox genes besides those in this cluster (and of other regulatory gene types) must have increased as well. At some point during this trend, bilaterians became able to displace sediment and thus to produce trace fossils, and they entered the fossil record of the Late Vendian. By the time that the last common ancestor of deuterostomes and protostomes evolved, presumably still during the Vendian, a large *Hox/HOM* cluster was present in that lineage. The Cambrian explosion, then, may have occurred largely or entirely within organisms that had a single *Hox/HOM* cluster. However, the major increase in the upper bound of complexity during the Phanerozoic may be associated with a series of duplications of the cluster.

The postulated branching of lineages at proto-molluscan grades that led eventually to a variety of descendants with distinctive body plans requires that the details of pattern formation responsible for the derived features of each of those body plans evolved independently. In this view the *Hox/HOM* cluster had a responsibility for anteroposterior differentiation in the proto-mollusk (and earlier) and retained that role, coming to be associated with the development of segmentation in arthropods, in annelids, and with anteroposterior structures in vertebrates, even though those features are quite different and evolved independently. Many early developmental steps also were greatly modified after the ancestral branching that led to separate phyla (Davidson, 1990, 1991). In other words, nearly all of the body architectures, and many of the developmental steps that characterize the phyla, evolved after the pattern formation and selector genes were assembled. The Cambrian explosion in metazoan body plans may not have involved any great expansion of the gene regulatory apparatus but probably could not have occurred if the regulatory systems were not already sophisticated, and thus able rapidly to create novel morphologies, as the morphological innovations were evidently achieved through modes of regulatory evolution (Britten and Davidson, 1971; Valentine and Campbell, 1975; Jacobs, 1990; Valentine, 1994).

External Factors. The abrupt appearance of higher taxa in the Cambrian has stimulated a search for possible environmental changes that might have produced this evolutionary outburst; there has been no shortage of suggestions (Valentine *et al.*, 1991; Signor and Lipps, 1992). However, the geological record has not revealed unequivocal evidence of unique Cambrian events that might be held accountable. One of the more common suggestions has been that oxygen levels rose in the late Precambrian to values that could sustain metazoans and thus permitted evolution to produce increasingly active organisms through time. Recent studies have indeed revealed series of late Precambrian–Cambrian geochemical excursions that can be plausibly interpreted as involving CO_2 and oxygen levels and as being associated with the biogeochemical cycling of the time (Knoll, 1994).

It is not clear that an external trigger was needed to produce the Cambrian explosion. A continuous expansion of an already moderately complex fauna as various lineages acquired evolutionary access to broadening arrays of marine habitats, with the resulting enhancement of ecological interactions within that fauna, may be all that was required to produce the record we have, insofar as the origin of phyla is concerned. However, continuing geochemical and other studies will surely lead to a better understanding of the nature of environmental change during the Cambrian explosion, and then we shall be better able to judge the possible role of physical triggers.

SUMMARY

A broad variety of body plans and subplans appear during a period of perhaps 8 million years (my) within the Early Cambrian, an unequaled explosion of morphological novelty, the ancestral lineages represented chiefly or entirely by trace fossils. Evidence from the fossil record can be combined with that from molecular phylogenetic trees to suggest that the last common ancestor of (*i*) protostomes and deuterostomes was a roundish worm with a blood vascular system and (*ii*) of arthropods and annelids was similar, with a hydrostatic hemocoel; these forms are probably among trace makers of the late Precambrian. Cell-phenotype numbers in living phyla, and a model of cell-phenotype number increase, suggest an origin of metazoans near 600 my ago, followed by a passive rise in body-plan complexity. Living phyla appearing during the Cambrian explosion have a *Hox/HOM* gene cluster, implying its presence in the common ancestral trace makers. The explosion required a repatterning of gene expression that mediated the development of novel body plans but evidently did not require an important, abrupt increase in genomic or morphologic complexity.

I thank Dave Jablonski (University of Chicago) and Allen Collins and Chris Meyer (University of California, Berkeley) for valuable discussions and reviews of the manuscript, and Clint Turbeville for insights into coelom development. This research was based on work supported by Grant EAR-9196068, National Science Foundation, and by Faculty Research grants, University of California, Berkeley.

REFERENCES

Akam, M. (1987) The molecular basis for metameric pattern in *Drosophila* embryos. *Development* **101,** 1–22.

Anderson, D. T. (1973) *Embryology and Phylogeny of Annelids and Arthropods* (Pergamon, Oxford).

Backeljau, T., Winnepenninckx, B. & De Bruyn, L. (1993) Cladistic analysis of metazoan relationships: a reappraisal. *Cladistics* **9,** 167–181.

Ballard, J. W. O., Olsen, G. J., Faith, D. P., Odgers, W. A., Rowell, D. M. & Atkinson, P. W. (1992) Evidence from 12S ribosomal RNA sequences that onychophores are modified arthropods. *Science* **258,** 1345–1348.

Barnes, R. S. K., Callow, P. & Olive, P. J. W. (1993) *The Invertebrates, 2nd. Ed.* (Blackwell Sci., Oxford).

Bartels, J. L., Murtha, M. T. & Ruddle, F. H. (1993) Multiple *Hox/HOM*-class homeobox genes in Platyhelminthes. *Mol. Phylog. Evol.* **2,** 143–151.

Bengtson, S. & Conway Morris, S. (1992) Early radiation of biomineralizing phyla. In *Origin and Early Evolution of the Metazoa*, eds. Lipps, J. H. & Signor, P. W. (Plenum, New York), pp. 448–481.

Bengtson, S., Matthews, S. C. & Missarzhevsky, V. V. (1986) The Cambrian netlike fossil *Microdictyon*. In *Problematic Fossil Taxa*, Hoffman, A. & Nitecki, M. H. (Oxford Univ. Press, Oxford), pp. 97–113.

Bergström, J. (1989) The origin of animal phyla and the new phylum Procoelomata. *Lethaia* **22,** 259–269.

Bonner, J. T. (1965) *Size and Cycle* (Princeton Univ. Press, Princeton, NJ).

Bowring, S. A., Grotzinger, J. P., Isachsen, C. E., Knoll, A. H., Pelechaty, S. M. & Kolosov, P. (1993) Calibrating rates of Early Cambrian evolution. *Science* **261,** 1293–1298.

Briggs, D. E. G. & Collins, D. (1988) A Middle Cambrian chelicerate from Mount Stephen, British Columbia. *Palaeontology* **31,** 779–798.

and a speculation on the origins of evolutionary novelty. *Quart. Rev. Biol.* **46,** 111–133.

Budd, G. (1993) A Cambrian gilled lobopod from Greenland. *Nature* **364,** 709–711.

Bürglin, T. R. & Ruvkun, G. (1993) The *Caenorhabditis elegans* homeobox gene cluster. *Curr. Opin. Genet. Develop.* **3,** 615–620.

Cartwright, P., Dick, M. & Buss, L. W. (1993) HOM/Hox type homeoboxes in the chelicerate *Limulus polyphemus. Mol. Phyl. Evol.* **2,** 185–192.

Chalfie, M. (1993) Homeobox genes in *Caeonorhabditis elegans. Curr. Opin. Genet. Develop.* **3,** 275–277.

Clark, R. B. (1964) *Dynamics in Metazoan Evolution* (Clarendon Press, Oxford).

Conway Morris, S. (1992) Burgess Shale-type faunas in the context of the "Cambrian explosion": a review. *Jour. Geol. Soc. London* **149,** 631–636.

Conway Morris, S. (1993) Ediacaran-like fossils in Cambrian Burgess Shale-type faunas of North America. *Palaeontology* **36,** 593–635.

Conway Morris, S. & Peel, J. S. (1990) Articulated halkieriids from the Lower Cambrian of north Greenland. *Nature* **345,** 802–805.

Conway Morris, S., Peel, J. S., Higgins, A. K., Soper, N. J. & Davis, N. C. (1987) A Burgess Shale-like fauna from the Lower Cambrian of north Greenland. *Nature* **326,** 181–183.

Crimes, T. P. (1989) Trace Fossils. In *The Precambrian-Cambrian Boundary,* eds. Cowie, J. W. & Brazier, M. D. (Clarendon Press, Oxford), pp. 166–185.

Crimes, T. P. (1992) Changes in the trace fossil biota across the Proterozoic-Phanerozoic boundary. *Jour. Geol. Soc. London* **149,** 637–646.

Crimes, T. P. & Droser, M. L. (1992) Trace fossils and bioturbation: the other fossil record. *Ann. Rev. Ecol. Syst.* **23,** 339–360.

Davidson, E. H. (1990) How embryos work: a comparative view of diverse modes of cell fate specification. *Development* **108,** 365–389.

Davidson, E. H. (1991) Spatial mechanisms of gene regulation in metazoan embryos. *Development* **113,** 1–26.

Droser, M. & Bottjer, D. J. (1993) Trends and patterns of Phanerozoic ichnofabrics. *Ann. Rev. Earth Planet. Sci.* **21,** 205–225.

Dzik, J. & Lendzion, K. (1988) The oldest arthropods of the East European Platform. *Lethaia* 29–38.

Echelard, Y., Epstein, D. J., St-Jacques, B., Shen, L., Mohler, J., McMahon, J. A. & McMahon, A. P. (1993) Sonic hedgehog, a member of a family of putative signalling molecules, is implicated in the regulation of CNS polarity. *Cell* **75,** 1417–1430.

Eernisse, D. J., Albert, J. S. & Anderson, F. E. (1992) Annelida and Arthropoda are not sister taxa: a phylogenetic analysis of spiralian metazoan morphology. *Syst. Biol.* **41,** 305–330.

Elder, H. Y. (1980) Peristaltic mechanisms. In *Aspects of Animal Movement,* eds. Elder, H. Y. and Trueman, E. R. (Cambridge Univ. Press, Cambridge), pp. 71–92.

Erwin, D. H., Valentine, J. W. & Sepkoski, J. J., Jr. (1987) A comparative study of diversification events: the early Paleozoic versus the Mesozoic. *Evolution* **41,** 1177–1186.

Fedonkin, M. A. (1994) Vendian body fossils and trace fossils. In *Early Life on Earth,* ed. Bengtson, S. (Columbia Univ. Press, New York), pp. 370–388.

Fedonkin, M. A. (1985a) Systematic description of Vendian Metazoa. In *The Vendian System* Vol. 1, eds. Sokolov, B. S. & Ivanovsky, A. B. (Nauka, Moscow), Vol. 1, pp. 70–106 (in Russian).

Fedonkin, M. A. (1985b) Systematic description of Vendian Metazoa. In *The Vendian*

Fedonkin, M. A. (1985b) Systematic description of Vendian Metazoa. In *The Vendian System* Vol. 1, eds. Sokolov, B. S. & Ivanovsky, A. B. (Nauka, Moscow), Vol. 1, pp. 112–117 (in Russian).

Fedonkin, M. A. & Runnegar, B. (1992) Proterozoic metazoan trace fossils. In *The Proterozoic Biosphere*, eds. Schopf, J. W. & Klein, C. (Cambridge Univ. Press, Cambridge), pp. 389–395.

Field, K. G., Olsen, G. J., Lane, D. J., Giovannoni, S. J., Ghiselin, M. G., Raff, E. C., Pace, N. R. & Raff, R. A. (1988) Molecular phylogeny of the animal kingdom. *Science* **239,** 748–753.

Fisher, D. C. (1986) Progress in organismal design. In *Patterns and Processes in the History of Life*, eds. Raup, D. M. & Jablonski, D. (Springer, Berlin), pp. 99–117.

Glaessner, M. (1969) Trace fossils from the Precambrian and basal Cambrian. *Lethaia* **2,** 369–393.

Glaessner, M. F. (1984) *The Dawn of Animal Life* (Cambridge Univ. Press, Cambridge).

Gould, S. J. (1988) Trends as changes in variance: a new slant on progress and directionality in evolution. *Paleontol.* **62,** 319–329.

Hinz, I., Kraft, P., Mergl, M. & Müller, K. J. (1990) The problematic *Hadimopanella, Kaimenella, Milaculum* and *Utahphospha* identified as sclerites of Paleoscolecida. *Lethaia* **23,** 217–221.

Hou, X.-G., Ramsköld, L. & Bergström, J. (1991) Composition and preservation of the Chengjiang fauna—a Lower Cambrian soft-bodied biota. *Zool. Scr.* **20,** 395–411.

Ingham, P. (1988) The molecular genetics of embryonic pattern formation in *Drosophila. Nature* **335,** 25–34.

Jacobs, D. K. (1990) Selector genes and the Cambrian radiation of the bilateria. *Proc. Natl. Acad. Sci. USA* **87,** 4406–4410.

Jefferies, R. P. S. (1986) *Ancestry of the Vertebrates* (Cambridge Univ. Press, Cambridge).

Kappen, C., Schughart, K. & Ruddle, F. H. (1989) Two steps in the evolution of Antennapedia-class vertebrate homeobox genes. *Proc. Natl. Acad. Sci. USA* **86,** 5459–5463.

Kenyon, C. & Wang, B. (1991) A cluster of *Antennapedia*-class homeobox genes in a nonsegmented animal. *Science* **253,** 516–517.

Knoll, A. H. (1994) Neoproterozoic evolution and environmental change. In *Early Life on Earth*, ed. Bengtson, S. (Columbia Univ. Press, New York), pp. 439–449.

Krauss, S., Concordet, J.-P., & Ingham, P. W. (1993) A functionally conserved homolog of the Drosophia segment polarity gene *hh* is expressed in tissues with polarizing activity in zebrafish embryos. *Cell* **75,** 1431–1444.

Lake, J. A. (1990) Origin of the multicellular animals. *Proc. Natl. Acad. Sci. USA* **87,** 763–766.

McGinnis, W. (1985) Homeo box sequences of the *Antennapedia* class are conserved only in higher animal genomes. *Cold Spring Harbor Symp. Quant. Biol.* **50,** 263–270.

McMenamin, M. (1993) Osmotrophy in fossil protoctists and early animals. *Invertebr. Repro. Develop.* **23,** 165–166.

McShea, D. W. (1991) Complexity and evolution: what everybody knows. *Biol. Philosph.* **6,** 303–324.

McShea, D. W. (1993) Evolutionary change in the morphological complexity of mammalian vertebral column. *Evolution* **47,** 730–740.

Narbonne, G. M., Myrow, P., Landing, E. & Anderson, M. M. (1987) A candidate stratotype for the Precambrian-Cambrian boundary, Fortune Head, Burin Peninsula, southeastern Newfoundland. *Can. Earth Sci.* **24,** 1277–1293.

Nursall, J. R. (1962) On the origins of the major groups of animals. *Evolution* **16,** 118–123.

Oliver, G., Vispo, M., Mailhos, A., Martinez, C., Sosa-Pineda, B., Fielitz, W. & Ehrlich, R. (1992) Homeoboxes in flatworms. *Gene* **121,** 337–342.

Patel, N. H., Martin-Blanco, E., Coleman, K. G., Poole, S. J., Ellis, M. C., Kornberg, T. B. & Goodman, C. S. (1989) Expression of *engrailed* proteins in arthropods, annelids and chordates. *Cell* **58,** 955–968.

Pendleton, J. W., Nagai, B. K., Murtha, M. T. & Ruddle, F. H. (1993) Expansion of the *Hox* gene family and the evolution of chordates. *Proc. Natl. Acad. Sci. USA* **90,** 6300–6304.

Raff, R. A. (1994) Developmental mechanisms in the evolution of animal form: origins and evolvability of body plans. In *Early Life on Earth,* ed. Bengtson, S. (Columbia Univ. Press, New York), pp. 489–500.

Raff, R. A. & Kaufman, T. C. (1983) *Embryos, Genes and Evolution* (Macmillan, New York).

Ramsköld, L. & Hou, X.-G. (1991) New Early Cambrian animal and onychophoran affinities of enigmatic metazoans. *Nature* **351,** 225–227.

Riddle, R. D., Johnson, R. L., Laufer, E. & Tabin, C. (1993) *Sonic hedgehog* mediates the polarizing activity of the ZPA. *Cell* **75,** 1401–1416.

Ruppert, E. E. & Balser, E. J. (1986) Nephridia in the larvae of hemichordates and echinoderms. *Biol. Bull.* **171,** 188–196.

Schierwater, B., Murtha, M., Dick, M., Ruddle, F. H. & Buss, L. W. (1991) Homeoboxes in cnidarians. *Exper. Zool.* **260,** 413–416.

Schummer, M., Scheurlen, I., Schaller, C. & Falliot, B. (1992) HOM/Hox homeobox genes are present in hydra (*Chlorohydra viridissima*) and are differentially expressed during regeneration. *EMBO* **11,** 1815–1823.

Seilacher, A. (1989) Vendozoa: organismic construction in the Proterozoic biosphere. *Lethaia* **22,** 229–239.

Sepkoski, J. J., Jr. (1984) A kinetic model of Phanerozoic taxonomic diversity. III. Post-Paleozoic families and mass extinctions. *Paleobiology* **10,** 246–267.

Sepkoski, J. J., Jr. (1992) Proterozoic-Early Cambrian diversification of metazoans and metaphytes. In *The Proterozoic Biosphere,* eds. Schopf, J. W. & Klein, C. (Cambridge Univ. Press, Cambridge, U.K.), pp. 553–564.

Sepkoski, J. J., Jr. (1993) Ten years in the library: new data confirm paleontological patterns. *Paleobiology* **19,** 43–51.

Shashikant, C. S., Utset, M. F., Violette, S. M., Wise, T. L., Einat, P., Einat, M., Pendleton, J. W., Schugart, K. & Ruddle, F. H. (1991) Homeobox genes in mouse development. *Crit. Rev. Eukaryotic Gene Expression* **1,** 207–245.

Signor, P. W. & Lipps, J. H. (1992) Origin and early radiation of the Metazoa. In *Origin and Early Evolution of the Metazoa,* eds. Lipps, J. H. & Signor, P. W. (Plenum, New York), pp. 3–23.

Sneath, P. H. A. (1964) Comparative biochemical genetics in bacterial taxonomy. In *Taxonomic Biochemistry and Serology,* ed. Leone, C. A. (Ronald, New York), pp. 565–583.

Turbeville, J. M. (1991) Nemertinea. In *Microscopic Anatomy of Invertebrates: Platyhelminthes and Nemertinea,* eds. Harrison, F. W. & Bogitsh, B. J. (Wiley-Liss, New York), Vol. 3, pp. 285–328.

Turbeville, J. M., Pfeifer, D. M., Field, K. G. & Raff, R. A. (1991) The phylogenetic status of arthropods, as inferred from 18S rRNA sequences. *Mol. Biol. Evol.* **8,** 669–686.

Turbeville, J. M., Field, K. G. & Raff, R. A. (1992) Phylogenetic position of Phylum Nemertini, inferred from 18S rRNA sequences: molecular data as a test of morphological character homology. *Mol. Biol. Evol.* **9,** 235–249.

Valentine, J. W. (1969) Patterns of taxonomic and ecological structure of the shelf benthos during Phanerozoic time. *Palaeontology* **12,** 684–709.

Valentine, J. W. (1989) Bilaterians of the Precambrian-Cambrian transition and the annelid-arthopod relationship. *Proc. Natl. Acad. Sci. USA* **86,** 2272–2275.

Valentine, J. W. (1990) Molecules and the early fossil record. *Paleobiology* **16,** 94–95.

Valentine, J. W. (1992) The macroevolution of phyla. In *Origin and Early Evolution of the Metazoa,* eds. Lipps, J. H. & Signor, P. W. (Plenum, New York), pp. 525–553.

Valentine, J. W. (1994) The Cambrian explosion. In *Early Life on Earth,* ed. Bengtson, S. (Columbia Univ. Press, New York), pp. 401–411.

Valentine, J. W. & Campbell, C. A. (1975) Genetic regulation and the fossil record. *Am. Sci.* **63,** 673–680.

Valentine, J. W., Collins, A. G. & Meyer, P. C. (1994) Morphological complexity increase in metazoans. *Paleobiology* **20,** 131–142.

Valentine, J. W., Awramik, S. M., Signor, P. W. & Sadler, P. M. (1991) The biological explosion at the Precambrian-Cambrian boundary. *Evol. Biol.* **25,** 279–356.

Valentine, J. W., Tiffney, B. & Sepkoski, J. J., Jr. (1991) Evolutionary dynamics of plants and animals: a comparative approach. *Palaios* **6,** 81–88.

Wainright, P. O., Hinkle, G., Sogin, M. L. & Stickel, S. K. (1993) Monophyletic origins of the Metazoa: an evolutionary link with fungi. *Science* **260,** 340–342.

Webster, P. J. & Mansour, T. E. (1992) Conserved classes of homeodomains in *Schistosoma mansoni,* an early bilateral metazoan. *Mech. Dev.* **38,** 25–32.

Wheeler, W. C., Cartwright, P. & Hayashi, C. Y. (1993) Arthropod phylogeny: a combined approach. *Cladistics* **9,** 1–39.

Whittington, H. B. (1985) *The Burgess Shale* (Yale Univ. Press, New Haven, CT).

Willmer, P. (1990) *Invertebrate Relationships* (Cambridge Univ. Press, Cambridge, U.K.).

Wright, S. (1949) Adaptation and selection. In *Genetics, Paleontology and Evolution,* eds. Jepson, G. L., Simpson, G. G. & Mayr, E. (Princeton Univ. Press, Princeton, NJ), pp. 365–389.

6
The Role of Extinction in Evolution

DAVID M. RAUP

The extinction of species is not normally considered an important element of Neodarwinian theory, in contrast to the opposite phenomenon, speciation. This is surprising in view of the special importance Darwin attached to extinction, and because the number of species extinctions in the history of life is almost the same as the number of originations; present-day biodiversity is the result of a trivial surplus of originations, cumulated over millions of years. For an evolutionary biologist to ignore extinction is probably as foolhardy as for a demographer to ignore mortality. The past decade has seen a resurgence of interest in extinction, yet research on the topic is still at a reconnaissance level, and our present understanding of its role in evolution is weak.

Charles Darwin on Extinction

In the *Origin* (1859), Darwin made his view of extinction, and its role in evolution, quite clear. He saw four essential features.

(*i*) Extinctions of species have occurred gradually and continuously throughout the history of life.

> . . . species and groups of species gradually disappear, one after another, first from one spot, then from another, and finally from the world. (Darwin, 1859, pp. 317–318)

David M. Raup is professor of geophysical sciences at the University of Chicago, Chicago, Illinois.

. . . the complete extinction of the species of a group is generally a slower process than their production: if the appearance and disappearance be represented . . . by a vertical line of varying thickness the line is found to taper more gradually at its upper end, which marks the progress of extermination. . . . (p. 218)

(*ii*) Sudden disappearances of many species, now called mass extinctions, did not actually occur. Although the Cretaceous–Tertiary (K–T) event was well known in Darwin's day (Lyell, 1833, p. 328), Darwin was convinced that sudden disappearances of species from the fossil record were due solely to unrecognized gaps in the temporal record.

With respect to the apparently sudden extermination of whole families or orders, as of Trilobites at the close of the palaeozoic period [Permian mass extinction] and of Ammonites at the close of the secondary period [K–T mass extinction], we must remember what has already been said on the probable wide intervals of time between our consecutive formations; and in these intervals there may have been much slower extermination. (Darwin, 1859, pp. 321–322)

Like his geologist colleague Charles Lyell, Darwin was contemptuous of those who thought extinctions were caused by great catastrophes.

. . . so profound is our ignorance, and so high our presumption, that we marvel when we hear of the extinction of an organic being; and as we do not see the cause, we invoke cataclysms to desolate the world, or invent laws on the duration of the forms of life! (p. 73)

(*iii*) Species extinction is usually, though not always, caused by the failure of a species in competition with other species. That is, causes of extinction are generally biological, not physical.

The inhabitants of each successive period in the world's history have beaten their predecessors in the race for life, and are, insofar, higher in the scale of nature. . . . (p. 345)

If . . . the eocene inhabitants . . . were put into competition with the existing inhabitants, . . . the eocene fauna or flora would certainly be beaten and exterminated; as would a secondary [Mesozoic] fauna by an eocene, and a palaeozoic fauna by a secondary fauna. (p. 337)

. . . each new variety, and ultimately each new species, is produced and maintained by having some advantage over those with which it comes into competition; and the consequent extinction of the less-favoured forms almost inevitably follows. (p. 320)

(*iv*) The extinction of species (and larger groups) is closely tied to the process of natural selection and is thus a major component of progressive evolution. In some passages of the *Origin*, Darwin seems to have seen extinction as part of natural selection; in others, as an inevitable outcome.

. . . extinction and natural selection . . . go hand in hand. (p. 172)

The extinction of species and of whole groups of species, which has played so conspicuous a part in the history of the organic world, almost inevitably follows on the principle of natural selection; for old forms will be supplanted by new and improved forms. (p. 475)

Thus, as it seems to me, the manner in which single species and whole groups of species become extinct accords well with the theory of natural selection. (p. 322)

In his final summary of the *Origin* (pp. 489–490), Darwin listed the fundamental components ("laws") of the evolutionary process: *reproduction, inheritance, variability, struggle for life,* and *natural selection,* with its "consequences" *divergence of character* and the *extinction of less-improved forms.* Despite Darwin's obvious concern for the role of extinction, the word does not appear in the index to the *Origin,* nor have biologists paid much attention to the phenomenon until the past decade. Mayr (1964) published an expanded and modernized index to the *Origin,* but even this contains only a small fraction of the possible citations to the word extinction. For some reason or reasons not entirely clear, extinction largely dropped out of the consciousness of evolutionary biologists and paleobiologists. Only with the advent of vigorous controversy (Alvarez *et al.*, 1980) over the causes of the K–T event, and with the development of concern for presently endangered species, has the role of extinction been confronted in modern terms.

George Gaylord Simpson on Extinction

In *Tempo and Mode* (1944), Simpson detailed what he considered to be the most important determinants of evolution. These were (Chapter II) *variability, mutation rate, character of mutations, generation length, population size,* and *natural selection.* But missing from this chapter is any indication that extinction plays an important role in evolution. To be sure, Chapter II includes occasional mention of specific extinctions, but not as significant drivers of evolution. For example, Simpson suggests that mammals with long generation times (equated with large body size) suffered greater extinction in the latest Pleistocene because natural selection could not operate quickly enough for adaptation to changing climatic conditions. But the implications of this are not developed, and Simpson clearly did not share Darwin's view that extinction is a vital part of the evolutionary process. Elsewhere in *Tempo and Mode,* however, Simpson noted that major extinctions provide opportunities (space, ecological niches, etc.) for later diversification by the survivors.

In sharp contrast to Darwin's view, Simpson saw interspecies competition as only rarely the cause of extinction of species or larger groups. He thought replacement of one group by another was generally passive.

In the history of life it is a striking fact that major changes in the taxonomic groups occupying various ecological positions *do not, as a rule, result from direct competition* of the groups concerned in each case and the survival of the fittest, as most students would assume a priori. On the contrary, the usual sequence is for one dominant group to die out, leaving the zone empty, before the other group becomes abundant. . . . (Simpson, 1944, p. 212; emphasis added)

As examples of passive replacement, Simpson lists the extinction of the ichthyosaurs millions of years before being replaced by cetaceans, the gap between the extinction of the pterodactyls and diversification of bats, and the fact that dinosaurs died out before the radiation of large terrestrial mammals. In the latter case, he notes the not uncommon observation that the successor group (diverse mammals) had been in existence, albeit not thriving, through much of the dinosaur reign.

Simpson has relatively little to say about whether extinction is sudden or gradual. Mass extinctions are acknowledged, but he follows Darwin in arguing that gaps, saltations, and other abrupt changes in the fossil record should not be taken at face value.

Probably there is always a considerable period of time corresponding with the gap in morphology, taxonomy, and phylogeny. It is impossible to prove that there are no exceptions to this generalization, so that there is some danger that it may represent the statement of an a priori postulate rather than evidence for the postulate; but I believe that this is a valid deduction from the facts. (p. 111)

Taken as a whole, Simpson's treatment of extinction is very different from Darwin's: they differ in their perceptions of the causes of extinction (but not the rates) and, above all, in the role of extinction in evolution.

Despite the foregoing, Simpson made important and lasting contributions to the study of extinction in the fossil record through his use of survivorship analysis above the species level. His now-classic comparison of the slopes of survivorship curves for bivalve mollusks and carnivorous mammals set the stage for development of the technique by Van Valen (1973) and many others. Much of our synoptic knowledge of extinction today relies on an expanded use of these techniques.

The Record of Extinction

The known fossil record contains roughly a quarter of a million species, most of which are extinct. Although fossils of the earliest forms are important to our knowledge of the history of life, the fossil record is dominated numerically by the remains of multicellular organisms from the last 600 million years (Myr). Fossil species are grouped into about 35,000 genera and 4000 families. About one-quarter of the families are still living.

Although the fossil record is ample for statistical purposes, it contains a very small fraction of all the species that have ever lived. Estimates of that fraction range from <1% to a few percent, depending on the organisms being considered and assumptions about past biodiversity and turnover rates. The probability of fossilization is strongly influenced by many biological and physical factors. Marine animals with hard skeletons are strongly favored, and as a result, the fossil record is dominated by these groups (e.g., mollusks, brachiopods, reef corals). Even for these organisms, however, biases in preservation abound.

The dinosaur fossil record illustrates some of the more extreme sampling problems. According to a review by Dodson (1990), 336 of the named species of dinosaur are taxonomically valid. Of these, 50% are known only from a single specimen, and about 80% are based on incomplete skeletons. The 336 species are grouped into 285 genera, and of these, 72% have been found only in the rock formation where they were first discovered, and 78% have been found in only one country. These numbers are astonishing if viewed as if the data were complete. The species/genus ratio being barely above unity is undoubtedly due to incomplete sampling, as is the apparent biogeographic restriction.

Incomplete sampling also influences our estimates of extinction rates. Lack of fossilization inevitably shortens the apparent life span of species, and this may explain why durations of dinosaur species are far shorter than is typical of other, better preserved organisms. On the other hand, short-lived, localized species have a low probability of appearing in the fossil record at all. The net effect of these biases is that statistical estimates of mean duration are almost certainly exaggerated. That is, the fossil record is biased in favor of successful species—successful in the sense of surviving for a long time and being ecologically and geographically widespread. Thus, analysis of past extinctions must operate in a sampling regime very different from that of present-day biodiversity.

Figure 1 shows the frequency distribution of recorded life spans of 17,500 genera of fossil marine animals. The distribution is highly skewed, with the mean (28 Myr) being the result of many short durations combining with a few very long ones. Survivorship analysis of the genus data indicates a mean species duration of 4 Myr (Raup, 1991a), although as indicated above, this is probably a high estimate because of the dominance of successful species in the sample. Regardless of the uncertainties, however, species and genus residence times on earth are very short on geologic time scales. The longest-lived genus in Figure 1 (160 Myr) lasted only about 5% of the history of life.

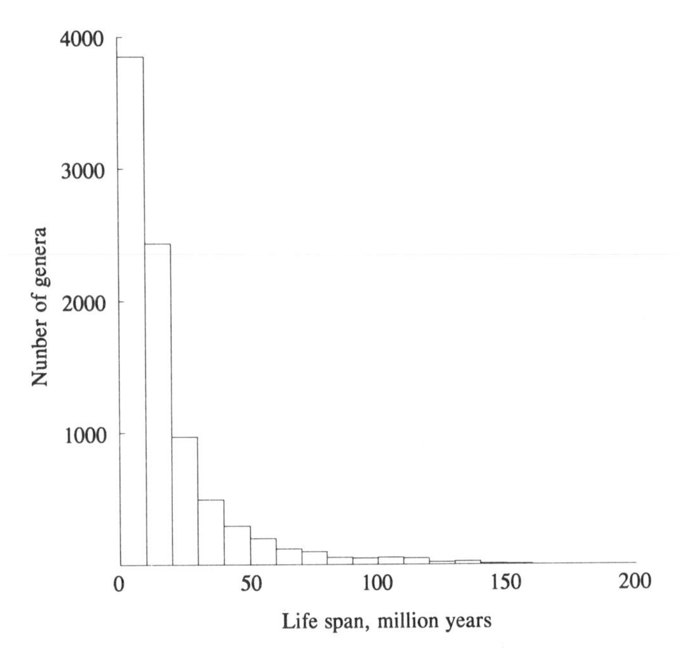

FIGURE 1 Life spans of about 17,500 extinct genera of marine animals (vertebrate, invertebrate, and microfossil) tabulated from data compiled by Sepkoski (1989).

Episodes of Extinction

It is conventional to divide extinctions into two distinct kinds: background and mass extinction. The term "mass extinction" is most commonly reserved for the so-called "Big Five" events: short intervals in which 75–95% of existing species were eliminated (Table 1). The K–T

TABLE 1 Comparison of species extinction levels for the Big Five mass extinctions

Extinction episode	Age, Myr before present	Percent extinction
Cretaceous (K–T)	65	76
Triassic	208	76
Permian	245	96
Devonian	367	82
Ordovician	439	85

Extinction data are from Jablonski (1991).

event, mentioned earlier, is one of the Big Five, but not the largest. Although the Big Five were important events, their combined species kill amounted to only about 4% of all extinctions in the past 600 Myr (Raup, 1993). The mass/background dichotomy is unfortunate because it implies two modes of extinction, yet there is no evidence for a discontinuity between them. Figure 2 shows variation in percent species kill in 1-Myr intervals for the past 600 Myr. The events called mass extinctions are concentrated in the right-hand tail, but there is no break between this tail and the main distribution. The data appear to produce a single, highly skewed distribution. Thus, segregating mass extinction from background has no more meaning than distinguishing hurricanes from other tropical cyclonic storms on the basis of some arbitrary wind speed (64-knot sustained surface winds). Continued use of the mass/background dichotomy serves only to hide interesting structure in extinction data.

Figure 3 shows a cumulative distribution of extinction frequency, the so-called "kill curve" for species of the past 600 Myr. The format is one used commonly to analyze severe storms, floods, earthquakes, and other natural phenomena where the larger the event, the rarer it is. The kill curve gives the average time interval (mean waiting time) between an extinction event and the next one of equal or greater magnitude. An "event" is defined as the species kill occurring in an arbitrarily short interval. Thus, 10% (or more) of the standing crop of species goes

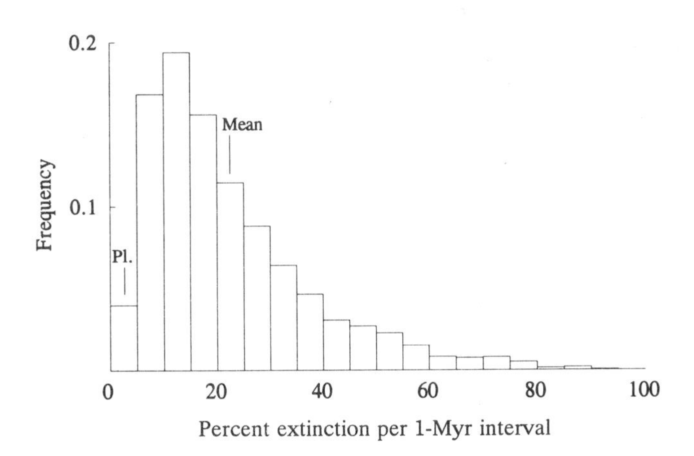

FIGURE 2 Variation in intensity of extinction for 1-Myr intervals during the past 600 Myr, based on the species kill curve (Raup, 1991a). "Pl." refers to extinction intensity of the Pleistocene glacial epoch. Mass extinctions occupy the right-hand tail of the distribution. The mean extinction rate, 25% extinction per 1 Myr, is the approximate reciprocal of the mean species duration (4 Myr).

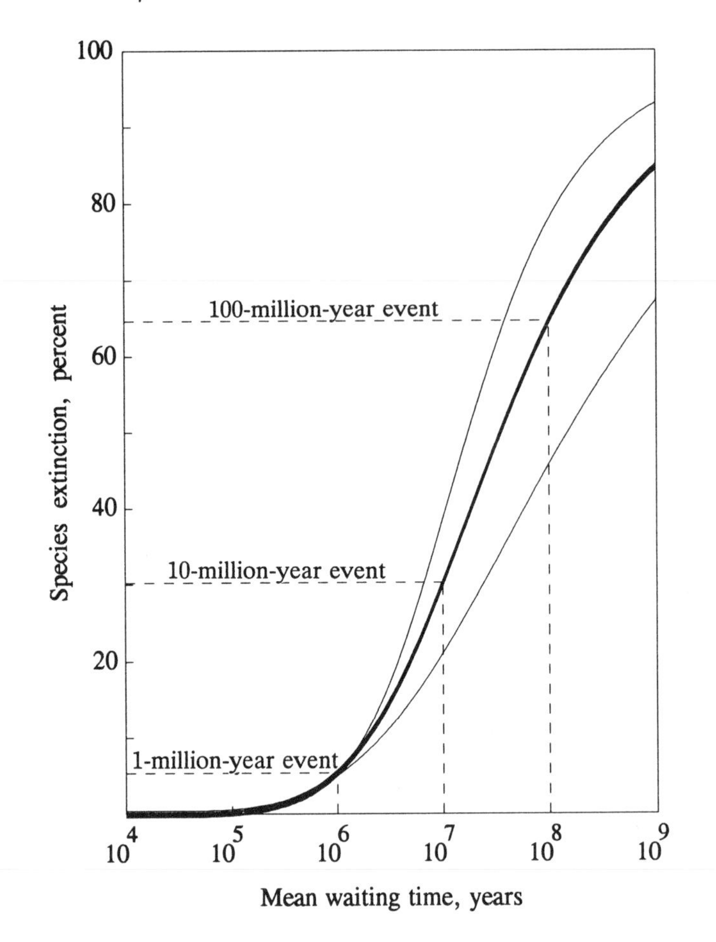

FIGURE 3 Kill curve (heavy line) for the past 600 Myr (Raup, 1991a). Waiting time is the average interval between events of a given extinction intensity. Thus, for example, a short episode of extinction which kills 30% of standing species diversity occurs on average every 10 Myr ("10-million-year event"), with no implication of uniform periodicity in the spacing of events. Light curves bound the uncertainty in placement of the kill curve from fossil data (Raup, 1992).

extinct, on average, every 1 Myr, 30% every 10 Myr, and 65% every 100 Myr (Raup, 1991a). The 100-Myr events include the Big Five mass extinctions.

Without further analysis, one could assert that the kill curve is a natural result of chance coincidence of independent events. That is, pure chance might produce an episode of nearly simultaneous extinctions if we wait long enough. This is emphatically not the case. For a random model that assumes that all species extinctions are independent of one other, the probability of a 10% extinction every 1 Myr (on average) is vanishingly small (Raup, 1991a). In fact, a kill curve based on this model could not be plotted at the scale of Figure 3: the curve would be indistinguishable from the horizontal axis. The only available conclusion is that extinctions are nonrandomly clustered in time, and this implies strongly that the K–T extinctions, for example, had a common cause.

Some of the clustering of extinction may be due to the removal of one or a few species that are crucial to the existence of other species. Or clusters may be due to destruction of one important ecosystem or habitat. However, for the larger events, at least, the extinctions are far more pervasive. At the end of the Cretaceous, high levels of species extinction (>50%) are found in all geographic areas and involve organisms as different as burrowing mollusks, planktonic microorganisms, land plants, and dinosaurs. This suggests, among other things, that the big mass extinctions cannot be explained by Darwin's species interactions unless one is willing to postulate an incredible degree of connectedness in the biosphere.

A striking effect of the typical mass extinction is its aftermath. For as long as 5–10 Myr, fossil faunas and floras are impoverished and are often dominated by only one or two species. The longest such interval followed the late Permian extinction (the largest of the Big Five): many major phyla and classes, known to have survived from later occurrences, are absent from the early Triassic assemblages. And about a third of the Triassic is characterized by what has been called the "coal gap," an interval where no coal deposits have been found—either of temperate or of tropical origin (A. M. Ziegler, University of Chicago, pers. comm., 1993).

When full diversity does return, it often has a strikingly different character. A classic example is the history of marine reefs. Reef communities have been wiped out several times in the past 600 Myr, coinciding in four cases with Big Five events. Each time reefs reappear, the principal framework organisms have changed, switching back and forth between calcareous algae, sponges, bryozoans, rudist mollusks, and various corals (Sheehan, 1985; Copper, 1988). The contemporary term "coral reef" describes only the current occupants of that adaptive zone.

Selectivity

Darwin argued that all extinction is selective: species not able to compete with other species die out. In one of the passages quoted above

(p. 337), Darwin expressed confidence that if ancient species could be re-created today and put in competition with their modern counterparts, the old species would be "beaten and exterminated." This is definitely not the current view, and major research programs are now being devoted to determining the extent, if any, of selectivity in past extinctions. A common (though by no means proven) view is that the victims of extinction are in no way different from the survivors, except for the fact of their extinction. Simpson (1944) was clearly moving in this direction when he suggested in *Tempo and Mode* that the mammals were the lucky recipients of space vacated by the dinosaurs.

Taxonomic Selectivity. Much of current extinction research attempts to identify taxonomic selectivity. Do some taxonomic groups suffer significantly more species extinction in an extinction episode than other groups? These studies can take advantage of the availability of taxonomic data bases, such as those compiled by Sepkoski for marine genera (Sepkoski, 1989) and families (Sepkoski, 1992), and thus have the benefit of large samples. The approach carries the tacit assumption that genealogical relatedness implies similarity of physiology, ecology, or other attributes that determine susceptibility to extinction.

Taxonomic selectivity has been documented, but the effect is generally quite small and requires massive samples for confirmation. For example, when extinction rates for several taxonomic groups are compared with the mean for all groups, about 10% differ from the mean at a 0.05 significance level, whereas 5% would be expected by chance. Similarly, about 2% of the tests are significant at the 0.01 level. Thus, taxonomic selectivity is present but minor.

Occasionally, pronounced taxonomic selectivity has been found. The dinosaur extinction is such a case. In the latest Cretaceous of western North America, Clemens (1986) tabulated 117 genera of fossil mammals, amphibians, reptiles, and fish; 50 of these (43%) died out at or near the end of the Cretaceous, including all 22 dinosaur genera. The null hypothesis that all genera shared the same probability of extinction (0.43) can be rejected easily, and this demonstrates a clear bias against dinosaur survival. But such cases are relatively rare. In the same data set, for example, 8 of 24 mammalian genera died out (33%), but because of the small sample size, it is impossible to demonstrate that this extinction rate is significantly lower than the mean for all groups.

Small sample sizes have plagued many studies of selectivity, giving rise to generalizations that are widely accepted but not supportable statistically. For example, it is often claimed that the amphibians survived the K–T event with little difficulty. In the Clemens data set, only one-third of the amphibian genera went extinct (equal to the mammalian rate), but the sample size (12 genera) is, again, much too small for statistical significance.

Selectivity for Specific Traits. Somewhat more success has been achieved by focusing directly on aspects of physiology, behavior, habitat, and biogeography. Jablonski (1986a) showed, for example, that marine mollusks with planktonic larvae survive longer during "background" times than those which develop directly from the egg. This result is reasonable because species with a planktonic stage have greater dispersal capabilities and can attain wider geographic distributions, and thus are more likely to survive stresses that eliminate species in small areas. Whereas Jablonski has confirmed that widespread species are significantly less likely to go extinct during most geologic intervals, he has also shown that this protection breaks down at times of severe mass extinction, when broad geographic range at the genus level becomes important (Jablonski, 1986b).

It is commonly thought that tropical organisms suffer more extinction than those in higher latitudes. Although this is supported by anecdotal data for several extinction events, a recent study based on a global data base (3514 occurrences of 340 genera) found no recognizable geographic pattern in extinction of bivalve mollusks at the end of the Cretaceous, once reef-dwelling rudists were omitted (Raup and Jablonski, 1993). Approximately 50% of all genera died out, regardless of geographic position. Unpublished follow-up studies show a lack of habitat selectivity for bivalves and gastropods of the Gulf Coast during the same interval (D. Jablonski, University of Chicago, pers. comm., 1993). It may be that extinction is selective when overall extinction rates are low (so-called background extinction), but not when rates are high.

Large body size is thought to increase the risk of extinction. Indeed, many apparently good examples exist, including the now-extinct dinosaurs, ammonites, eurypterids, mammoths and mastodons, and rudist clams (LaBarbera, 1986). In the terrestrial realm, at least, decreased survival can be related easily to body size through demographic considerations (small populations, large home range, low birth rates, etc.). But the issue is clouded by the lack of rigorously controlled statistical analysis and by the fact that in several cases (e.g., eurypterids and ammonites), the largest species did not exist late in the group's range.

Species Selection. A special case of selective extinction involves differential origination and extinction of species in an evolving clade. In theory, species carrying a favored trait should survive longer and thus have greater opportunity for speciation than less well-adapted species.

The expected result is an increase in frequency of species carrying the favored trait. The species-selection idea is close to Darwin's view of selective extinction among closely related species. Unfortunately, well-documented cases of species selection are few. Also, there has been vigorous disagreement among evolutionary biologists and paleobiologists on whether species selection could ever be an important force in evolution. On the one hand, it is argued that species selection can alter frequencies of alternative traits, even eliminating some, but cannot be responsible for complex adaptations such as eyes and limbs (Levinton *et al.*, 1986). The counterargument is that adaptations, originating at the population level by natural selection, may be sustained, even "cloned," by species selection, thus making possible further improvements in the trait by natural selection operating within the descendent species. By the latter scenario, species selection plays a useful and possibly indispensable part in the evolution of those adaptations (including complex organs) that require more time than is available during the life span of a single species. Again, it must be emphasized that there are too few authenticated cases of species selection to build a strong case for or against its role in evolution.

Units of Selectivity. To the extent that selective extinction occurs, it may operate at several alternative hierarchical levels. In the dinosaur extinction, all species of two major orders (Saurischia and Ornithischia) were eliminated. Presumably these species had something in common that made them all susceptible to the environmental stresses of the terminal Cretaceous. Selectivity at this high level is the only way by which highly diverse groups of species (classes or orders) have a measurable probability of being eliminated completely. It can be shown easily that if the determinants of extinction were at the species level, independent of membership in a larger group, diverse clades would never die out (Raup, 1981). But the fossil record contains ample instances of large, successful clades going extinct, either gradually (e.g., trilobites) or suddenly (e.g., ammonites).

Summary. Extinction is evidently selective at certain times and places but the effects tend to be subtle and require careful analysis of large data bases. Darwin's contention that all extinction is selective cannot be sustained, although this may reflect only our inability to recognize complex patterns in an imperfect fossil record.

Causes of Extinction

A remarkable feature of the history of life is that so many successful species have died out. Many of the extinctions recorded in the fossil record are of species or large groups of species that were ecologically tolerant and occurred in great numbers in all parts of the world. If these extinctions were caused by slow declines over long periods of time, as Darwin thought, they might be explicable in terms of the cumulative effect of very slight deficiencies or disadvantages. But it is becoming increasingly clear that successful species often die out quickly. This is best documented for the K–T mass extinction because of the extensive field work inspired by the controversy over the cause of that event. Several important biologic groups, including the ammonites and dinosaurs, now appear to have existed at full diversity right up to the K–T boundary (Ward, 1990; Sheehan *et al.*, 1991).

For a species to survive for several million years, as many do, it must be well adapted to the physical and biological stresses normal in its environment. Tree species, for example, that can withstand, or even benefit from, forest fires have presumably evolved this ability because forest fires are common in their environment. It may well be that most species have evolved ways of surviving anything that their environment can throw at them, as long as the stress occurs frequently enough for natural selection to operate. This implies, in turn, that likely causes of extinction of successful species are to be found among stresses that are *not* experienced on time scales short enough for natural selection to act.

The recent Pleistocene glaciation produced very few complete extinctions of species. To be sure, extinction rates during the last deglaciation were high among large mammals and some bird groups but overall, global data (including marine organisms) show the Pleistocene to be on the left-hand tail of the distribution in Figure 2. A reasonable explanation is that although the glaciation was associated with marked shifts in climatic regimes, most species were already equipped to cope with the changes by natural physiological tolerance, by having populations in refugia, or by having the ability to migrate to more favorable areas. This appears to be especially true in the marine realm (Clarke, 1993).

In view of the foregoing, recent hypotheses for extinction caused by the catastrophic effects of extremely rare physical events (e.g., asteroid or comet impact, global volcanism) have great appeal.

The Role of Extinction

Despite many uncertainties, we can formulate a reasonable statement of the probable role of extinction, containing the following elements.

(*i*) Extinction of a widespread species, or a widespread group of species, requires an environmental shock (physical or biological) which is not normally encountered during the geological life spans of such species or groups, and the shock must be applied rapidly enough over a broad geographic area to prevent adaptation by natural selection or escape by migration. If the most effective extinction mechanisms are beyond the experience of the victims, a high degree of apparent randomness should be expected. Survivors are most likely to be those organisms which are fortuitously preadapted to an "unexpected" stress (Raup, 1991b).

(*ii*) The most intense episodes of extinction, like the Big Five, produce major restructuring of the biosphere. Three-quarters, or more, of the standing diversity is removed, and diversification of the surviving lineages yields a global biosphere very different from that before the extinctions. Previously successful clades are lost, and unlikely survivors expand. Although the extinction does not, by itself, make a creative contribution to the evolution of complex structures such as wings or limbs, it may be decisive in sustaining or eliminating such structures.

The pterosaurs died out in the latest Cretaceous, and reptiles never again achieved powered flight. Did this foster the Tertiary radiation of bats? What further adaptations might pterosaurs have evolved had they survived? Seen in this light, the major extinctions have a profound influence on the future course of evolution, sometimes constructive and sometimes destructive.

(*iii*) At lower levels of extinction intensity, Darwin-style selectivity may be relatively common, but except for a few spectacular cases, including Jablonski's studies of the effects of larval development and geographic range (Jablonski, 1986a), we do not have enough hard evidence to claim that low-level extinction has anything approaching the importance given it by Darwin. Further studies in this area, under the rubric of species selection, are sorely needed.

SUMMARY

The extinction of species is not normally considered an important element of neodarwinian theory, in contrast to the opposite phenomenon, speciation. This is surprising in view of the special importance Darwin attached to extinction, and because the number of species extinctions in the history of life is almost the same as the number of originations; present-day biodiversity is the result of a trivial surplus of originations, cumulated over millions of years. For an evolutionary biologist to ignore extinction is probably as foolhardy as for a demographer to ignore mortality. The past decade has seen a resurgence of interest in extinction, yet research on the topic is still at a reconnaissance level, and our present understanding of its role in evolution is weak. Despite uncertainties, extinction probably contains three important elements. (*i*) For geographically widespread species, extinction is likely only if the killing stress is one so rare as to be beyond the experience of the species, and thus outside the reach of natural selection. (*ii*) The largest mass extinctions produce major restructuring of the biosphere wherein some successful groups are eliminated, allowing previously minor groups to expand and diversify. (*iii*) Except for a few cases, there is little evidence that extinction is selective in the positive sense argued by Darwin. It has generally been impossible to predict, before the fact, which species will be victims of an extinction event.

I thank David Jablonski for many helpful discussions and for his helpful comments on the manuscript. This research was supported by National Aeronautics and Space Administration Grants NAGW-1508 and NAGW-1527.

REFERENCES

Alvarez, L. W., Alvarez, W., Asaro, F. & Michel, H. V. (1980) Extraterrestrial cause for the Cretaceous-Tertiary extinction. *Science* **205,** 1095–1108.

Clarke, A. (1993) Temperature and extinction in the sea: a physiologist's view. *Paleobiology* **19,** 499–518.

Clemens, W. A. (1986) Evolution of the vertebrate fauna during the Cretaceous-Tertiary transition, in *Dynamics of Extinction,* ed. Elliot, D. K. (Wiley-Interscience, New York), pp. 63–85.

Copper, P. (1988) Ecological succession in Phanerozoic reef ecosystemns: Is it real? *Palaios* **3,** 136–152.

Darwin, C. (1859) *On the Origin of Species* (Murray, London).

Dodson, P. (1990) Counting dinosaurs: How many kinds were there? *Proc. Natl. Acad. Sci. USA* **87,** 7608–7612.

Jablonski, D. (1986a) Causes and consequences of mass extinctions: a comparative approach, in *Dynamics of Extinction,* ed. Elliot, D. K. (Wiley-Interscience, New York), pp. 183–229.

Jablonski, D. (1986b) Mass and background extinctions: the alternation of macroevolutionary regimens. *Science* **231,** 129–133.

Jablonski, D. (1991) Extinctions: a paleontological perspective. *Science* **253,** 754–757.

LaBarbera, M. (1986) The evolution and ecology of body size, in *Patterns and Processes in the History of Life,* eds. Raup, D. M. & Jablonski, D. (Springer-Verlag, Berlin), pp. 69–98.

Levinton, J. S. et al. (1986) Organismic evolution: the interaction of microevolutionary and macroevolutionary processes, in *Patterns and Processes in the History of Life,* eds. Raup, D. M. & Jablonski, D. (Springer-Verlag, Berlin), pp. 167–182. [other authors: Bandel, K., Charlesworth, B., Muller, G., Nagl, W., Runnegar, B., Selander, R. K., Stearns, S. C., Turner, J. R. G., Urbanek, A. J. & Valentine, J. W.]

Lyell, C. (1833) *Principles of Geology* (Murray, London), Vol. 3.

Mayr, E. (1964) *Index to Facsimile Edition of Darwin's Origin of Species* (Harvard Univ. Press, Cambridge, MA), pp. 497–513.

Raup, D. M. (1981) Extinction: bad genes or bad luck? *Acta Geol. Hispan.* **16,** 25–33.

Raup, D. M. (1991a) A kill curve for Phanerozoic marine species. *Paleobiology* **17,** 37–48.

Raup, D. M. (1991b) *Extinction: Bad Genes or Bad Luck?* (Norton, New York).

Raup, D. M. (1992) Large-body impact and extinction in the Phanerozoic. *Paleobiology* **18,** 80–88.

Raup, D. M. (1993) Extinction from a paleontological perspective. *Eur. Rev.* **1,** 207–216.

Raup, D. M. & Jablonski, D. (1993) Geography of end-Cretaceous marine bivalve extinctions. *Science* **260,** 971–973.

Sepkoski, J. J., Jr. (1989) Periodicity in extinction and the problem of catastrophism in the history of life. *J. Geol. Soc. London* **146,** 7–19.

Sepkoski, J. J., Jr. (1992) A compendium of fossil marine families. *Milwaukee Contrib. Biol. Geol.* **83,** 1–156.

Sheehan, P. M. (1985) Reefs are not so different—they follow the evolutionary pattern of level-bottom communities. *Geology* **13,** 46–49.

Sheehan, P. M., Fastovsky, D. E., Hoffmann, R. G., Berghaus, C. B. & Gabriel, D. L. (1991) Sudden extinction of the Dinosaurs: Latest Cretaceous, Upper Great Plains, U.S.A. *Science* **254,** 835–839.

Simpson, G. G. (1944) *Tempo and Mode in Evolution* (Columbia Univ. Press, New York).

Van Valen, L. (1973) A new evolutionary law. *Evol. Theory* **1,** 1–30.

Ward, P. D. (1990) A review of Maastrichtian ammonite ranges. *Geol. Soc. Am. Spec. Pap.* **247,** 519–530.

7

Tempo and Mode in the Macroevolutionary Reconstruction of Darwinism

STEPHEN JAY GOULD

We yearn to capture the essence of complexity in a line. Rabbi Hillel (*ca.* 30 B.C.–A.D. 10) wrote: "What is hateful to you do not do to your neighbor. That is the whole Torah. The rest is commentary." And Marcus Aurelius, a century later and a culture apart, stated: "Look to the essence of a thing, whether it be a point of doctrine, of practice, or of interpretation."

But conceptual complexity is not reducible to a formula or epigram (as we taxonomists of life's diversity should know better than most). Too much ink has been wasted in vain attempts to define the essence of Darwin's ideas, or Darwinism itself. Mayr (1991) has correctly emphasized that many different, if related, Darwinisms exist, both in the thought of the eponym himself, and in the subsequent history of evolutionary biology—ranging from natural selection, to genealogical connection of all living beings, to gradualism of change.

It would therefore be fatuous to claim that any one legitimate "essence" can be more basic or important than another. Yet I wish to focus on a Darwinism that is more pervasive than some of the other meanings—a status won by its role as the fundamental operational, or methodological postulate of all Darwin's theorizing and experimentation.

Stephen Jay Gould is professor of geology in the Museum of Comparative Zoology at Harvard University, Cambridge, Massachusetts.

Darwin's Uniformitarianism and the Downgrading of Macroevolution

Charles Lyell was Darwin's guru and intellectual father figure. Darwin commented, in a statement that (for once in his writing) does not reek of false modesty in proper Victorian taste, "I always feel as if my books came half out of Lyell's brain" (in F. Darwin, 1903, p. 117). Much of Lyell's thinking did not contribute to Darwin's evolutionism and may have acted as an impediment to transmutation—in particular, Lyell's steady-state vision of change without direction. But we can scarcely doubt that Lyell's major working postulate and philosophical premise—his uniformitarian vision—became just as firmly embedded in Darwin's thought and scientific action.

Lyell's uniformitarianism held that the full panoply of past events, even those of greatest extent and apparent effect, must be explained as extrapolations from causes now operating at their current observable rates and intensities. In other words, and invariably, the small and immediate may be extended and smoothly accumulated—drop by drop and grain by grain—through time's immensity to produce all scales of historical events. Time is the great enabler. No uniquenesses should be attributed to events of large scale and long times; no principles need be established for the great and the lengthy; all causality resides in the smallness of the observable present, and all magnitudes may be explained by extrapolation.

Darwin accepted and promulgated Lyell's uniformitarian vision in all its uncompromising intensity. Extrapolationism (the methodological side of uniformity) underlies and unites the otherwise disparate pieces and opinions in the *Origin of Species.* What other principle could coordinate, for example, Darwin's hostility to mass extinction (1859, pp. 317–329), his brilliant section on graded structural transition in the evolution of complex and "perfect" organs like the eye (pp. 186–189), his initial case of pigeon breeding as a model for change at all scales (pp. 20–28), and even his choice of the phrase "natural selection" as an analogy to small-scale changes produced by breeders and called "artificial selection."

Consider just two statements from the *Origin of Species* on the power of geological time to build small and present changes into any observed or desired effect. First, on nature's greater power based on time and fuller scrutiny:

> As man can produce and certainly has produced a great result by his methodical and unconscious means of selection, what may not nature effect? Man can act only on external and visible characters: nature cares nothing for appearances. . . She can act on every internal organ, on every shade of constitutional difference,

on the whole machinery of life. . . How fleeting are the wishes and efforts of man! how short his time! and consequently how poor will his products be, compared with those accumulated by nature during whole geological periods. (1859, p. 83)

Second, on time's promotion of the infinitesimal to great magnitude:

It may be said that natural selection is daily and hourly scrutinising, throughout the world, every variation, even the slightest; rejecting that which is bad, preserving and adding up all that is good; silently and insensibly working, whenever and wherever opportunity offers, at the improvement of each organic being in relation to its organic and inorganic conditions of life. We see nothing of these slow changes in progress, until the hand of time has marked the long lapse of ages. (1859, p. 84)

The pure extrapolationism of Darwin's uniformitarian perspective creates an enormous, if not fatal, problem for paleontology. We would like to be a source of meaningful evolutionary theory, for this discipline explains the patterning of the objects we study. But if every event at our scale may be built by extrapolation from a present that contains all causes, then we have no theoretical contribution to make. We are still needed in a lesser role, of course, for history is massively contingent, as Darwin well knew, and theory must therefore underdetermine actual events. But paleontology, in this status, only provides phenomenology—a descriptive accounting, dedicated to documenting that life followed this particular pathway, rather than another route equally plausible in theory. Moreover, paleontology, in Darwin's view, cannot even provide particularly good phenomenology (however honored *faute de mieux*) because an imperfect fossil record so blots, confuses, and distorts the pathway. Remember that Darwin's first geological chapter bears no triumphant title, but rather the apologetic: "On the Imperfection of the Geological Record."

The demotion imposed by pure extrapolationism—to description devoid of theory—must be the chief source of paleontology's curiously low and almost ironic reputation: to be beloved and glamorized by the public (with a series of images from *Indiana Jones* to *Jurassic Park*), and almost invisible within professional halls of status and funding. Consider two assessments of our absent contribution to evolutionary theory. Sadly, as Julian Huxley notes in beginning the first quote, paleontologists have often defended their own debasement—an all too common phenomenon noted among slaves, hostages, and other oppressed people who adopt the assessments of their captors (psychologists even have a label for it, as the Patty Hearst syndrome). Huxley wrote in the book that gave our theory its name (1942, p. 38):

As admitted by various paleontologists, a study of the course of evolution cannot be decisive in regard to the method of evolution. All that paleontology can do in this latter field is to assert that, as regards the type of organisms which it studies, the evolutionary methods suggested by the geneticists and evolutionists shall not contradict its data.

And even so iconoclastic a morphologist as D. Dwight Davis stated for the Princeton meeting on genetics, paleontology, and evolution (1949, p. 77)—the gathering that oversaw the foundation of our major professional society and its journal, *Evolution*:

Paleontology supplies factual data on the actual rates of change in the skeleton and the patterns of phyletic change in the skeleton. Because of the inherent limitations of paleontological data, however, it cannot perceive the factors producing such changes. Attempts to do so merely represent a superimposition of neobiological concepts on paleontological data.

Such invalid statements in professional publications often follow an unfortunate path towards inclusion in basic textbooks—and errors in this particular medium are almost immune to natural selection, as extinction-proof as a living fossil in the deep ocean. One major, and very fine, introductory text (Allen and Baker, 1971, p. 524) states:

Evolution can be studied on the population level only with living organisms. The fossil record provides too few data to allow such treatment; it merely allows paleontologists to reconstruct the history of animal and plant groups [the restriction of our efforts to descriptive phenomenology]. The population approach makes it possible to ask such questions as: What is the rate of evolution in a given species? What factors influence the course or rate of evolution? What conditions are necessary for evolution to begin or cease?

Funny. I would include these three questions within a set most amenable to resolution by the data of fossils and their temporal distribution!

As a final illustration of the reductionistic biases that still beset this most comprehensive of fields, and of the usual tendency to ignore or devalue theory based on whole organisms or long times, the assigned reporter for *Science* magazine presented a remarkably skewed and parochial view of the conference that honored Simpson's *Tempo and Mode* at its half-century, and formed the basis for this published symposium (Cohen, 1994). The meeting itself was broad and comprehensive, with talks spanning a full range of levels and durations, from molecules at moments to faunas over geological periods. Yet the reporter ignored about two-thirds of the presentations, including all from Simpson's own professional domain, and focused entirely upon molecular insights—a central issue to be sure, but surely not the exclusive or even the primary theme of a meeting called to honor Simpson's work and its sequelae. Under the headline "Will Molecular Data Set the Stage for a Synthesis," *Science*'s one-dimensional reviewer got Simpson's title wrong and then stated:

tradition for denying such a claim, and asserting the need for principles contrary to Darwinism in explaining evolution in the fullness of time (orthogenesis, various forms of vitalism and finalism), Simpson felt especially compelled to argue that the entire past, in all magnitude and duration, could be fully encompassed by extrapolation from microevolutionary principles of the moment—Darwinian uniformitarianism in its purest form.

Thus, although Simpson did enunciate a methodology—modes from tempos—for discovering uniquely macroevolutionary theory, he applied the procedure to deny this possible outcome. In other words, he developed a method that might have yielded theory, and then claimed that none was to be found. And this conclusion was no passive or subsidiary result of other purposes, but the central goal—and, in Simpson's view, the intellectual triumph—of his work. Paleontology became a dutiful son to the synthesis, and no longer an unruly child. Simpson concluded, with evident satisfaction (1944, p. 124):

> The materials for evolution and the factors inducing and directing it are also believed to be the same at all levels and to differ in mega-evolution only in combination and in intensity. From another point of view mega-evolution is, according to this theory, only the sum of a long, continuous series of changes that can be divided taxonomically into horizontal phyletic subdivisions of any size, including subspecies.

2. Simpson's later moves to greater conventionality. Of his three evolutionary modes, Simpson always emphasized the one—phyletic evolution—most supportive of extrapolationism, for trends in the phyletic mode work by pure, step-by-step anagenetic accumulation, the "march of frequency distributions" through time. He exalted the phyletic mode as primary by two strategies. First, by asserting the predominant relative frequency of this maximally extrapolationist mode—the 9/10 figure previously cited: "Nine-tenths of the pertinent data of paleontology fall into patterns in the phyletic mode" (1944, p. 203).

As a second strategy, he downplayed the other two modes. He saw speciation as a low-level process, capable only of producing iterated variety (and perhaps of protecting adaptations by sorting them into several lineages), but not as participating in sustained evolutionary trends: "This sort of differentiation draws mainly on the store of preexisting variability in the population. The group variability is parceled out among subgroups. . . . The phenotypic differences involved in this mode of evolution are likely to be of a minor sort or degree. They are mostly shifting averages of color patterns and scale counts, small changes in sizes and proportions, and analogous modifications" (1944, p. 201).

But quantum evolution posed a different challenge to the dominance of phyletic extrapolationism. Simpson had never granted quantum evolution a high relative frequency, but he did regard this mode as responsible for some of the most profound anatomical transitions in life's history. In *Tempo and Mode,* Simpson did present quantum evolution as an alternative to the phyletic mode, with different primary causes (though still tolerably uniformitarian in invoking Wright's genetic drift). But Simpson radically changed his view in his larger, and far more conservative, later book, *The Major Features of Evolution* (1953). He now demoted quantum evolution from a separate mode to merely an extreme value in the phyletic spectrum. He began by denying any efficacy to Wright's process: "Genetic drift is certainly not involved in all or in most origins of higher categories, even of very high categories such as classes or phyla" (1953, p. 355). He then redefined quantum evolution as one among four styles of phyletic evolution, all characterized by "the continuous maintenance of adaptation" (1953, p. 385). Quantum evolution was therefore transmogrified from a distinct mode to extrapolative accumulation of adaptive change at fastest rates: Quantum evolution, he now claimed, "is not a different sort of evolution from phyletic evolution, or even a distinctly different element of the total phylogenetic pattern. It is a special, more or less extreme and limiting case of phyletic evolution" (1953, p. 389).

I see a kind of supreme irony in Simpson's argument and its ontogenetic development. He made a brilliant and expansive move in recognizing that paleontology had access to theory through the quantification of tempos and inference of modes. But he then found no theory where it might have resided, and he became ever more wedded to the synthetic proposition that all in time's vastness could be rendered by extrapolation from Darwinian processes seen in the genetics of modern populations. Paleontology therefore remained the subsidiary playing field for a game with rules fully specified elsewhere. Simpson hoped to win respect for paleontology by defining his field as an ally to the synthesis but, as in politics and war, faithfulness without independence will be used to the utmost, but never really honored with equality.

A Solution in Bonded Independence

Dichotomy is both our preferred mental mode, perhaps intrinsically so, and our worst enemy in parsing a complex and massively multivariate world (both conceptual and empirical). Simpson, in discussing "the old but still vital problem of micro-evolution as opposed to macroevolution" (1944, p. 97), correctly caught the dilemma of dichotomy by writing (1944, p. 97): "If the two proved to be basically different, the innumerable studies of micro-evolution would become relatively unimportant and would have minor value for the study of evolution as a whole."

Faced with elegant and overwhelming documentation of microevolution, and following the synthesists' program of theoretical reduction to a core of population genetics, Simpson opted for denying any distinctive macroevolutionary theory and encompassing all the vastness of time by extrapolation. But if we drop the model of dichotomous polarization, then other, more fruitful, solutions become available.

The Synthesis arose in a reductionistic age, as best evidenced by the contemporary "unity of science" movement initiated by philosophers of the Vienna Circle (see Smocovitis, 1992, for a fascinating account of these links), and by the general intellectual context now called modernism, and then so dominant in a variety of fields from architecture to classical music. Modernism's emphasis on the abstract, the simplified, the fully universal, the underlying principles that build the unique and complex from the small and general, all fueled the preference within evolutionary biology for a comprehensive micro-level theory that could build all scales and sizes by smooth extrapolation. Theory introduced at the macro-level seemed antithetical—a true dichotomous contrary—to such a program.

We now live in an age of self-styled "postmodern" reformation—and though this movement has engendered silliness in architecture and incomprehensibility in literature, postmodernism has also greatly benefited intellectual life by stressing themes of pluralism, multi-level causality, virtues of complexity, individuality, and, yes, even a bit of playfulness. Modernism's hegemonic idea of universal reduction to lower-level principles and causes has been replaced by respect for the legitimacy of multiple levels and perspectives and for their causal mechanisms and insights.

In this postmodern context, it should be easy to grasp a stunningly simple and utterly unprofound solution to "the old but still vital problem of microevolution. . . [and] macroevolution." (But you do need the context to see the "obvious," hence the unavailability of this solution under modernism.) I put an ellipsis in Simpson's statement to eliminate the three words that cause all the trouble—"as opposed to." Micro- and macroevolution are not opposed, but neither does one follow by extrapolation from (and therefore become intellectually subservient to) the other. The existence of genuinely independent macroevolutionary theory does not imply that "the innumerable studies of micro-evolution would become relatively unimportant." These studies are vitally important both as controlling in their own domain, and powerfully contributory to macroevolution as well. Contributory, but neither exclusive nor decisive. No dichotomy exists. There is no single pathway of reductive explanation. Our evolutionary world is a hierarchy of levels, each of legitimacy and irreducible worth. I propose no California love-fest of "I'm OK, you're OK." Genuine pluralism is tough minded and rigorous in trying to map theoretical complexity upon our hierarchical world. Empiricism adjudicates, and some levels may turn out to be unimportant in nature, though plausible in theory. But we must entertain the legitimacy of all logically coherent levels in order to find out.

In seeking an independent body of macroevolutionary theory, not construed as contrary to microevolutionary knowledge, but viewed as truly complementary in bonding to produce a more satisfying total explanation, I would focus upon two themes that share the common feature of rejecting Darwin's uniformitarian extrapolationism, not his natural selection (or other major meanings of Darwinism).

1. Causal boundaries between levels, breaking the possibility of smooth upward extrapolation. Darwin's uniformity requires isotropy in extension, all the way from low causal to high phenomenological; nothing in the structure of causation may break the ever-growing inclusion. But if, on the other hand, important new causes arise at higher phenomenological levels of long time or great magnitude—even if most of the results be complementary to those produced by lower-level causes (though they need not be congruent, and may well be contrary or orthogonal)—then the extrapolationist paradigm is invalid.

I believe that nature is so hierarchically ordered in a causal sense and that distinct processes emerge at a series of ascending breakpoints in time and magnitude. To mention the two themes that have been most widely discussed in paleontological literature during the past twenty years:

(i) Punctuated equilibrium and trends within clades. Trends in the anagenetic mode may be understood as pure extrapolation and accumulation by selection (or other processes) operating at sequential moments in populations. But if species tend to be stable after geologically momentary origins, as punctuated equilibrium proposes (see Cheetham, 1986, for a best case; Gould and Eldredge, 1993, for a compendium of support; and Levinton, 1987, and Hoffman, 1989, for opposition), then trends must be described as the differential success of certain species within a clade (as a result of greater longevity, higher propensity to speciate, or biased direction of speciation)—and the reasons for geological success of species are both intrinsically macroevolutionary, and distinct from accumulation by natural selection within a continuously evolving population. Moreover, if the characters causing differential species success are emergent properties of species themselves (Vrba and Gould, 1986), then the reasons for macroevolutionary change by species selection within clades are formally irreducible to conventional Darwinian selection upon organisms within populations.

(ii) Mass extinction and patterns of waxing and waning among clades. Darwin, as noted above, feared and rejected mass extinction (see Raup's article in this symposium)—not because such coordinated dying is inconsistent with natural selection (for nothing in this form of Darwinism guarantees that organisms can adapt to environmental change of such magnitude and rapidity), but because mass extinction breaks the extrapolative causal continuum that the uniformitarian meaning of Darwinism requires. Mass extinctions are not random, but survival through them works by different rules (see Jablonski, 1986, for a general argument; Kitchell, Clark and Gombos, 1986, for an intriguing example) from those that regulate success in Darwinian struggles of normal times. Darwinian accumulation through normal times cannot, therefore, encompass the history of life. If mass extinctions only accelerated, but otherwise coincided in causal direction with events of normal times (the "turning up the gain" model in my terminology—Gould, 1985), or if mass extinctions were only minor patterning agents, then the extrapolative Darwinism of normal times would still rule. But mass extinctions are not coincident, and they are truly massive (up to 96% species death of marine invertebrates in a well-known estimate for the largest, late Permian great dying—Raup, 1979). They are, therefore, causal patterning agents separate from the daily Darwinism of normal times.

2. The hierarchical reconstruction of the theory of natural selection. Darwin's key notion, that natural selection works almost entirely upon individual organisms as primary units, arises from several aspects of his thinking—from, for example, his uniformitarianism (for organisms are the noticeable biological objects of moments), and his overthrow of Paleyan teleology. (What a delicious irony—to claim that good organic design and ecological harmony, once seen as proof of God's wise benevolence, truly arise only as the side consequence of a process with apparently opposite ethical meaning—organisms struggling for their own benefits alone, defined as individual reproductive success.) Classical Darwinism, as a single-level theory causally focused upon organisms, makes sense in traditional terms (while the attempt of Dawkins, 1976, and others to reduce the level of causality even further to genes can only be called hyperdarwinism, or more of the same). Williams (1992, p. 6) correctly identifies conventional Darwinian methodology: "In practice, higher levels of selection are seldom invoked, and biologists routinely predict and find that the properties of organisms are

those expected if selection operates mainly on the varying capabilities of individuals."

In this context, I believe that the most portentous and far-ranging reform and expansion of Darwinism in our generation has been the growing (Lewontin, 1970; Wade, 1978; Gould, 1982), if so far ill-coordinated, attempt to reconstruct the theory of natural selection as a more general process, working simultaneously on biological objects at many levels of a genealogical hierarchy. The revised theory is in no way antithetical to Darwinian natural selection and should be read as an extension rather than a replacement. But the hierarchical theory has a structure very different from conventional, single-level Darwinism working on individual organisms—so the revised theory is a fascinating novelty, not a more inclusive extrapolation. After all, there is a world of difference between the claim that nature's momentarily stable objects are optima or maxima set by one canonical form of selection and the statement that such stabilities are balances among distinct levels of selection that may work coincidentally, orthogonally, or contrarily. Since most of these newly recognized levels are intrinsically macroevolutionary (species selection, clade selection, and some forms of interdemic selection), and since their ways and modes are distinct from conventional natural selection on individuals, the hierarchical theory also affirms a substantial theoretical space for macroevolution and its paleontological basis.

To be a unit of selection, biological objects must embody five basic properties: birth points, death points, sufficient stability through their existence, reproduction, and inheritance of parental traits by offspring. (The first three properties are required to individuate any named item as a distinct entity rather than an arbitrary segment of a continuum; the last two are prerequisites for agents of Darwinian selection, defined as differential reproductive success.) Organisms are the quintessential biological objects endowed with these five properties, hence their role as canonical Darwinian individuals in the basic theory.

But many other kinds of biological objects maintain these five properties, and can therefore act as causal agents of selection. The hierarchical theory is therefore explicitly causal, and not merely phenomenological. We may start with gene selection—not the false Dawkinsian version, which tries to break all higher-level processes down to this supposedly universal locus of causality, but the proper form of genes acting "for" themselves, as in the badly named phenomenon of "selfish DNA" (Doolittle and Sapienza, 1980; Orgel and Crick, 1980). (In the general theory of selection, all objects work for themselves by struggling for differential reproductive success at their own level; multiply replicating DNA, producing no benefits to organisms thereby, can only be viewed as selfish if all evolutionary change be judged by impact upon organisms—the very Darwinian parochialism now superseded by the hierarchical theory!)

Moving up a level, Buss (1987) has made a fascinating case for a distinct form of cell-lineage selection, with cancer as one mark of its pyrrhic victory over conventional selection on organisms. We next encounter ordinary Darwinian selection on organisms, a powerful mode surely responsible for adaptive design of bodies [but not, therefore, *pace* Dawkins (1986), more intrinsic or more important than other evolutionary phenomena, like waxing and waning dominance among clades through geological time—a phenomenon that surely cannot be fully rendered by differential merits of adaptive design among organisms].

Moving to levels above organisms, we first encounter the confusing field of selection among groups or demes within species—a theme once infused with woolly thinking (see Williams, 1966, for historically needed correction) that gave the entire subject a bad name, but now being treated more rigorously and surely containing much of enormous value in various modes termed interdemic, trait-group, etc. (Wright, 1980; Wilson, 1983). Above this complex field, we encounter the two clear levels of truly macroevolutionary selection, largely based upon paleontological data, and capable of producing important phenomena of evolutionary pattern not fully rendered by causes at lower levels—species selection (Stanley, 1975; Jablonski, 1987; Vrba, 1989) for trends within clades and clade selection (Williams, 1992) for differential waxing and waning of monophyletic groups.

The developing field of hierarchical selection theory is beset with conceptual difficulties so thorny that I sometime wonder if our innately dichotomizing minds are sufficiently well constructed for thinking about simultaneous levels interacting in all possible modes (or perhaps I'm just stupid, although the issues seem to beset others as well). Two problems have been paramount in the developing discussion.

1. How shall selection itself be identified and defined? Since we desire an explanatory theory, we must clearly distinguish (see Vrba and Gould, 1986) the causal process of selection (differential survival based on active and intrinsic properties of the biological objects under review), from the descriptive phenomenon of sorting (differential survival that might be causally based upon selection at lower or higher levels, yielding sorting as an effect). Even with this proviso, several partly contradictory criteria for the definition of selection as a causal process at higher levels have been proposed. Most firm and unambiguous, but most elusive and hard to document, is the "emergent trait" approach (Vrba and Gould, 1986; Vrba, 1989), where selection is only identified if explicit features responsible for sorting can be specified as emergent in the objects being sorted.

Differential success based on emergent traits is surely selection by anyone's definition and permits us to speak of genuine adaptation at higher levels—but emergent traits may be rare, and are surely hard to define, often demanding narrative knowledge of selective processes not available from data of fossils.

The "emergent fitness" approach (Lloyd and Gould, 1993) is more general and operational (through use of ANOVA-type models applicable to quantitative data of fossils), but fitnesses are not traits, and the analog to adaptation is thereby lost, along with clear correspondence to vernacular notions of "advantageous." This approach does, however, provide the enormous benefit of including selection upon variability as a legitimate form of causality at higher levels.

Williams (1992) has proposed an even more inclusive definition for clade selection, an interesting position for a man who formerly criticized all proposed forms of group selection so brilliantly, and who became identified thereby as a champion of lower-level selection (Williams, 1966). Williams seems to define as higher-level selection any form of sorting between groups that can be described as nonconcordant with any simultaneously observed mode of sorting at the organism level (see his interesting hypothetical example in 1992, pp. 50–52).

2. How shall the items and units of selection be identified and defined? Two major contributors to this debate on hierarchical selection—Eldredge (1989) and Williams (1992)—have tried to establish parallel hierarchies of equal causal import: genealogical and ecological for Eldredge, material and codical for Williams. I believe that these efforts are ill-advised and that only the genealogical and material sequences should be viewed as causal units participating in Darwinian selection.

Williams makes his distinction between entities and information, speaking (1992, p. 10) of "two mutually exclusive domains of selection, one that deals with material entities and another that deals with information and might be termed the codical domain." But I do not think that the codical domain has meaning or existence as a locus for causal units of selection, for two reasons:

(i) Odd mapping upon legitimate intuitions. Williams uses a criterion of selection that arises from an important literature developed by Hull (1988), Sober (1984), and others on replicators and interactors—and that requires relatively faithful replication across generations in order to qualify an entity as a unit of selection. (Sexual organisms, dispersed and degraded by half in each offspring of the next generation, do not qualify on this criterion—a major argument advanced by gene selectionists for locating causality instead at the lower level of faithfully replicating sequences of DNA.) Williams accepts this criterion for his codical domain, thus leading to the following peculiar position: *genes* are units of selection (as the replicating consequence in the codical domain of selection upon organisms in the material domain); *gene pools* are also units of selection (as replicating consequences of higher-level selection upon groups to clades); whereas *genotypes*, in an intermediate category, are *not* units of selection (except in asexual organisms, where replication is faithful). Thus the codical domain skips a space in the hierarchy, and contains no organismic level of selection (except for asexual forms) because the corresponding codex is impersistent.

This linkage of selective agency to faithful replication has been so often repeated in the past decade that the statement has almost achieved status as dogma in evolutionary theory. Yet I think that this criterion is entirely wrong. Selection isn't about unitary persistence—never was, and never should have been so formulated. Selection is about concentration—that is, the differential passage of more of "youness" into the next generation, an increase in relative representation of the heritable part of whatever you are (whether you pass yourself on as a whole, or in disaggregated form into the future of your lineage). Consider the standard 19th Century metaphor for selection: a sieve. The sieve is shaken, and particles of a certain size become concentrated, as others pass through the net (lost by selection). Integral "you" may be disaggregated in this process, but so long as the next generation contains a relative increase in your particles, and so long as you qualified as an active causal agent of the Darwinian struggle while you lived, then you are a unit of selection (and a winning unit in this case).

(ii) The codical domain as bookkeeping only. We may indeed, and legitimately as a practical measure, choose to keep track of an organism's success in selection by counting the relative representation of its genes in future generations (because the organism does not replicate faithfully and therefore cannot be traced as a discrete entity). But this practical decision for counting does not deprive the organism of status as a causal agent, nor does it grant causality to the objects counted.

The listing of accounts is bookkeeping—a vitally important subject in evolutionary biology, but not a form of causality. I think that Williams's codical domain is not an alternative realm of causality, as he claims, but just a fancy name for the necessary bookkeeping function of evolutionary calculation. Williams almost seems to admit as much in two passages (1992, pp. 13 and 38):

> For natural selection to occur and be a factor in evolution, replicators must manifest themselves in interactors, the concrete realities that confront a biologist. The truth and usefulness of a biological theory must be evaluated on the basis of its success in explaining and predicting material phenomena. It is

equally true that replicators (codices) are a concept of great interest and usefulness and must be considered with great care for any formal theory of evolution. (1992, p. 13)

Fine. But codices are units of information useful in bookkeeping, not material entities "out there" in the Darwinian struggle—and bookkeeping is not causality.

However we ultimately define the levels in a genealogical hierarchy of effective selection upon each, and however we decide to codify the criteria for identifying selection at these levels, the hierarchical, multilevel theory of natural selection should put an end to an unhappy and unhelpful conflict rooted in the false mental tactic of dichotomization: the modes of macro- and microevolution as intrinsically opposed and in battle for a common turf. This model led the Synthesis to deny any theoretical status to macroevolution at all—thus preserving hegemony for a microevolutionary theory that could supposedly encompass all scales by smooth extrapolation. But macroevolution is complementary, not oppositional—and each domain holds unique turf (while maintaining a rich and fascinating interaction with all other realms). A grant of independence and theoretical space to a previously rejected domain does not mark a retreat or a submission, but rather a commitment to probe all the richness of nature with all the mental equipment that our limited faculties can muster. For a fine poet once stated this "Happy Thought" in *A Child's Garden of Verses*:

The world is so full of a number of things,
I'm sure we should all be as happy as kings.

SUMMARY

Among the several central meanings of Darwinism, his version of Lyellian uniformitarianism—the extrapolationist commitment to viewing causes of small-scale, observable change in modern populations as the complete source, by smooth extension through geological time, of all magnitudes and sequences in evolution—has most contributed to the causal hegemony of microevolution and the assumption that paleontology can document the contingent history of life but cannot act as a domain of novel evolutionary theory. G. G. Simpson tried to combat this view of paleontology as theoretically inert in his classic work, *Tempo and Mode in Evolution* (1944), with a brilliant argument that the two subjects of his title fall into a unique paleontological domain and that modes (processes and causes) can be inferred from the quantitative study of tempos (pattern). Nonetheless, Simpson did not cash out his insight to paleontology's theoretical benefit because he followed the strict doctrine of the Modern Synthesis. He studied his domain of *potential* theory and concluded that no *actual* theory could be found—and that a full account of causes could therefore be located in the microevolutionary realm after all. I argue that Simpson was unduly pessimistic and that modernism's belief in reductionistic unification (the conventional view of Western intellectuals from the 1920s to the 1950s) needs to be supplanted by a postmodernist commitment to pluralism and multiple levels of causation. Macro- and microevolution should not be viewed as opposed, but as truly complementary. I describe the two major domains where a helpful macroevolutionary theory may be sought—unsmooth causal boundaries between levels (as illustrated by punctuated equilibrium and mass extinction) and hierarchical expansion of the theory of natural selection to levels both below (gene and cell-line) and above organisms (demes, species, and clades). Problems remain in operationally defining selection at non-organismic levels (emergent traits vs. emergent fitness approaches, for example) and in specifying the nature and basis of levels, but this subject should be the central focus in formulating a more ample and satisfactory general theory of evolution on extended Darwinian principles.

REFERENCES

Allen, G. E. & Baker, J. J. W. (1971) *The Study of Biology* (Addison–Wesley, Menlo Park, CA).

Buss, L. W. (1987) *The Evolution of Individuality* (Princeton Univ. Press, Princeton, NJ).

Cheetham, A. H. (1986) Tempo of evolution in a Neogene bryozoan: Rates of morphological change within and across species boundaries. *Paleobiology* **12,** 190–202.

Cohen, J. (1994) Will molecular data set the stage for a synthesis? *Science* **263,** 758.

Darwin, C. (1859) *On the Origin of Species* (J. Murray, London).

Darwin, F. (1903) *More Letters of Charles Darwin* (Appleton, New York), p. 117.

Davis, D. D. (1949) Comparative anatomy and the evolution of vertebrates. In *Genetics, Paleontology and Evolution,* eds. Jepsen, G. L., Mayr, E., & Simpson, G. G. (Princeton Univ. Press, Princeton, NJ), pp. 64–89.

Dawkins, R. (1976) *The Selfish Gene* (Oxford Univ. Press, New York).

Dawkins, R. (1986) *The Blind Watchmaker* (Norton, New York).

Dobzhansky, T. (1937) *Genetics and the Origin of Species* (Columbia Univ. Press, New York).

Doolittle, W. F. & Sapienza, C. (1980) Selfish genes, the phenotype paradigm and genome evolution. *Nature (London)* **284,** 601–603.

Eldredge, N. (1989) *Macroevolutionary Dynamics: Species, Niches and Adaptive Peaks* (McGraw–Hill, New York).

Gould, S. J. (1982) Darwinism and the expansion of evolutionary theory. *Science* **216,** 380–387.

Gould, S. J. (1985) The paradox of the first tier: An agenda for paleobiology. *Paleobiology* **11,** 2–12.

Gould, S. J. & Eldredge, N. (1993) Punctuated equilibrium comes of age. *Nature (London)* **366,** 223–227.

Hoffman, A. (1989) *Arguments on Evolution* (Oxford Univ. Press, New York).

Hull, D. L. (1988) Interactors versus vehicles. In *The Role of Behavior in Evolution*, ed. Plotkin, H. C. (MIT Press, Cambridge, MA).

Huxley, J. S. (1942) *Evolution. The Modern Synthesis* (George Allen & Unwin, London).

Jablonski, D. (1986) Background and mass extinctions: The alternation of macroevolutionary regimes. *Science* **231,** 129–133.

Jablonski, D. (1987) Heritability at the species level: Analysis of geographic ranges of Cretaceous mollusks. *Science* **238,** 360–363.

Kitchell, J. A., Clark, D. L. & Gombos, A. M., Jr. (1986) Biological selectivity of extinction: A link between background and mass extinction. *Palaios* **1,** 504–511.

Levinton, J. S. (1987) *Genetics, Paleontology and Macroevolution* (Cambridge Univ. Press, New York).

Lewontin, R. C. (1970) The units of selection. *Annu. Rev. Ecol. Syst.* **1,** 1–18.

Lloyd, E. A. & Gould, S. J. (1993) Species selection on variability. *Proc. Natl. Acad. Sci. USA* **90,** 595–599.

Mayr, E. (1991) *One Long Argument: Charles Darwin and the Genesis of Modern Evolutionary Thought* (Harvard Univ. Press, Cambridge, MA).

Mayr, E. (1942) *Systematics and the Origin of Species* (Columbia Univ. Press, New York).

Orgel, L. E. & Crick, F. H. C. (1980) Selfish DNA: The ultimate parasite. *Nature (London)* **284,** 604–607.

Raup, D. M. (1979) Size of the Permo-Triassic bottleneck and its evolutionary implications. *Science* **206,** 217–218.

Simpson, G. G. (1944) *Tempo and Mode in Evolution* (Columbia Univ. Press, New York).

Simpson, G. G. (1953) *The Major Features of Evolution* (Columbia Univ. Press, New York).

Smocovitis, V. B. (1992) Unifying biology: The evolutionary synthesis and evolutionary biology. *J. Hist. Biol.* **25,** 1–65.

Sober, E. (1984) *The Nature of Selection* (MIT Press, Cambridge, MA).

Stanley, S. M. (1975) A theory of evolution above the species level. *Proc. Natl. Acad. Sci. USA* **72,** 646–650.

Stebbins, G. L. (1950) *Variation and Evolution in Plants* (Columbia Univ. Press, New York).

Vrba, E. S. (1989) Levels of selection and sorting with special reference to the species level. *Oxford Surv. Evol. Biol.* **6,** 111–168.

Vrba, E. & Gould, S. J. (1986) The hierarchical expansion of sorting and selection: Sorting and selection cannot be equated. *Paleobiology* **12,** 217–228.

Wade, M. J. (1978) A critical review of the models of group selection. *Quart. Rev. Biol.* **53,** 101–114.

Williams, G. C. (1966) *Adaptation and Natural Selection* (Princeton Univ. Press, Princeton, NJ).

Williams, G. C. (1992) *Natural Selection: Domains, Levels, and Challenges* (Oxford Univ. Press, New York).

Wilson, D. S. (1983) The group selection controversy: History and current status. *Annu. Rev. Ecol. Syst.* **14,** 159–187.

Wright, S. (1980) Genetic and organismic selection. *Evolution* **34,** 825–843.

8

Morphological Evolution Through Complex Domains of Fitness

KARL J. NIKLAS

. . . the central problem of evolution . . . is that of a mechanism by which the species may continually find its way from lower to higher peaks.

Sewall Wright

The history of life is to be studied by a great variety of means, among which special importance attaches to the actual historical record in rocks and the fossils contained in them.

George Gaylord Simpson

A powerful metaphor, proposed by Sewall Wright (Wright, 1931, 1932; Provine, 1986), conceives of evolution as a "local search" for "adaptive peaks" by progressively fitter mutants. This image of a walk over a fitness-landscape forcefully draws attention to the relation between the number and location of fitness peaks, on the one hand, and the number and magnitude of phenotypic transformations among neighboring variants required to increase fitness, on the other (Kauffman, 1993; Maynard Smith, 1970). However, comparatively few attempts have been made to quantify the relation between the topology of landscapes and the dynamics of walks (Eigen, 1987; Gillespie, 1983; Kauffman and Levin, 1987). Among the numerous obstacles to quantitative analyses of Wright's metaphor are (i) the *de minimus* requirement for mapping all possible genotypes onto their corresponding pheno-

Karl J. Niklas is professor of botany at Cornell University, Ithaca, New York.

types for a complete analysis of phenotypic variation (Scharloo, 1991), (ii) the possibility that the fitness contributed by performing one task depends upon the ability to perform other tasks simultaneously (i.e., epistatic fitness-contributions; Ewens, 1979; Franklin and Lewontin, 1970; Lewontin, 1974), (iii) the likelihood that the dynamics of a walk depend upon the walk's point of origin on the landscape in addition to intrinsic genetic or developmental barriers to phenotypic transformations, and (iv) the requirement to treat biologically realistic temporal and spatial variations in the topology of fitness-landscapes, as well as (v) the complex interactions among a panoply of physical and biological variables that collectively define fitness (Kauffman, 1993; Gould, 1980; Levin, 1978).

Nonetheless, the image of the fitness-landscape continues to inspire questions about the tempo and mode of evolution—for example, what is the relation between the number of fitness peaks and the number of functional tasks that an organism must simultaneously perform to grow, survive, and reproduce? Although there is no *a priori* reason to assume that the number and location of phenotypic optima depend upon the number of tasks an organism must perform, there are good reasons to believe that manifold functional obligations author "course-grained" landscapes with many phenotypic optima. For example, engineering theory shows that the number of equally efficient designs for an artifact generally is proportional to both the number and the complexity of the tasks than an artifact must perform (Meredith *et al.*, 1973) because the efficiency with which each of many tasks is performed must be relaxed due to unavoidable conflicting design specifications for individual tasks (Gill *et al.*, 1981), and, as the number of tasks increases, the number of configurations that achieve equivalent or nearly equivalent performance levels increases (Brent, 1973). If such relationships hold true for organisms, these relations may account for the morphological and anatomical diversity seen among even closely related species. Indeed, the sharp logical distinction between "optima" and "maxima," on the one hand, and the observation that a multi-task artifact may assume diverse appearances, on the other, suggest the hypothesis that the imposition of manifold obligations increases the number of equally fit phenotypes.

Although the problematic analogy between engineered and biological systems speaks to the topology of fitness-landscapes, it sheds no light on questions related to the dynamics of walks—for example, what is the relation between the number of tasks that an organism must perform and the magnitude of the morphological transformations between neighboring variants required to reach fitness optima? Are walks confined to nearest-neighbor variants or are they free to reach comparatively distant morphologies? At some level, the number and magnitude of phenotypic transformations comprising a walk must depend upon both the location of fitness peaks and the extent to which the fitness of neighboring variants are correlated. However, it is evident also that the extent to which a walk proceeds depends upon the ability of an organism to alter its phenotype. Although the developmental repertoire of most organisms permits some latitude in external shape and internal structure, walks undoubtedly are governed by genetic or developmental mechanisms that establish barriers to transformations among neighboring variants on the landscape (Alberch, 1980, 1981, 1989; Odell *et al.*, 1981; Oster *et al.*, 1980). Thus, morphological transformations among phenotypes are not equiprobable, and walks cannot be governed exclusively by the topology of the fitness-landscape.

Plants as a Venue for Simulated Walks

The extent to which walks are genetically or developmentally unfettered is a matter of relative rather than absolute degree because it undoubtedly varies among organisms and changes over evolutionary time. For example, the developmental "plasticity" of plants appears extremely high in comparison with most animals (Schmid, 1993; Sultan, 1987; Van Tienderen, 1990). By the same token, certain periods of evolutionary time are characterized by exceptionally high rates of phenotypic innovation as, for example, the colonization of the terrestrial landscape by the first vascular plants (Figure 1). Indeed, one is left with the impression that the walks of plants, in general, and those of early tracheophytes, in particular, feature phenotypic transformations sufficient to achieve many, perhaps most, of the morphological optima widely scattered over their fitness-landscapes.

An added advantage to dealing with plants is that their fitness calibrates closely with the operation of physical laws and processes governing the exchange of mass and energy between the plant body and the external environment, which have remained constant over evolutionary time (Gates, 1965; Brent, 1973; Alberch, 1980, 1981, 1989; Oster *et al.*, 1980; Gill *et al.*, 1981; Odell *et al.*, 1981; Nobel, 1983; Sultan, 1987; Van Tienderen, 1990; Niklas, 1992; Schmid, 1993). Thus, the broad outlines of the fitness-landscape for plants likely have remained comparatively constant. If so, then plants may be the ideal venue for examining the relation between the topology of fitness-landscapes and the dynamics of more or less unrestricted walks.

The assumption that walks are unimpeded over stable fitness-landscapes greatly simplifies attempts to explore the relation between

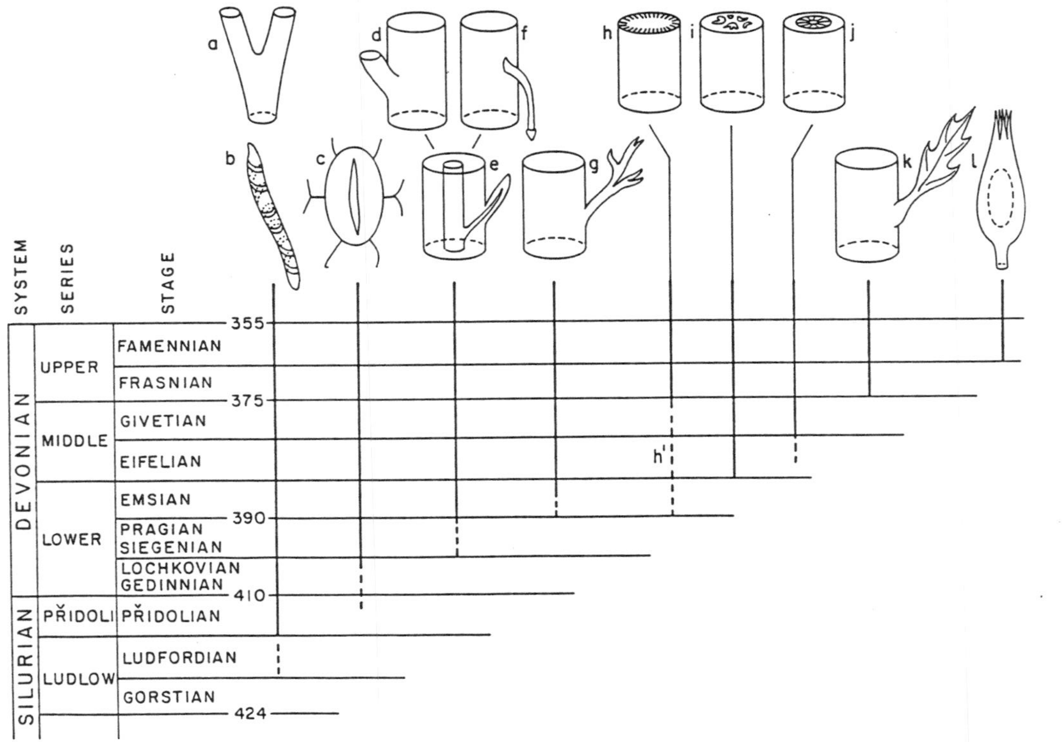

FIGURE 1 First occurrences of some morphological and anatomical features in the fossil record of Upper Silurian and Devonian Period vascular land plants (Niklas, 1992): (*a*) equal branching, (*b*) tracheids, (*c*) stomata, (*d*) unequal branching, (*e*) vascularized appendages, (*f*) adventitious roots, (*g*) planated lateral branches, (*h*) continuous periderm, (*h′*) patches of periderm, (*i*) dissected steles, (*j*) secondary xylem, (*k*) planated, webbed leaves, (*l*) ovules.

landscape topology and the dynamics of walks, particularly in terms of computer simulations. The first step is to simulate a multidimensional domain of all conceivable phenotypes—a "morphospace" (*sensu* Thomas and Reif, 1993). The next step is to determine the ability of every hypothetical phenotype to perform each of a few biologically realistic tasks, in addition to its ability to simultaneously perform various combinations of these tasks—that is, the fitness of phenotypes must be mapped to establish and quantify the topology of the fitness-landscape. Then, beginning with the same ancestral phenotype, a computer algorithm can be used to search the morphospace for successively more fit phenotypes. Simulations of this sort are brought to closure when each phenotypic maximum or optimum within the morphospace is reached by a walk, after which the number and magnitude of the phenotypic transformations in a walk, as well as the number of phenotypic maxima or optima within different fitness-landscapes, are computed and compared. Clearly, to be useful, this heuristic protocol requires nonarbitrary definitions for "morphology," "function," and "ancestor." It also must be cast in terms of a real evolutionary episode against which simulated walks and predicted phenotypic maxima or optima can be compared and contrasted with the actual morphological trends established by the fossil record.

The early evolution of vascular land plants is a case in point. The most ancient tracheophytes had cylindrical, bifurcating axes that lacked leaves and roots (Banks, 1975; Edwards *et al.*, 1992; Stewart and Rothwell, 1993; Taylor and Taylor, 1993). These morphologies are easily simulated by means of a computer with only six variables (Niklas and Kerchner, 1984; Niklas, 1988). Referring to Figure 2, in which each axis of a bifurcate pair is distinguished by the subscript 1 or 2, the six variables are the probabilities of branching P_1 and P_2, the rotation angles subtended between each axis and the horizontal plane γ_1 and γ_2, and the bifurcation angles subtended between the longitudinal axis of each axial member and the longitudinal axis of the subtending member ϕ_1 and ϕ_2. Indeed, a morphospace containing 200,000 phenotypes, encompassing virtually the entire spectrum of early vascular land-plant morphology, can be simulated by establishing the limiting conditions (and increments) for these six variables: $0 \leq P \leq 1$ (in increments of 0.01), and $0° \leq \gamma \leq 180°$ and $0° \leq \phi \leq 180°$ (both in increments of 1°). Within this morphospace, the simplest phenotype (i.e., a Y-shaped plant) results when $P_1 = P_2 = 0$ and $\phi_1 = \phi_2$. Higher values of P produce more complex, highly branched morphologies. Morphologies with equal (isometric) branching are simulated when $P_1 = P_2$. Plants with anisometric (unequal) branching, very much like those that appear in the Devonian Period, are obtained when $P_1 \neq P_2$. And phenotypes with

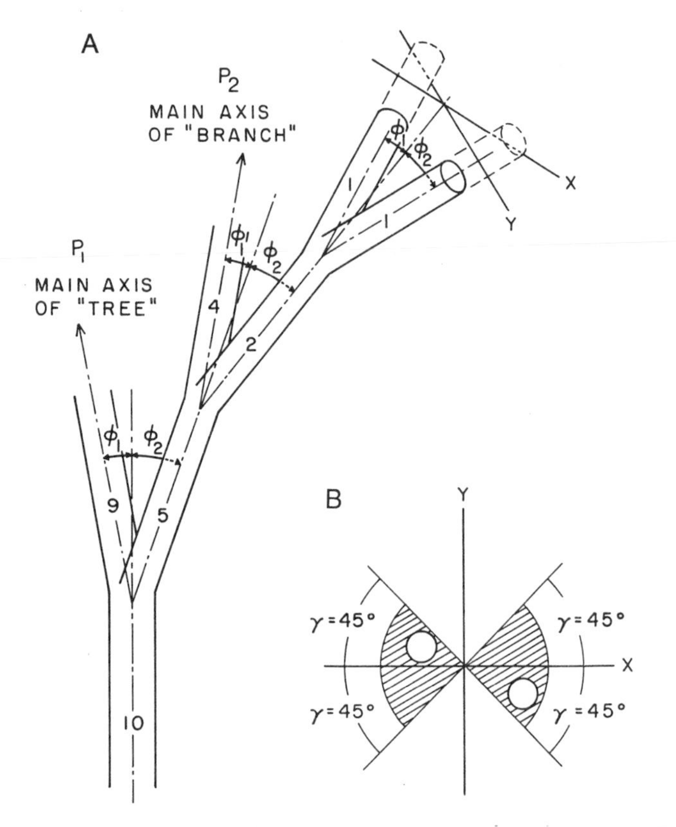

FIGURE 2 Variables used to simulate branched morphology of early vascular land plants: *P*, probability of branching; ϕ, bifurcation angle; γ, rotation angle (subscripts 1 and 2 distinguish parameters for each axis in a branch-pair). (*A*) Three-dimensional bifurcate pattern of branching for which the orientation of each axis is prescribed by ϕ and γ. (*B*) Transverse view of terminal branch-pair in *A* showing the geometric affect of γ.

horizontally flattened (planated) branching systems are simulated when $\gamma_1 = \gamma_2 = 0°$.

Because six variables are required to simulate ancient tracheophytes, hypothetical phenotypes occupy a multidimensional space. Although this makes the graphic display of simulated walks somewhat difficult, the situation is greatly simplified by initiating simulations of walks in the isometric domain of the morphospace (i.e., $P_1 = P_2$) and permitting optima within this domain to enter the anisometric domain of the morphospace (i.e., $P_1 \neq P_2$). Conceptually, this simulation is illustrated in Figure 3*A*.

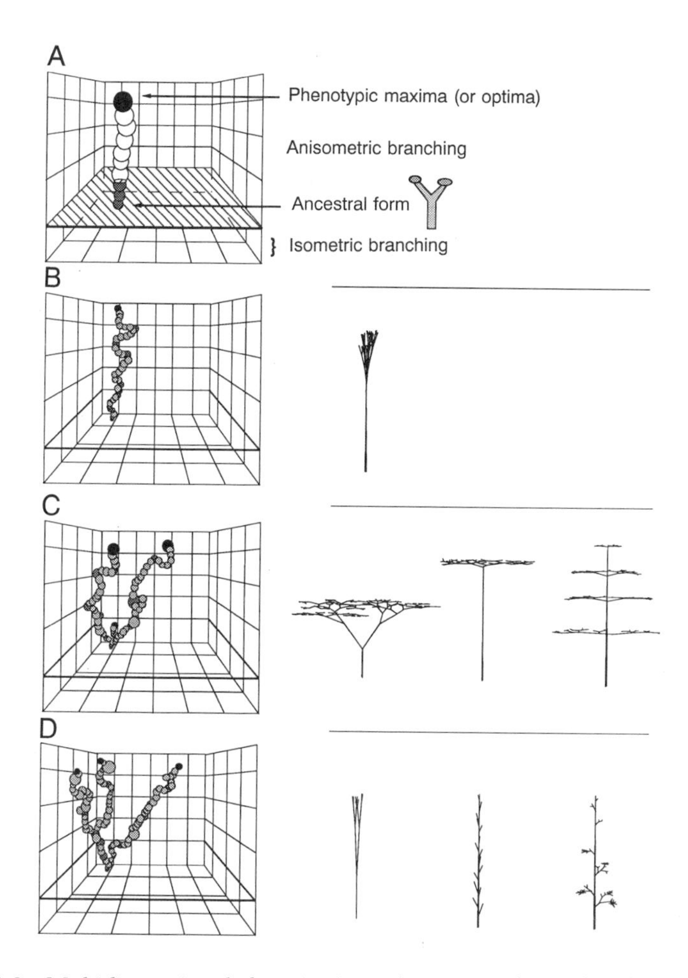

FIGURE 3 Multidimensional domain (morphospace) of simulated branching patterns (*A*) and simulated walks through three single-task fitness-landscapes (*B–D*). (*A*) Simplified version of the morphospace represented in Cartesian space (x, y, z) consisting of two subdomains of shape (isometric and anisometric branching). Each walk is "seeded" with the simplest shape in the isometric domain (i.e., a Y-shaped ancestral form) and represents a series of genetically and developmentally unfettered steps that obtain phenotypic maxima or optima (shown at right). (*B*) A walk through a reproductive-success landscape. (*C*) A walk through a light interception landscape. (*D*) A walk through a mechanical stability landscape (see Table 1).

Turning attention to the topology of the fitness-landscape, the functional obligations assuring growth, survival, and reproduction must be known and quantified. For early vascular land plants, these obligations can be inferred from living tracheophytes and undoubtedly include the requirement to intercept sunlight, to mechanically sustain the weight of aerial organs, and to be able to produce and disperse diaspores some distance from parental plants. Fortunately, each of these tasks can be quantified by means of closed-form equations derived from biophysics or biomechanics. For example, the efficiency E of a computer-simulated phenotype to intercept solar irradiance (of intensity I measured perpendicular to its surfaces) is given by the formula

$$E = \int_{0^\circ}^{180^\circ} \left(\frac{S_P}{S}\right) I \cdot d\theta, \tag{1a}$$

where S_p is the total unshaded surface area of the phenotype projected toward incident light, S is the total surface area of the phenotype, and θ is the incident solar angle, which varies between 0° and 180° in each diurnal cycle (Niklas and Kerchner, 1984). Because the magnitude of I is independent of θ (assuming atmospheric conditions are clear), Eq. **1a** reduces to

$$E = \lim_{\Delta\theta \to 0^\circ} \sum_{\theta=0^\circ}^{\theta=180^\circ} \left(\frac{S_P}{S}\right) \theta \cdot \Delta\theta. \tag{1b}$$

Although the total projected area S_p of each morphology varies as a function of θ, it also depends upon the orientation of axes in addition to the extent to which neighboring axes shade one another. All of these variables can be dealt with by even the most simple computer.

In terms of mechanical stability, the maximum bending stresses σ_{max} that develop in a cylindrical plant axis may be computed from the formula

$$\sigma_{max} = \pm\frac{32M}{\pi d^3}, \tag{2a}$$

where M is the bending moment, which has units of force times length, and + and − denote tensile and compressive stresses, respectively (Niklas, 1992). For any value of d, the bending stresses are directly proportional to the bending moment that, when expressed in terms of ϕ and γ, is given by the formula

$$M = \frac{\pi}{\gamma} d^2 L^2 \rho g \sin\phi, \tag{2b}$$

where ρ is the bulk density of tissues used to fabricate a plant and g is gravitational acceleration. Assuming that the bulk density varies little from one plant to another, $M \propto \gamma^{-1} \sin \phi$. Thus, the bending moment and therefore the maximum bending stresses are minimized when $\phi =$ 90° (i.e., a vertically oriented axis).

In terms of spore dispersal, an elementary ballistic model suffices:

$$x = \frac{HU_{\mathrm{h}}}{U_{\mathrm{T}}},$$

where x is the lateral distance of transport, H is the height at which spores are released from the parent plant, U_{T} is the average settling velocity of spores, and U_{h} is the horizontal wind speed averaged between H and ground level, which is assumed to parabolically diminish from the top to the base of the plant (Okubo and Levin, 1989). Assuming that U_{T} is independent of H and that U_{h} is proportional to H, the maximum lateral transport distance for spores is proportional to the square of plant height, $x \propto H^2$, indicating that even a small increase in plant height confers a selective advantage to spore dispersal. However, the number of spores a plant produces is as important to its reproductive success as the distance spores are transported (Niklas, 1986). Thus, the fitness contribution of the number of spores produced per plant must be considered in addition to the fitness contributed by elevating spores above ground level. Assuming that spores are produced at the tips of branches and that the number of spores per sporangium varies little among hypothetical phenotypes, reproductive fitness R is maximized by maximizing both the number n and height of branch tips: $R = f(n, H)$.

The foregoing implicitly assumes that fitness is proportional to E and R and inversely proportional to M. Assuming that each of these three functional obligations contributes equally and independently to fitness, the most parsimonious mathematical expression for the total fitness F of a phenotype is the geometric mean of E and R divided by M—i.e., $F = [(E)(R)]^{1/2}M^{-1}$. Importantly, the topology of the fitness-landscape for each of the three functional tasks differs because different functional obligations can have different phenotypic requirements. For example, phenotypes that maximize light interception also maximize their bending moments. However, a phenotype with a high fitness in terms of its ability to garner irradiate energy for photosynthesis will have a low fitness in terms of its high probability for mechanical failure. Consequently, when both of these functions are considered simultaneously, the objective of a walk is to optimize E/M. Although E cannot be maximized without maximizing M, some functional obligations reinforce the same morphological solution because their phenotypic requirements are very similar. Phenotypes that maximize the potential for long-distance spore dispersal tend to minimize the bending moment on their vertical axes. Theoretically, therefore, walks that optimize R/M are comparatively direct and simple. From the formula for total phenotypic fitness F, the topology of the fitness-landscape resulting when all three tasks are considered simultaneously is more complex than those resulting when only one or two tasks are considered. Specifically, phenotypes that maximize long-distance spore dispersal (high fitness) also minimize their bending moments (high fitness) but minimize their ability to intercept sunlight (low fitness) because most of their branches are vertically oriented and therefore bunched together.

Once the topology of the fitness-landscape has been quantified, walks must be "seeded"—that is, they must be assigned a nonarbitrary point of origin. Once again, the fossil record for early tracheophytes is invaluable in this regard. The simplest and most ancient phenotype known for vascular land plants is epitomized by the Silurian fossil remains of *Cooksonia*. The sporophyte of this genus consisted of one, or more than one, short-branched cylindrical axes that terminated in sporangia and lacked leaves or roots (Banks, 1975; Edwards *et al.*, 1992). This morphology can be taken as the point of origin for each walk, regardless how fitness is mathematically defined. Each walk proceeds as a sequence of N number of steps, each representing a morphological transformation to a more fit phenotype from the preceding phenotype. The sequence of steps in a walk, therefore, serves to identify more fit phenotypes. The magnitude of each phenotypic transformation can be depicted as the volume of the morphospace that must be searched by a computer-driven algorithm until the next more fit phenotype is reached. The volume searched by each step in a walk can be quantified by its diameter D. Each walk is permitted to branch when two or more phenotypes with equivalent fitness are identified by the algorithm. Each walk is brought to closure when it obtains all the phenotypic maxima within a single-function landscape or all the phenotypic optima within multi-function landscape. The mean diameter $\bar{D}$ and the SE of D for all the steps in a walk quantify the mean variation in the phenotypic transformations required to achieve all maxima or optima within a particular fitness-landscape. The volume fraction VF of the entire morphospace occupied by a walk can be computed from the formula $VF = [(\Sigma D_i)V_T^{-1}] \times 100\%$, where V_T is the total volume of the morphospace. Because the sample statistics obtained from single- and multiple-function fitness-landscapes (i.e., N, n, $\bar{D}$, VF) have unequal variances, the approximate t test, $|t'_s|$, can be used to test for equality of sample means (Snedecor and Cochran, 1980; Sokal and Rohlf, 1981).

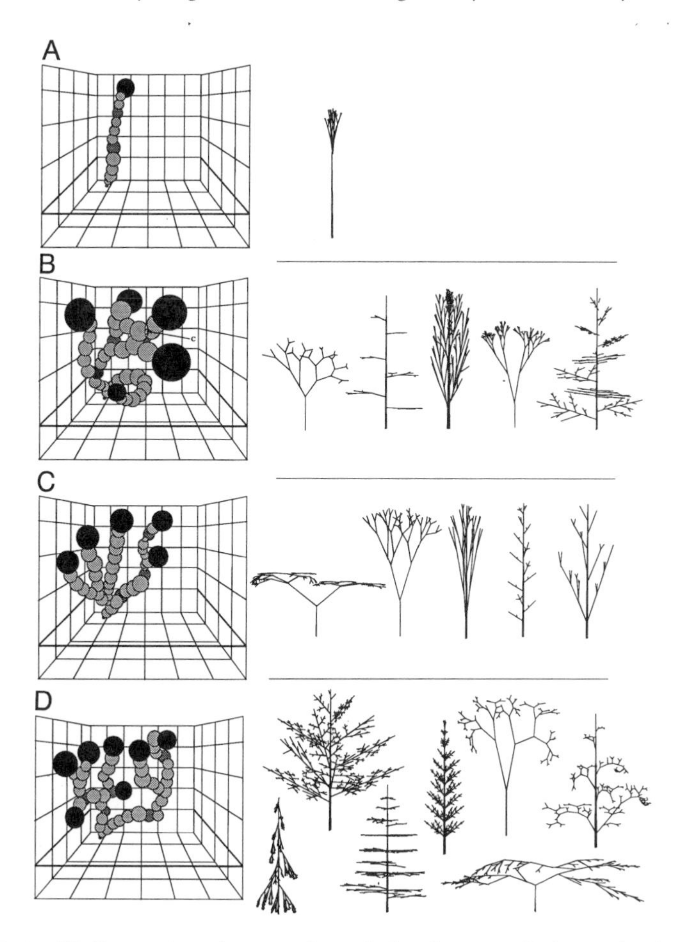

FIGURE 4 Walks within four multi-task landscapes (*Left*) and phenotypic optima (*Right*). (*A*) A walk through a mechanical stability and reproductive success landscape. (*B*) A walk through a light interception and mechanical stability landscape. (*C*) A walk through a light interception and reproductive success landscape. (*D*) A walk through a light interception, mechanical stability, and reproductive success landscape (see Table 1).

Single- Versus Multi-task Walks

Unfettered walks over stable fitness-landscapes are illustrated in Figs. 3 and 4 for a morphospace containing 200,000 phenotypes mimicking early vascular land plants. As noted, the morphospace is multidimensional, consisting of a domain occupied by "ancient" isometrically branched phenotypes and another occupied by more "derived" anisometrically branched morphologies (Figure 3*A*). Consequently, the plots

TABLE 1 Sample statistics for simulated walks in Figs. 3 and 4

	Sample statistics*			
	N	n	$\overline{D}$ (± SE)	VF
Single-task walks				
Light interception, E	116	3	5.81 ± 0.152	0.568
Mechanical stability, M	81	3	5.49 ± 0.186	0.335
Reproductive success, R	56	1	4.37 ± 0.130	0.117
$\overline{Y}$ ± SE	84.3 ± 17.4	2.33 ± 1.15	5.22 ± 0.44	0.34 ± 0.13
Multi-task walks				
E-M-R	49	7	14.4 ± 0.58	3.65
E-R	41	5	15.1 ± 1.02	3.53
E-M	40	5	21.6 ± 1.26	10.1
M-R	13	1	13.7 ± 1.63	0.84
$\overline{Y}$ ± SE	35.8 ± 7.85	4.50 ± 1.26	16.2 ± 1.82	4.52 ± 2.26

*N, number of steps in walk; n, number of phenotypic maxima or optima in landscape; D, mean diameter of steps in walk; VF, volume fraction of morphospace occupied by walk.

of walks are graphic redactions of mathematically more complex features. Every walk proceeds through the same morphospace; walks differ solely as a consequence of differences in fitness topologies. Specifically, walks are shown for three single-task and four multi-task-defined fitness-landscapes. The three individual tasks are light interception E, mechanical stability M, and reproduction R (Figure 3). In turn, these tasks have four combinatorial permutations (Figure 4), three of which give two-task landscapes (i.e., E-M, E-R, M-R) and one of which is a three-task landscape (i.e., E-M-R).

The most apparent differences between simulated single- and multi-task walks are the number and magnitude of their phenotypic transformations, on the one hand, and the number of phenotypic maxima and optima they reach within landscapes, on the other (Table 1). Single-task walks have many small phenotypic transformations within landscapes that contain what appear to be comparatively few phenotypic maxima. By contrast, multi-task walks have few, but large, transformations within landscapes that, at first glance, appear to contain comparatively many phenotypic optima. Also, the mean morphospace volume occupied by multi-task walks is greater than that occupied by single-task walks (Table 1). Statistical comparisons indicate that the hypothesis that multi-task walks require significantly larger phenotypic transformations than those of single-task walks can be accepted (Table 2). However, statistical comparisons show that the mean number of phenotypic maxima in single-task landscapes does not significantly differ from that of optima in multi-task landscapes. In part, this is due to the small sample sizes for each of these two categories of fitness-landscape and to

TABLE 2. Summary of t'_s tests for the equality of $\bar{Y}$ values (see Table 1)

	Parameter*			
	N	n	$\bar{D}$	VF
$\lvert t'_s \rvert$	1.55	2.17	5.87	9.46
$t'_{0.05}$	3.66	3.69	3.25	3.19

*$\lvert t'_s \rvert$, Absolute value of the t test for sample mean; $t'_{0.05}$, approximate critical value of t distribution; $\lvert t'_s \rvert > t'_{0.05}$ indicates that mean values significantly differ at the 5% level.

the fact that the topology of fitness resulting from performing two tasks, mechanical stability and reproduction, contains a single phenotypic optimum (Figure 4*A*) that dramatically depresses the mean and inflates the SE for the mean number of optima. Nonetheless, the largest number of optima observed among all simulated landscapes is attained when fitness is quantified in terms of all three tasks, suggesting that the hypothesis would have been accepted had walks been simulated in landscapes for which fitness was defined in terms of other biological tasks in addition to light interception, mechanical stability, and reproduction.

It is instructive to compare the fitness of phenotypic maxima with the fitness of phenotypic optima. Although every simulated walk is permitted to reach all the maxima or optima in a particular landscape, the elevation of peaks (maxima or optima) differs from one landscape to another because fitness is defined in different terms in each landscape. Because phenotypic maxima and optima occupy fitness peaks, their absolute fitnesses define a landscape's elevation and, therefore, the gradient of the phenotypic transformations attending a walk. In theory, the magnitudes of the fitness of phenotypic maxima are greater than the magnitudes of the fitness of phenotypic optima. Therefore, the observation that the fitness of phenotypic maxima is greater than that of phenotypic optima is somewhat trivial. What is not unimportant, however, is that the "currency" in which fitness is measured differs among landscapes—that is, fitness was measured in different units (e.g., quanta of light absorbed, probability of mechanical failure, distance of spore dispersal). Therefore, comparisons among the elevations of different landscapes can be made only in relative, rather than absolute, terms. The ratio of the fitness of a phenotypic maximum (or optimum) and the fitness of the ancestral phenotype is useful because it normalizes the elevation of peaks and can be used to crudely compare the topologies of very different landscapes.

Noting that the normalized fitness values are ratios, a comparison for the fitness-landscapes shown in Figure 3 with those shown in Figure 4 indicates that the topologic relief of single-task landscapes is, on the average, 10 times greater than that of multi-task landscapes—that is, the fitnesses of phenotypic maxima with respect to the fitnesses of their ancestral condition are 10 times greater than the fitnesses of phenotypic optima with respect to their ancestral condition. Thus, the fitness of phenotypic optima apparently falls closer to the mean fitness of all the phenotypes within a landscape as the functional complexity of the phenotypes under selection increases. As noted, this result is consistent with engineering theory that indicates that the performance levels of artifacts designed to perform individual functional tasks are higher than those of artifacts designed to simultaneously perform two or more of the same tasks. Additionally, within both categories of fitness-landscapes, the relative fitness values of phenotypes occupying fitness peaks decreases as the number of peaks increases ($r = 0.82$; $N = 25$ maxima and optima)—that is, the number of phenotypic maxima (or optima) increases as the elevation of a landscape declines.

These observations are crudely summarized in Figure 5; they suggest that both the number and the accessibility of phenotypic optima increase as the number of functional obligations contributing to total fitness increases. Put differently, as the complexity of optimal phenotypes increases, the fitness of these optima falls closer to the mean fitness of all the phenotypes under selection. One implication of this conclusion is that the majority of walks over complex fitness-landscapes occurs over fitness plateaus and, therefore, is largely undirected by gradients of fitness until walks approach the foothills of fitness peaks.

Weaknesses and Strengths of Simulated Walks

However entertaining they may be, computer-simulated walks have four obvious weaknesses. First, "fitness" was measured in terms of comparatively few biological tasks that were further assumed to contribute to fitness in an independent and equal manner. The obvious epistatic relation between photosynthesis and reproduction (Gates, 1965; Nobel, 1983), therefore, was entirely neglected, as was the possibility that some tasks are more important to fitness than others (Franklin and Lewontin, 1970; Lewontin, 1974; Ewens, 1979). Second, walks were simulated as continuous processions among more fit phenotypes. Alternatively, plausible types of walks were not considered (Gillespie, 1983, 1984; Kauffman, 1993). Third, fitness-landscapes were assumed to be spatially stable in pointed neglect of evident changes in the environment that are predicted to shift the location of fitness optima

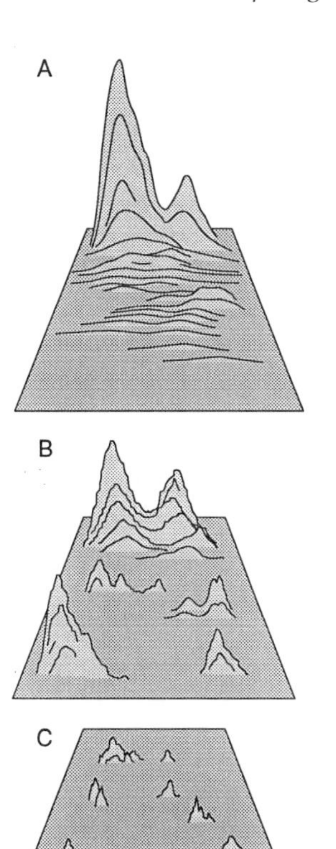

FIGURE 5 Diagrammatic relations among relative phenotypic fitness, the number of phenotypic maxima or optima, and the coarseness of fitness-landscapes. As phenotypic complexity (i.e., the number of biological tasks to be simultaneously performed by organisms) increases, the relative elevation of fitness peaks declines, but the number of phenotypic maxima (or optima) and the coarseness of the landscape increase (*A*–*C*).

and accelerate evolution, particularly in very large panmictic populations (Wright, 1932). And fourth, all walks were assumed to be unfettered by genetic or development constraints. Even for plants, which arguably may be more phenotypically "plastic" than animals, this is a naive expectation (Maynard Smith *et al.*, 1985).

On the other hand, the approach taken here has some obvious strengths. First, walks were simulated over a dimensionally complex morphospace containing phenotypes representative of the entire spectrum of vascular land-plant morphology. Second, although only three were considered, the functional obligations elected to define and quantify fitness are biologically realistic for the majority of past and present terrestrial plant species. Third, although the environment, living and nonliving, is in constant flux, the particular episode of plant evolution focused upon here most likely was dominated by the operation of physical laws and processes. Metaphorically, the fitness-landscape of the first occupants of the terrestrial landscape was painted in the primary colors of biophysics rather than the subtle hues of complex biotic interactions characterizing subsequent plant history. And fourth, the morphological trends predicted by the phenotypic transformations attending simulated walks are, in very broad terms, compatible with those actually seen in the fossil record of early tracheophytes. This correspondence is important on two accounts. First, it suggests that the developmental repertoire of early vascular plants was capable of phenotypic transformations sufficiently dramatic to warrant the assumption that the walks entertained by these organisms were largely developmentally unfettered. Second, if the hypothetical relations between fitness topologies and the dynamics of walks forecast by computer simulations have any relevancy, then they must, at the very least, mimic trends evinced in the fossil record of plants.

There can be little doubt that the phenotypic optima reached by simulated walks, particularly those over the tripartite fitness-landscape, are morphologically complex (Figure 4*D*) nor that all phenotypic maxima or optima are attainable by walks proceeding essentially from the archetypal vascular land plant (i.e., *Cooksonia*). Fossil remains from the Late Silurian Period indicate that the first tracheophytes were comparatively short, consisting of equally branched, naked axes that simultaneously functioned as photosynthetic and reproductive organs. During much of the Devonian Period, the maximum stature of sequentially younger plant taxa steadily increased. This intertaxonomic trend in plant size was attended by significant evolutionary changes in branching morphology, among which unequal branching and horizontally planated lateral branching systems are notable. Importantly, unequal branching is a requisite for and, therefore, prefigures the elaboration of a main vertical axis for mechanical support. By the same token, the planation of lateral branches, which facilitates light interception, has been traditionally interpreted as a precursor to the evolution of megaphylls (Stewart and Rothwell, 1993). Therefore, it is not unreasonable to suppose that the transition from equal to unequal branching, which was evolutionarily rapid and invariably adopted by simulated walks, was a requisite for subsequent morphological trends and likely positioned derived phenotypes at the foothills near (rather than within the valleys between) fitness peaks (Figure 6).

By the latest Devonian Period arborescent species bearing planated lateral branching systems and true leaves were not uncommon (Stewart and Rothwell, 1993; Taylor and Taylor, 1993). This fact is consistent with the observation that the majority of the phenotypic maxima in multi-task landscapes have horizontally planated lateral branching systems, as do the phenotypic maximum in the fitness-landscape defined by light interception. Virtually every major vascular land-plant lineage evolved arborescent, leaf-bearing species, although these species obviously

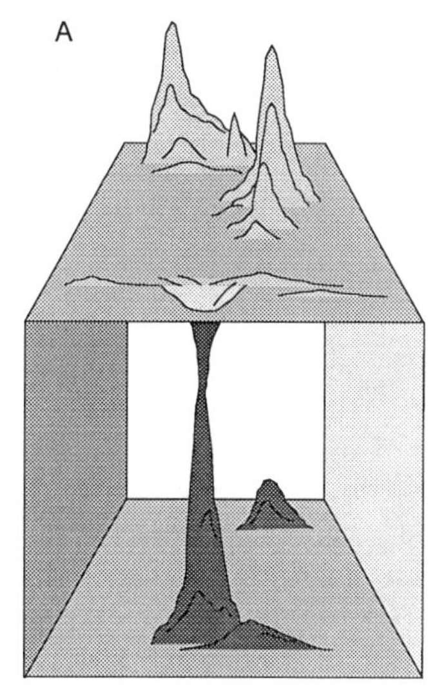

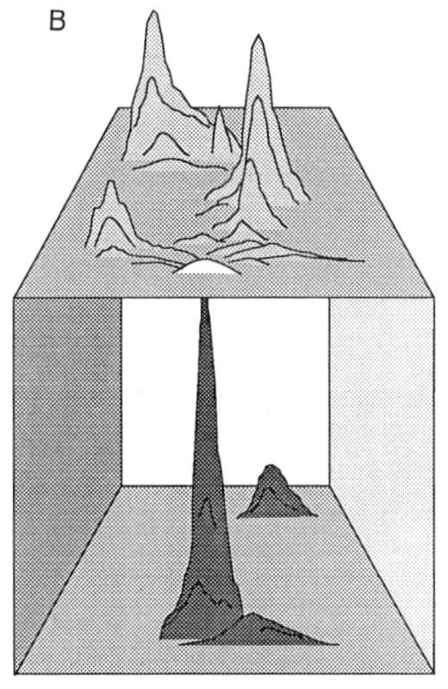

FIGURE 6 Alternative hypothetical topologies of a complex fitness-landscape. In both cases, the ancestral phenotype resides within the lower (boxed) domain, and descendent phenotypic variants must ascend the prominent (darkly shaded) fitness peak in this domain before a walk can proceed over the upper (unboxed) domain. (*A*) The first phenotypic variant entering the upper domain occupies a fitness minimum (valley) distant from fitness peaks. (*B*) The first phenotypic variant entering the upper domain resides on a local fitness-maximum (foothill) near fitness peaks.

differed in morphological, as well as anatomical, details. In very broad terms, the appearance of ancient tracheophyte lineages evincing parallel or convergent phenotypic evolution is a feature mimicked by simulated walks. In all but two fitness-landscapes, walks repeatedly branched to obtain many phenotypic maxima or optima, most of which have a tree-like appearance (i.e., unequally branched with a main vertical "stem") with many lateral, planated "branches" or "leaf-like appendages." Some walks even converge on and cross through the same regions in the hypothetical morphospace of early vascular land plants (see Figure 4*B*).

Another, although highly problematic, parallel that can be drawn between simulated walks and the early evolution of tracheophytes relates to the "dilated" terminal steps of each branch in the walks through multi-task landscapes. These terminal steps indicate that the last phenotypic transformations required to reach optima within a landscape are more sensational compared with prior phenotypic transformations. This hypothesis is consistent with the fact that the highest-per-taxon origination rates as well as the highest rates of appearance of morphological (as well as anatomical) characters occur toward the end of the Devonian Period, which marks the closure of early land-plant evolution (Niklas *et al.*, 1980; Knoll *et al.*, 1984). Interestingly, a significant temporal lag between the first appearance and rapid taxonomic diversification of animal, as well as plant, lineages is not atypical (Sepkoski, 1979; Tiffney, 1981). It should be noted, however, that simulated walks have no temporal component—steps in walks are vectors whose magnitudes are measured in space, not time. Although the volume of the morphospace occupied by a step denotes the phenotypic variance required to reach the next more fit phenotype, the "time" required to achieve this variance cannot be specified.

However tantalizing the similarities between simulated walks and the broad morphological trends seen in early land-plant evolution, they cannot be taken as *prima facie* evidence that the mathematical and statistical properties of simulated walks reflect reality, nor can they be taken as evidence that the early evolution of vascular plant shape was governed by the biological obligations to intercept sunlight, remain mechanically stable, or to disperse large numbers of spores over great distances. However important these biological tasks may be to plant growth, survival, and reproductive success, the correspondence between simulated and empirically determined morphological trends for early tracheophytes may be simply fortuitous. And, under any circumstances, correlation can never be taken as evidence for a cause–effect relationship.

With these caveats clearly in mind, the following hypotheses are tentatively proposed for future study: (1) The fitness-landscapes defined by a single biological task require comparatively small, but numerous, morphological transformations to reach phenotypic maxima (i.e., single-task landscapes are "fine-grained," yet steep), whereas manifold functional obligations obtain "course-grained," but less steep, fitness-landscapes; (2) course-grained landscapes contain a greater number of phenotypic optima than the number of phenotypic maxima in fine-grained landscapes; (3) multi-task-driven walks occupy a greater volume of the morphospace than do single-task-driven walks; (4) multi-task optima have a lower overall fitness than single-task phenotypic maxima; (5) organisms evincing developmental plasticity (and therefore the potential for significant phenotypic variation) (i) evolutionarily benefit by their ability to walk "rapidly" through and occupy a greater volume of the theoretically available morphospace, (ii) will be at increasing advantage as the number of manifold functional obligations increases, but (iii) are at a disadvantage whenever circumstances abruptly change to favor a single functional obligation rather than the full complement of functions.

Wright's metaphor suggests a "trial and error mechanism on a grand scale by which the species may explore the region surrounding the small portion of the fitness field which it occupies" (Wright, 1932). Attempts to cast this and other evolutionary mechanisms in terms of biologically realistic models by means of computer simulations are still very much in their infancy. As heuristic tools, however, simulations designed to forecast opportunistic phenotypic transformations over the topologies of fitness-landscapes illustrate some of the initial steps required to adapt Wright's metaphor to understanding the evolution of plant morphology.

SUMMARY

Computer simulated phenotypic walks through multi-dimensional fitness-landscapes indicate that (1) the number of phenotypes capable of reconciling conflicting morphological requirements increases in proportion to the number of manifold functional obligations an organism must perform to grow, survive, and reproduce, and (2) walks over multi-task fitness-landscapes require fewer but larger phenotypic transformations than those through single-task landscapes. These results were determined by (1) simulating a "morphospace" containing 200,000 phenotypes reminiscent of early Paleozoic vascular sporophytes, (2) evaluating the capacity of each morphology to perform each of three tasks (light interception, mechanical support, and reproduction) as well as the ability to reconcile the conflicting morphological requirements for the four combinatorial permutations of these tasks, (3) simulating the walks obtaining all phenotypic maxima or optima within the seven "fitness-landscapes," and (4) computing the mean morphological variation attending these walks. The results of these simulations, whose credibility is discussed in the context of early vascular land-plant evolution, suggest that both the number and the accessibility of phenotypic optima increase as the number of functional obligations contributing to total fitness increases (i.e., as the complexity of optimal phenotypes increases, the fitnesses of optima fall closer to the mean fitness of all the phenotypes under selection).

REFERENCES

Alberch, P. (1980) Ontogenesis and morphological differentiation. *Am. Zool.* **20,** 653–667.

Alberch, P. (1981) Convergence and parallelism in foot morphology in the neotropical salamander *Bolitoglossa*. I. Function. *Evolution* **35,** 84–100.

Alberch, P. (1989) The logic of monsters: Evidence for internal constraints in development and evolution. In *Ontogenèse et Evolution*, eds. David, B., Dommergues, J. L., Chaline, J. & Laurin, B. (Geobios, Lyon, France), Vol. 12, pp. 21–57.

Banks, H. P. (1975) Early vascular land plants: proof and conjecture. *BioScience* **25,** 730–737.

Brent, R. P. (1973) *Algorithms for Minimization Without Derivatives* (Prentice–Hall, Englewood Cliffs, NJ), p. 195.

Edwards, D., Davies, K. L. & Axe, L. (1992) A vascular conducting strand in the early land plant *Cooksonia*. *Nature (London)* **357,** 683–685.

Eigen, M. (1987) New concepts for dealing with the evolution of nucleic acids. In *Cold Spring Harbor Symp. Quant. Biol.* **52,** 307–320.

Ewens, W. (1979) *Mathematical Population Genetics* (Springer, New York), p. 325.

Franklin, I. & Lewontin, R. C. (1970) Is the gene the unit of selection? *Genetics* **65,** 707–717.

Gates, D. M. (1965) Energy, plants, and ecology. *Ecology* **46,** 1–16.

Gill, P. E., Murray, W. & Wright, M. H. (1981) *Practical Optimisation* (Academic, London), p. 401.

Gillespie, J. H. (1983) A simple Stochastic gene substitution model. *Theor. Popul. Biol.* **23,** 202–215.

Gillespie, J. H. (1984) Molecular evolution over the mutational landscape. *Evolution* **38,** 1116–1129.

Gould, S. J. (1980) The evolutionary biology of constraint. *Daedalus* **109,** 39–52.

Kauffman, S. A. (1993) *The Origins of Order* (Oxford Univ. Press, Oxford), p. 709.

Kauffman, S. A. & Levin, S. A. (1987) Towards a general theory of adaptive walks on rugged landscapes. *J. Theor. Biol.* **128,** 11–45.

Knoll, A. H., Niklas, K. J., Gensel, P. G. & Tiffney, B. H. (1984) Character diversification and patterns of evolution in early vascular plants. *Paleobiology* **10,** 34–47.

Levin, S. A. (1978) On the evolution of ecological parameters. In *Ecological Genetics: The Interface*, ed. Brussard, P. F. (Springer, New York), pp. 3–26.

Lewontin, R. C. (1974) *The Genetic Basis of Evolutionary Change* (Columbia Univ. Press, New York), p. 346.

Maynard Smith, J. (1970) Natural selection and the concept of protein space. *Nature (London)* **225,** 563–564.

Maynard Smith, J., Burian, R., Kauffman, S., Alberch, P., Campbell, J., Goodwin, B., Lande, R., Raup, D. & Wolpert, L. (1985) Developmental contraints and evolution. *Q. Rev.* **60,** 265–287.

Meredith, D. D., Wong, K. W., Woodhead, R. W. & Wortman, R. H. (1973) *Design and Planning of Engineering Systems*, eds. Newmark N. M. & Hall, W. J. (Prentice–Hall, Englewood Cliffs, NJ), p. 393.

Niklas, K. J. (1992) *Plant Biomechanics: An Engineering Approach to Plant Form and Function* (Univ. of Chicago Press, Chicago), p. 607.

Niklas, K. J. (1988) Biophysical limitations on plant form and evolution. In *Plant Evolutionary Biology*, eds. Gottlieb, L. D. & Jain, S. K. (Chapman and Hall, London), pp. 185–220.
Niklas, K. J. (1986) Computer-simulated plant evolution. *Sci. Am.* **254,** 78–86.
Niklas, K. J. & Kerchner, V. (1984) Mechanical and photosynthetic constraints on the evolution of plant shape. *Paleobiology* **10,** 79–101.
Niklas, K. J., Tiffney, B. H. & Knoll, A. H. (1980) Apparent changes in the diversity of fossil plants: A preliminary assessment. In *Evolutionary Biology*, eds. Hecht, M., Steere, W. & Wallace, B. (Plenum, New York), Vol. 12, pp. 1–89.
Nobel, P. S. (1983) *Biophysical Plant Physiology and Ecology* (Freeman, New York), p. 608.
Odell, G. M., Oster, G., Alberch, P. & Burnside, B. (1981) The mechanical basis of morphogenesis. I. Epithelial folding and invagination. *Dev. Biol.* **85,** 446–462.
Okubo, A. & Levin, S. A. (1989) A theoretical framework for data analysis of wind dispersal of seeds and pollen. *Ecology* **70,** 329–338.
Oster, G. F., Odell, G. & Alberch, P. (1980) Mechanics, morphogenesis and evolution. In *Lectures on Mathematics in the Life Sciences*, ed. Oster, G. (Am. Math. Soc., Providence, RI), pp. 165–255.
Provine, W. B. (1986) *Sewall Wright and Evolutionary Biology* (Univ. of Chicago Press, Chicago), p. 545.
Scharloo, W. (1991) Canalization, genetic and developmental aspects. *Annu. Rev. Ecol. Syst.* **22,** 265–294.
Schmid, B. (1993) Phenotypic variation in plants. *Evol. Trends Plants* **6,** 45–60.
Sepkoski, J. J. (1979) A kinetic model of Phanerozoic taxonomic diversty. II. Early Phanerozoic families and multiple equilibria. *Paleobiology* **5,** 222–251.
Snedecor, G. W. & Cochran, W. G. (1980) *Statistical Methods* (Iowa State Univ. Press, Ames), 7th Ed., p. 507.
Sokal, R. R. & Rohlf, F. J. (1981) *Biometry: The Principles and Practice of Statistics in Biological Research* (Freeman, San Francisco), p. 859.
Stewart, W. N. & Rothwell, G. W. (1993) *Paleobotany and the Evolution of Plants* (Cambridge Univ. Press, Cambridge, U.K.), p. 521.
Sultan, S. E. (1987) Evolutionary implications of phenotypic plasticity in plants. *Evol. Ecol.* **21,** 127–178.
Taylor, T. N. & Taylor, E. L. (1993) *The Biology and Evolution of Fossil Plants* (Prentice–Hall, Englewood Cliffs, NJ), p. 982.
Thomas, R. D. K. & Reif, W.-E. (1993) The skeleton space: A finite set of organic designs. *Evolution* **47,** 341–360.
Tiffney, B. H. (1981) Diversity and major events in the evolution of land plants. In *Paleobotany, Paleoecology, and Evolution*, ed. Niklas, K. J. (Praeger, New York), Vol. 2, pp. 193–230.
Van Tienderen, P. H. (1990) Morphological variation in *Plantago lanceolata*: limits of plasticity. *Evol. Trends Plants* **4,** 35–43.
Wright, S. (1932) The role of mutation, inbreeding, crossbreeding, and selection in evolution. *Proc. Int. Congr. Genet. 6th* **1,** 356–366.
Wright, S. (1931) Evolution in Mendelian population. *Genetics* **16,** 97–159.

Part III

HUMAN EVOLUTION

Human evolution figures remarkably in *Tempo and Mode* by its complete absence. The paleontological record of human evolution illuminates general issues of rate and pattern of evolution, and human evolution was a subject about which Simpson had much to say in later years. But the paleontological record of mankind's history was much too scanty at the time of *Tempo and Mode*. Not so at present. Henry M. McHenry, in Chapter 9, elucidates that human morphological evolution was mosaic. Bipedalism appeared early; the enlargement of the brain, much later. Some locomotor features changed only well after our ancestors had evolved bipedal gait. Dentition and face remained quite primitive for some time after the evolution of a distinctively hominid cranium. McHenry projects the haphazard pattern of brain-size increases over a reconstruction of the phylogenetic relationships among the, at least, eight known hominid species, from *Australopithecus africanus* to *Homo sapiens*.

The genetic diversity of the human histocompatibility complex is wondrous. At least 41 alleles are known at the *B* locus, 60 at *C*, 38 at *DPB1*, 58 at *DRB1*, and more than a dozen at each of three other loci. This gene complex serves to differentiate self from nonself and in the defense against parasites and other foreign invaders. The alleles at any one locus are quite divergent, the living descendants of lineages that recede separately for millions of years into the past. Francisco J. Ayala and colleagues rely on the theory of gene coalescence to conclude, in Chapter 10, that our ancestral lineage has been at least 100,000 individuals strong, on the average, for the last 30 million years. If a population retrenchment occurred at any time, the bottleneck could not have been smaller than a few thousand individuals, a conclusion that is also buttressed by computer simulations. These results contradict the claim propagated by the media that all modern humans descend from a single woman or very few women that lived 200,000 years ago.

9

Tempo and Mode in Human Evolution

HENRY M. McHENRY

In the 50 years since the publication of Simpson's *Tempo and Mode in Evolution* (Simpson, 1944) the paleontological record of Hominidae has improved more than a 100-fold. The improvements include precise geological dating and rich collections of well-preserved fossil hominids. Particularly valuable are newly discovered postcranial remains of early species that permit body-size estimation (McHenry, 1992; McHenry 1994b; Ruff and Walker, 1993). These new data show that the pattern of morphological change in the hominid lineage was mosaic. Different parts of the body evolved at different times and at various rates. This report focuses on hominid phylogeny and the tempo and mode of evolution of bipedalism, the hominid dental configuration, and encephalization.

Species, Clades, and Phylogeny

Views differ on the definitions of fossil hominid species and their phylogenetic relationships for many reasons but especially because of 1) the difficulty in identifying paleospecies (Kimbel and Martin, 1993; Tobias, 1991a; Walker and Leakey, 1993; Wood, 1991) and 2) the pervasiveness of homoplasy (Skelton and McHenry, 1992). One view (Skelton and McHenry, 1992) consists of five species of *Australopithecus*

Henry M. McHenry is professor of anthropology at the University of California, Davis.

(*A. afarensis, A. aethiopicus, A. africanus, A. boisei,* and *A. robustus*) and three of *Homo* (*H. habilis, H. erectus,* and *H. sapiens*). Table 1 presents the geological dates and the estimated body, brain, and tooth sizes of these species and apes.

Analysis of the states of 77 craniodental characters in these species of *Australopithecus* and *H. habilis* (Skelton and McHenry, 1992) reveals that the cladogram in Figure 1*A* is the most parsimonious (tree length = 12,796, consistency index = 0.72). The two late "robust" australopithecines, *A. robustus* and *A. boisei* are the most highly derived and form a sister group with early *Homo*. This branch links with *A. africanus* to form a clade containing *A. africanus, A. robustus, A. boisei,* and early *Homo*. *A. aethiopicus* branches from this clade next with *A. afarensis* as a sister species to all later hominids.

Figure 1*B* displays the phylogenetic tree implied by the most parsimonious cladogram. This phylogeny implies that *A. afarensis* is the most primitive hominid and that all later hominids shared a common ancestor that was more derived than *A. afarensis*. This post-*afarensis* hypothetical ancestor may someday be discovered. Its morphology can be reconstructed by observing the many ways *A. aethiopicus* resembles later hominids (especially *A. africanus*) and not *A. afarensis*. For example, the canine eminences of the face are prominent in the outgroup and in *A. afarensis* but are reduced or absent in all other species of hominid, which implies that the common ancestor of all post-*afarensis* species had canine eminences that were also reduced. This hypothetical ancestor would have a strongly developed metaconid on the lower first premolar. It would not, however, resemble *A. aethiopicus* in traits related to masticatory hypertrophy (heavy chewing), nor would it resemble any other post-*afarensis* species because they are all too derived in flexion of the base of the skull, orthognathism (flat faced), and encephalization to have been the ancestor of *A. aethiopicus*. After the divergence of *A. aethiopicus*, this phylogeny depicts a common ancestor of *A. africanus, A. robustus, A. boisei,* and *Homo* that resembled *A. africanus* in its development of anterior dentition, basicranial flexion, orthognathism, and encephalization. A second hypothetical common ancestor appears in Figure 1*B* to account for the numerous derived traits shared by *A. robustus, A. boisei,* and early *Homo* that are not seen in *A. africanus*. This ancestor would have the degree of basicranial flexion and orthognathism seen in early *Homo* and the amount of encephalization seen in *A. robustus* and *boisei*. This phylogeny proposes a third hypothetical ancestor that would be at the root of the lineage leading to *A. robustus* and *A. boisei*. This ancestor probably resembled *A. robustus* in traits related to heavy chewing.

Although the most parsimonious cladogram implies this phylogeny, other cladograms are possible but less probable. A cladogram linking *A. aethiopicus* to *A. boisei* and *robustus* as one branch and *A. africanus*/early *Homo* as another requires more evolutionary steps (tree length = 13332;

TABLE 1 Species of *Australopithecus*, *Homo*, and modern African apes with geological ages, estimated body weights, brain volumes, relative brain sizes (EQ), cheek-tooth area, and relative cheek-tooth area (MQ)

Species	Dates, (My)	Body weight, kg[a]		Brain volume,[b] (cc)	EQ[c]	Tooth area,[d] mm^2	MQ[e]
		Male	Female				
A. afarensis	4–2.8	45	29	384	2.2	460	1.7
A. africanus	3–2.3	41	30	420	2.5	516	2.0
A. aethiopicus	2.7–2.3			399		688	
A. boisei	2.1–1.3	49	34	488	2.6	756	2.5
A. robustus	1.8–1.0	40	32	502	2.9	588	2.2
H. habilis[f]	2.4–1.6	52	32	597	3.1	502	1.7
Early *H. erectus*	1.8–1.5	58	52	804	3.3	377	1.0
Late *H. erectus*	0.5–0.3	60	55	980	4.0	390	1.0
H. sapiens	0.4–0	58	49	1350	5.8	334	0.9
Pan paniscus	0	38	32	343	2.0	227	0.9
Pan troglodytes	0	49	41	395	2.0	294	0.9
Gorilla gorilla	0	140	70	505	1.7	654	1.0

[a]See refs. (McHenry, 1992 and 1994).
[b]Endocranial volume is transformed into brain volume by formula 4 in ref (Aiello and Dunbar, 1993).
[c]EQ is the ratio of brain volume and expected volume. Expected brain volume is 0.0589 (species body weight in $g)^{0.76}$; see ref. (Martin, 1981).
[d]Tooth area is the sum of the md × bl diameters of P_4, M_1, and M_2; see ref. (McHenry, 1984).
[e]MQ is the ratio of observed tooth area and expected area; expected area is 12.15 (species body weight in $kg)^{0.86}$. See ref. (McHenry, 1988).
[f]Two species may be represented in this sample. Using Wood's 1988 classification, I calculate the values for *H. habilis sensu stricto* and *Homo rudolfensis* as follows: male body weight, 37 and 60 kg; female body weight, 32 and 51 kg; brain volume, 579 and 709 cm^3; EQ, 3.5 and 3.0; tooth area, 478 and 570 mm^2; MQ, 1.9 and 1.5 kg; see ref. (McHenry, 1994b).

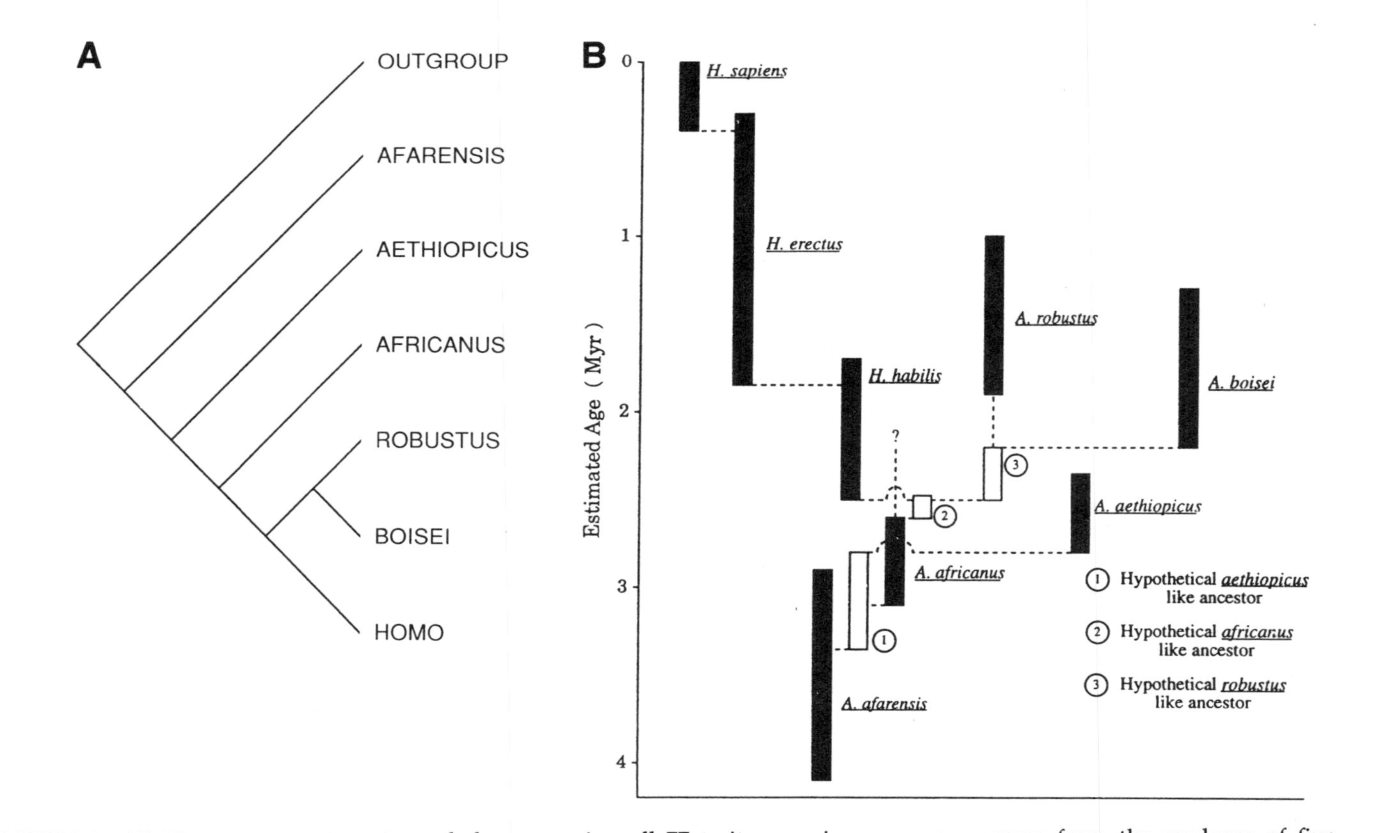

FIGURE 1 (*A*) The most parsimonious cladogram using all 77 traits or using summary scores from the analyses of five functional complexes or seven anatomical regions. Tree length is 12,796 and consistency index is 0.722. (*B*) The phylogeny implied by the most parsimonious cladogram. Three hypothetical ancestors are predicted.

consistency index = 0.69) because the later "robusts" resemble early *Homo* in so many features. These features include many aspects of basicranial flexion, loss of prognathism (muzzle), changes in the anterior dentition, and encephalization. The postcrania, although not included in this analysis, support the view that at least *A. robustus* and early *Homo* are monophyletic relative to other species of early hominid.

Whatever the true phylogeny is, and there can be only one, the fact remains that homoplasy is commonplace. Some resemblances appeared independently and not because of evolution from a common ancestor that possessed the same feature. Either adaptations for heavy chewing evolved twice or basicranial flexion, orthognathism, reduced anterior dentition, and encephalization each evolved more than once.

Bipedalism and the Postcranium

However the specific phylogeny of Hominidae is reconstructed, the important point is that these species are closely related to *H. sapiens*, and, in general, the more recent in time the species is, the more derived features it shares with our species. The earliest species, *A. afarensis*, is the most primitive in the sense that it shares the fewest of these derived traits and retains a remarkable resemblance to the common ancestor of African apes and people in many craniodental features. Its postcranium, however, is highly derived (McHenry, 1994a). It is fundamentally reorganized from the form typical of apes to that specific to Hominidae (McHenry, 1982, 1986, 1991, 1994a; Berge, 1993; Johanson *et al.*, 1982; Latimer, 1991; Latimer and Lovejoy, 1989, 1990a, 1990b; Lovejoy, 1978, 1988).

Figure 2A presents features in which the postcranium of *A. afarensis* differs from African apes and approaches the condition characteristic of humans. The most significant features for bipedalism include shortened iliac blades, lumbar curve, knees approaching midline, distal articular surface of tibia nearly perpendicular to the shaft, robust metatarsal I with expanded head, convergent hallux (big toe), and proximal foot phalanges with dorsally oriented proximal articular surfaces. A commitment to bipedalism in *A. afarensis* is also shown by the 3.5 million year (Myr) Laetoli footprints, which show very human-like proportions, arches, heel strike, and convergent big toes (McHenry, 1991; Leakey and Hay, 1979; Tuttle, 1987; White, 1980).

The nature of *A. afarensis* implies that bipedalism evolved well before the appearance of most other hominid characteristics. The appearance of bipedalism is sudden in the sense that it involved a complex alteration of structure in a relatively short period of time. Unfortunately, the fossil record does not yet include hominid postcrania predating

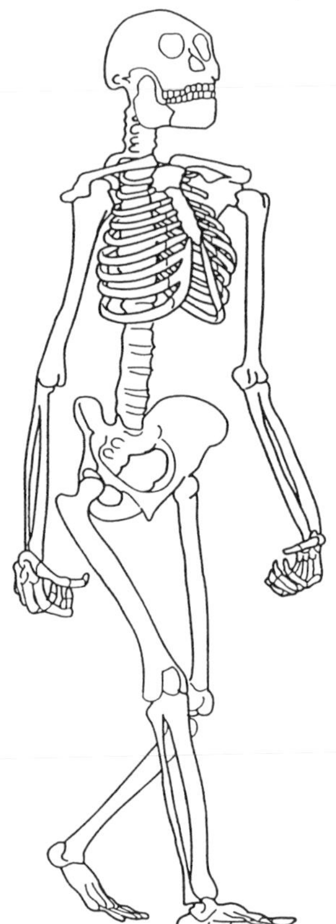

FIGURE 2A Derived postcranial traits shared by *A. afarensis* and *H. sapiens*. Abbreviations: MC, metacarpal. MT, metatarsal

4.0 Myr that would document the transition from ape-like to hominid locomotion. The fundamental changes had already taken place in *A. afarensis*.

These bipedal alterations seen in *A. afarensis* are incomplete relative to modern *H. sapiens*, however (McHenry, 1983, 1986; Deloison, 1991; Jungers, 1982, 1988; Senut, 1981, 1991; Schmid, 1983, 1991; Senut and Tardieu, 1985; Stern and Susman, 1983; Susman, *et al.*, 1984; Tardieu, 1983; Tuttle, 1981). Figure 2B presents traits in which this species differs in its postcranium from later species of Hominidae. These plesiomorphies probably imply that the bipedalism of *A. afarensis* was kinematically and energetically different from modern humans and may imply that they were more efficient tree climbers than modern humans. This arborealism would have been different from ape-like tree climbing, however, because the hindlimb was specialized for bipedality and had lost essential climbing adaptations such as hallucial divergence.

The pattern of change in these traits in later species of Hominidae is complex. Most of the postcranial elements that can be directly compared reveal a period of stasis with no change between *A. afarensis* and *A. africanus* (McHenry, 1986, 1983). This is particularly striking in the capitate bone in the wrist and pelvis. Both have the identical combination of modern pongid, modern human, and unique characteristics. In the metacarpals and hand phalanges, however, *A. africanus* has some *Homo*-like features absent in *A. afarensis* (Ricklan, 1987, 1990). The distal thumb phalanx of *A. africanus*, for example, is very human-like with its broad apical tuft that contrasts sharply with the relatively narrow, chimp-like tufts of the distal phalanges of *A. afarensis*. Limb proportions remain similar to *A. afarensis* in all species until the appearance of *H. erectus* at 1.7 Myr (McHenry, 1992). Even *H. erectus* retains some primitive characteristics relative to *H. sapiens* (Walker and Leakey, 1993). The most conspicuous of these is the relatively small cross-sectional area of the lumbar and sacral bodies (Latimer and Ward, 1993). Narrow pelvic inlets and long femoral necks are characteristic of *A. afarensis*, *A. africanus*, and *H. erectus* and are probably related to parturition of smaller-head neonates (Berge *et al.*, 1984; Lovejoy, 1978, 1988; McHenry, 1975a, 1975b; McHenry, 1975a, 1975b; Tague and Lovejoy, 1986; Walker, 1993; Walker and Ruff, 1993).

Body size remains relatively small in all species of *Australopithecus*, including the surprisingly petite bodies of the "robust" australopithecines (McHenry, 1992, 1994b; Ruff and Walker, 1993, Table 1, column 3). Sexual dimorphism in body size decreases from *A. afarensis* to *A. africanus* to *A. robustus*. Specimens attributed to *H. habilis* vary enormously in size and may imply (with other evidence) the existence of two species (McHenry, 1994a, 1994b, n.d.; Wood, 1992). A sudden change

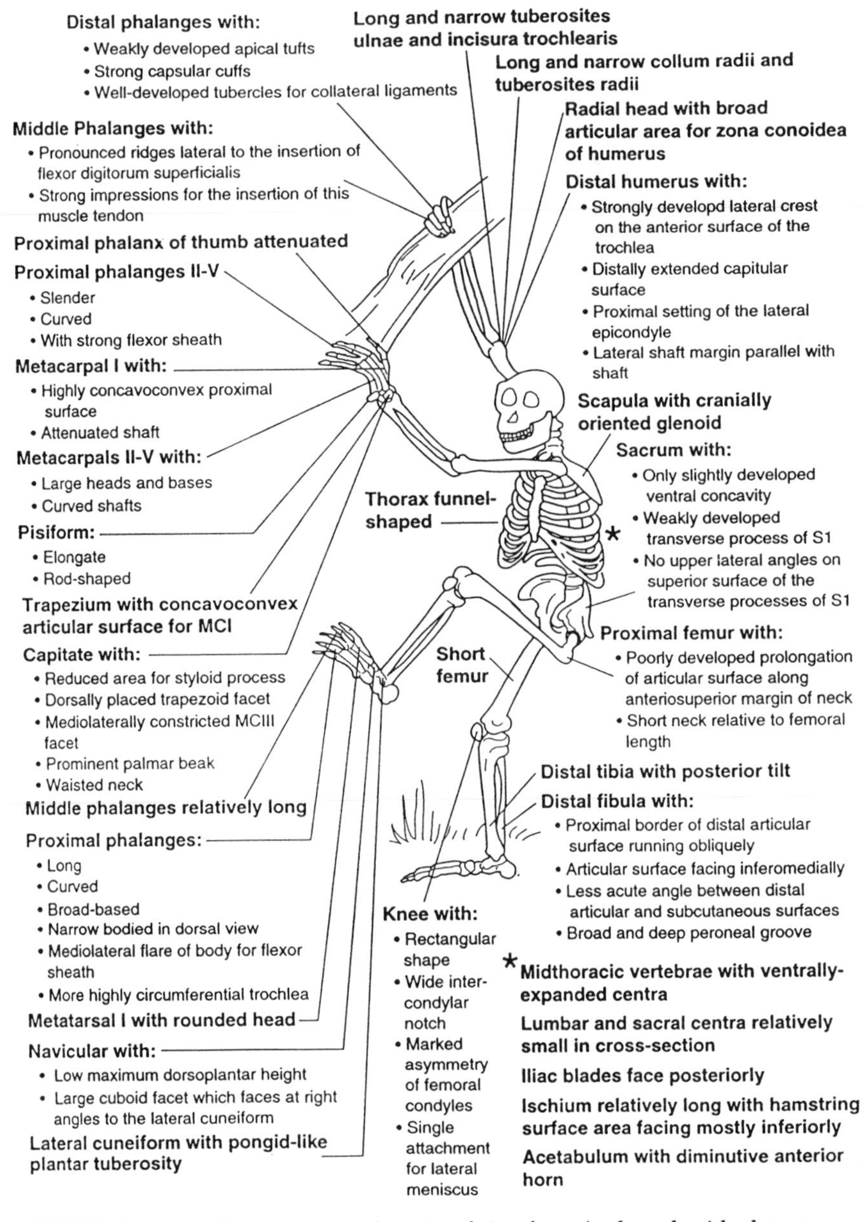

FIGURE 2B Primitive postcranial traits of *A. afarensis* shared with the reconstructed common ancestor of African apes and humans. Abbreviation: MC, metacarpal.

occurs at 1.8 Myr with the appearance of *H. erectus* with body weights as high as 68 kg and a substantial reduction in sexual dimorphism. There is no evidence of a gradual trend of increased body weight through time, as might be expected from Cope's law.

Mastication

The distinction between the hominid and pongid dental pattern was sharply delineated before the discovery of *A. afarensis* (Clark, 1967), but that species bridged the gap (Johanson and White, 1979; White *et al.*, 1981). Overall, the dentition of the earliest species of hominid is more similar to the inferred last common ancestor than it is to *H. sapiens*. Most notable primitive traits include large central and small lateral upper incisors, projecting upper canine with marginal attrition facets, small metaconid of the lower first premolar and parallel or convergent tooth rows. The positions of the masticatory muscles are also primitive, particularly the posterior placement of the main fibers of the temporalis. But there are numerous derived features shared with later hominids as well. The most conspicuous of these is the reduced canines with apical wear.

Hominid species postdating *A. afarensis* lose this species' primitive dental characteristics. *A. africanus* is variable in size and shape of its anterior teeth, but some specimens are more *Homo*-like (Kimbel and Martin, 1993; White *et al.*, 1981). Its lower first premolar is decidedly bicuspid. The mass of the temporalis muscle has moved forward into a more *Homo*-like position. Prognathism is reduced. The primitive dental features of *A. afarensis* are lost in hominid species postdating the appearance of *A. africanus*.

One unexpected characteristic of all early hominid species is postcanine megadontia and associated features related to heavy chewing (Skelton and McHenry, 1992; Aiello & Dean, 1990; Grine, 1988; McHenry, 1984; Rak, 1983; Tobias, 1967, 1991b; Turner and Wood, 1993; Walker *et al.*, 1986; Wood, 1988; Wood and Chamberlain, 1987). Relative to body size, the cheek-teeth of *A. afarensis* are 1.7 times larger than expected from that seen in modern species of Hominoidea (Table 1, column 8). Relative cheek-tooth size is higher in *A. africanus* (2.0) and higher still in *A. robustus* (2.2) and *A. boisei* (2.5). The appearance of *Homo* is marked by a reduction to 1.7. From the earliest *Homo* species to *H. erectus* to *H. sapiens* there has been dental reduction. Presumably the masticatory hypertrophy within species of *Australopithecus* is related to diet and to the amount of grit entering the mouth. Reduction of tooth size in *Homo* may reflect dietary change, but also it is probably related to the use of tools in preparing food.

The phylogeny presented in Figure 1*B* implies traits related to heavy chewing evolved by parallel evolution in two lineages. One of these is the lineage from *A. afarensis* to *A. aethiopicus*. The second is the lineage from *A. afarensis* to *A. africanus* to the late "robust" australopithecines, *A. robustus* and *A. boisei*. This is a surprising result because *A. aethiopicus* and *A. boisei* share a suite of unique character states such as extreme anterior projection of the zygomatic bone, huge cheek teeth, enormous mandibular robusticity, a heart-shaped foramen magnum, and temporoparietal overlap of the occipital at asterion (at least in males).

All of these traits, except for the heart-shaped foramen magnum, are related to the functional complex of heavy chewing. The huge cheek-teeth and robust mandibles of both species are obviously part of masticatory hypertrophy. The anterior projection of the zygomatic bones brings the masseter muscles into a position of maximum power. The encroachment by the root of the zygomaticoalveolar crest obscures the expression of the anterior pillars and upper canine jugae. Even the morphology of the temporoparietal overlap with occipital is related to the function of the forces generated by the chewing muscles (Skelton and McHenry, 1992).

Theoretically, it is understandable how such detailed similarity could be due to parallel evolution. These species are closely related and share ". . . so much in common in their constitution" (Darwin, 1872) that similar selective forces produce similar morphologies. The selective forces in this case are related to a feeding adaptation that is associated with a specialized ecological niche. As Mayr (1969, p. 125) points out ". . . most adaptations for special niches are far less revealing taxonomically than they are conspicuous. Occupation of a special food niche and the correlated adaptations have a particularly low taxonomic value." In fact, many of the same traits characteristic of *A. aethiopicus* and the other "robust" australopithecines reappear in distantly related species adapted to heavy chewing. Expansion of the cheek-teeth, shortening of the muzzle, and anterior migration of the attachment areas of the chewing muscles are seen in other primates whose diet requires heavy chewing (e.g., *Hadropithecus, Theropithecus,* probably *Gigantopithecus,* and *Ekmowehashala*).

Encephalization

Table 1, column 5 presents brain sizes in species of Hominidae. Absolute brain volume has more than tripled from *A. afarensis* to *H. sapiens*, and relative size has more than doubled (Tobias, 1991a; Wood, 1991; Walker, 1993; Aiello and Dunbar, 1993; Begun and Walker, 1993; Blumenberg, 1983; Falk, 1987; Foley, 1992; Godfrey and Jacobs, 1981;

Gould, 1975; Hofman, 1983; Holloway and Post, 1982; Holloway, 1983; Jerison, 1973; Leigh, 1992; Martin, 1981, 1983; McHenry, 1974, 1975a, 1975b, 1982; Parker, 1990; Passingham, 1985; Pilbeam and Gould, 1974; Shea, 1987; Tobias, 1971). Given the very human-like postcranium of *A. afarensis*, it is interesting that this species has a relative brain size very close to that of modern chimpanzees. Lamarck, Huxley, Haeckel, and Darwin speculated that bipedalism preceded encephalization, but they had no fossil proof (McHenry, 1975b). The early species of *Australopithecus* confirm their prediction.

Both absolute and relative brain size increase through time in the series from *A. afarensis* [384 cc, 2.2 ratio of brain volume and expected volume (EQ)] to *A. africanus* (420 cc, 2.5 EQ) to *A. boisei* (488 cc, 2.6 EQ) to *A. robustus* (502 cc, 2.9 EQ). Superficially, this increase through time appears to be by gradual increments, but samples are small and body weight determinations are inexact (McHenry, 1992). The sample of endocasts of *A. afarensis* consists of three specimens and of these, all are fragmentary, and one is the estimated adult size from a 2.5-yr-old child (Falk, 1987). Although there are six endocasts of *A. africanus*, three of these needed substantial reconstruction (Holloway, 1983). There is only one endocast of *A. robustus*, four of *A. boisei*, seven of *H. habilis*, five for early *H. erectus*, and five for late *H. erectus*. Body weight estimates may be off the mark, but the sample of postcranial specimens is sufficient to show that body weight remained at about the same relatively small size in all species of *Australopithecus*. This result implies that the apparent increase in brain size through time in species of *Australopithecus* is not due merely to an increase in body size. Body size and brain size are variable in specimens attributed to *H. habilis* with individuals as small as 32 kg and 484 cc and others as large as 57 kg and 709 cc. Although there are reasons to keep *H. habilis* as a single species (Tobias, 1991a), dividing the sample into two species is justifiable (Wood, 1991, 1992). With either taxonomy, the absolute brain sizes of these early *Homo* specimens lie between *Australopithecus* and *H. erectus*, although relative brain sizes of early members of *H. erectus* overlap the range of the smaller-bodied specimens of *H. habilis*. The relative brain size of early *H. erectus* is surprisingly small because body size is so large. By 1.7 Myr, individuals attributed to *H. erectus* grew to >180 cm, and by 1.5 Myr one individual (KNM-WT 15000) may have stood 185 cm and weighed 68 kg as an adult (Ruff and Walker, 1993). Despite the fact that the average early *H. erectus* brain was >200 cc larger than the average brain of *H. habilis*, the relative brain sizes are only slightly different (EQ = 3.1 and 3.3).

The pattern of encephalization since early *H. erectus* is difficult to interpret because geological dates are less accurate, variability is high, and body weights are difficult to establish. Figure 3 plots brain size

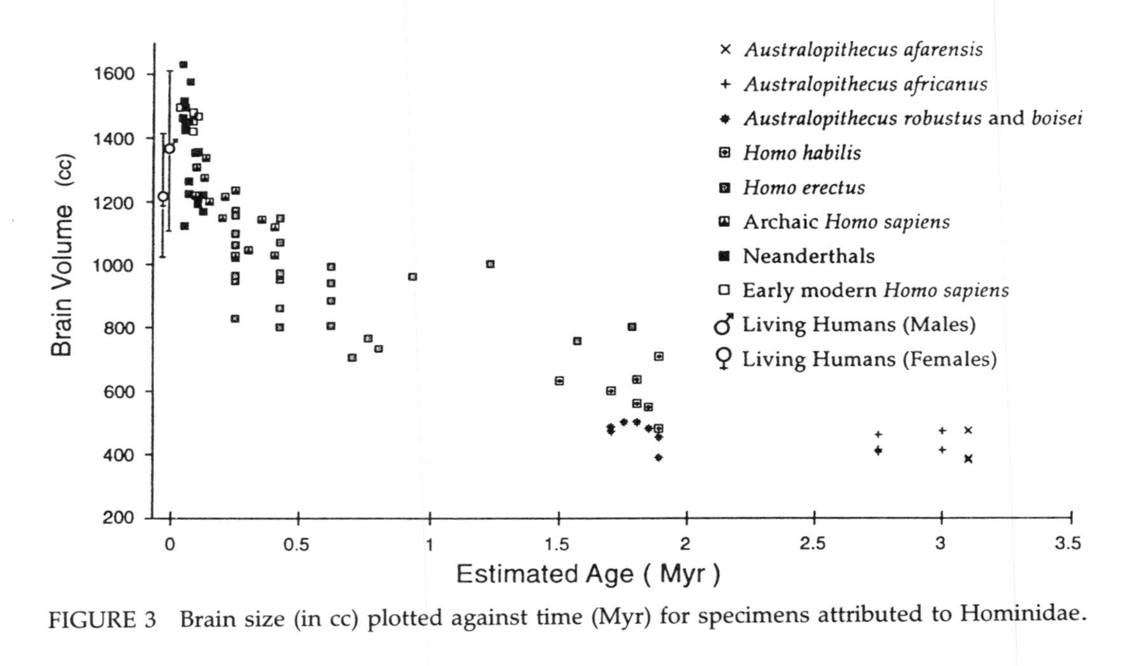

FIGURE 3 Brain size (in cc) plotted against time (Myr) for specimens attributed to Hominidae.

against time. For its first million years, *H. erectus* has absolute brain volumes that do not increase through time and therefore represent a period of stasis (Rightmire, 1990). It is difficult to establish whether relative brain sizes increased because there are very few postcranial fossils of *H. erectus* after 1.5 Myr from which to estimate body size. The few femora that are known are similar in size to those from early *H. erectus*. When taken over its entire range, the current sample of *H. erectus* does show a weak, but significant, positive increase in brain size through time (Leigh, 1992). The sample of archaic *H. sapiens* (0.4–0.125 Myr) shows a strong positive trend (Leigh, 1992). Variability is high. Many specimens as old as 0.4 Myr are within the modern human range of variation, and after 0.25 Myr all specimens are within this range. The average for the Neanderthals is 1369 cc compared with 1462 cc for early modern *H. sapiens*.

Stasis, Punctuation, and Trends

It is useful to regard evolutionary change in the hominid lineage from the point of view of Mayr's peripatric theory of speciation (Gould and Eldredge, 1993). Presumably, most of our samples derive from central populations of species and not from the small, isolated, and peripheral groups that are the most likely source of new species. When one of these peripheral isolates becomes reproductively isolated from the central species and its geographical range expands, it may overlap with the parent species, resulting in the coexistence of ancestral and descendant species. As depicted in Figure 1*B*, ancestral species overlap in time with descendants in most cases in hominid evolution, which is not what would be expected from gradual transformations by anagenesis (MacFadden, 1992). Trends through time observed in the fossil record are not necessarily the result of gradual change but rather ". . . an accumulation of discrete speciation events" (Gould and Eldredge, 1993; p. 223).

These events can be obscured by defining paleospecies too broadly, however. For example, it is conventional to define *H. erectus* as including specimens from deposits as old as 1.8 Myr and as young as 0.2 Myr (Rightmire, 1990). There is a slight trend in brain-size increase in this series (Leigh, 1992), but the earliest and smallest brained specimens are regarded by some as a separate species, *Homo ergaster* (Wood, 1992; Groves, 1989; Wood, 1993). Another example is the inclusion of specimens into *H. sapiens* that date back to perhaps 0.5 Myr, despite their decidedly archaic features. By this attribution, there is a strong positive trend in brain size through time (Leigh, 1992). An argument can be made, however, that this sample consists of several species (Stringer, 1994).

This view does not exclude the presence of change through time within species, however. As the original proponents of the theory of punctuated equilibrium point out (Gould and Eldredge, 1993), this view concerns the relative frequency of stasis, punctuation, and phyletic gradualism. Even within the multiple-species hypothesis of Middle to Late Pleistocene *Homo* (Stringer, 1994), all change through time does not occur at speciation events. For example, brain size and cranial morphology change from early to late specimens referred to *Homo neanderthalensis*. It is interesting, however, how little change occurs within most hominid species through time.

SUMMARY

The quickening pace of paleontological discovery is matched by rapid developments in geochronology. These new data show that the pattern of morphological change in the hominid lineage was mosaic. Adaptations essential to bipedalism appeared early, but some locomotor features changed much later. Relative to the highly derived postcrania of the earliest hominids, the craniodental complex was quite primitive (i.e., like the reconstructed last common ancestor with the African great apes). The pattern of craniodental change among successively younger species of Hominidae implies extensive parallel evolution between at least two lineages in features related to mastication. Relative brain size increased slightly among successively younger species of *Australopithecus*, expanded significantly with the appearance of *Homo*, but within early *Homo* remained at about half the size of *Homo sapiens* for almost a million years. Many apparent trends in human evolution may actually be due to the accumulation of relatively rapid shifts in successive species.

I thank the organizers of this symposium and particularly Francisco Ayala and Walter Fitch for the invitation to contribute this paper. I am indebted to my colleague, R. R. Skelton, with whom I did the phylogenetic analysis reported here. I thank all those whose work led to the discovery of the fossils and especially M. D. Leakey, R. E. Leakey, F. C. Howell, D. C. Johanson, Tadessa Terfa, Mammo Tessema, C. K. Brain, P. V. Tobias, the late A. R. Hughes, and T. White for many kindnesses and permission to study the original fossil material. I thank the curators of the comparative samples used in this study. Partial funding was provided by the Committee on Research of the University of California, Davis.

REFERENCES

Aiello, L. & Dean, C. (1990) *An Introduction to Human Evolutionary Anatomy* (Academic, London).

Aiello, L. C. & Dunbar, R. I. M. (1993) Neocortex size, group size, and the evolution of language. *Curr. Anthropol.* **34,** 184–193.

Begun, D. & Walker, A. (1993) The endocast. In *The Nariokotome Homo erectus Skeleton,* eds. Walker, A. & Leakey, R. (Harvard Univ. Press, Cambridge, MA), pp. 326–358.

Berge, C. (1993) *L'Évolution de la Hanche et du Pelvis des Hominidés: Bipedie, Parturition, Croissance, Allometrie* (Presses du CNRS, Paris).

Berge, C., Orban-Segebarth, R., & Schmid, P. (1984) Obstetrical interpretation of the australopithecine pelvic cavity. *J. Hum. Evol.* **13,** 573–587.

Blumenberg, B. (1983) The evolution of the advanced hominid brain. *Curr. Anthropol.* **24,** 589–623.

Clark, W. E. L. (1967) *Man-Apes or Ape-Men* (Holt, Rinehart & Winston, New York).

Darwin, C. (1872) *The Origin of Species* (Random House, New York), 6th Ed.

Deloison, Y. (1991) Les Australopitheques marchaient-ils comme nous? In *Origine(s) de la Bipédie chez les Hominidés,* eds. Coppens, Y. & Senut, B. (Presses du CNRS, Paris), pp. 177–186.

Falk, D. (1987) Hominid Paleoneurology. *Annu. Rev. Anthropol.* **16,** pp. 13–30.

Foley, R. A. (1992) Evolutionary ecology of fossil hominids. In *Evolutionary Ecology and Human Behaviour,* eds. Smith, E. A. & Winterhalder, B. (de Gruyter, New York), pp. 131–164.

Godfrey, L. & Jacobs, K. H. (1981) Gradual, autocatalytic and punctuational models of hominid brain evolution. *J. Hum. Evol.* **10,** 255–272.

Gould, S. J. (1975) Allometry in primates, with emphasis on scaling and the evolution of the brain. *Contrib. Primatol.* **5,** 244–292.

Gould, S. J. & Eldredge, N. (1993) Punctuated equilibrium comes of age. *Nature (London)* **366,** 223–227.

Grine, F. E., ed. (1988) Evolutionary history of the robust australopithecines: a summary and the historical perspective. In *Evolutionary History of the Robust Australopithecines* (de Gruyter, New York), 509–510.

Groves, C. P. (1989) *A Theory of Human and Primate Evolution* (Clarendon, Oxford).

Hofman, M. A. (1983) Encephalization in hominids: evidence for the model of punctuationalism. *Brain Behav. Evol.* **22,** 102–177.

Holloway, R. L. (1983) Human paleontological evidence relevant to language behavior. *Hum. Neurobiol.* **2,** 105–114.

Holloway, R. L. & Post, D. G. (1982) The relativity of relative brain measures and hominid mosaic evolution. In *Primate Brain Evolution: Methods and Concepts,* eds. Armstrong, E. & Falk, E. (Plenum, New York), pp. 57–76.

Jerison, H. (1973) *Evolution of the Brain and Intelligence* (Academic, New York).

Johanson, D. C. & White, T. D. (1979) A systematic assessment of early African hominids. *Science* **203,** 321–330.

Johanson, D. C., Taieb, M. & Coppens, Y. (1982) Pliocene hominids from the Hadar Formation, Ethiopia (1973–1977): Stratigraphic, chronologic, and paleoenvironmental contexts, with notes on hominid morphology and systematics. *Am. J. Phys. Anthropol.* **57,** 373–402.

Jungers, W. L. (1982) Lucy's limbs: skeletal allometry and locomotion in *Australopithecus afarensis. Nature (London)* **297,** 676–678.

Jungers, W. L. (1988) Relative joint size and hominoid locomotor adaptations with implications for the evolution of hominoid bipedalism. *J. Hum. Evol.* **17,** 247–266.

Kimbel, W. H. & Martin, L. B. (1993) in *Species, Species Concepts, and Primate Evolution,* eds. Kimbel, W. H. & Martin, L. B. (Plenum, New York), pp. 539–553.

Latimer, B. (1991) Locomotor adaptations in *Australopithecus afarensis*: the issue of arboreality. In *Origine(s) de la Bipédie chez les Hominidés*,

Latimer, B. & Lovejoy, C. O. (1989) The calcaneus of *Australopithecus afarensis* and its implications for the evolution of bipedality. *Am. J. Phys. Anthropol.* **78**, 369–386.

Latimer, B. & Lovejoy, C. O. (1990a) Hallucal tarsometatarsal joint in *Australopithecus afarensis*. *Am. J. Phys. Anthropol.* **82,** 125–134.

Latimer, B. & Lovejoy, C. O. (1990b) Metatarsophalangeal joint of *Australopithecus afarensis*. *Am. J. Phys. Anthropol.* **83,** 13–23.

Latimer, B. & Ward, C. V. (1993) The thoracic and lumbar vertebrae. In *The Nariokotome Homo erectus Skeleton*, eds. Walker, A. & Leakey, R. (Harvard Univ. Press, Cambridge, MA), pp. 266–293.

Leakey, M. D. & Hay, R. L. (1979) Pliocene footprints in the Laetolil beds at Laetoli, Northern Tanzania. *Science* **278,** 317–323.

Leigh, S. R. (1992) Cranial capacity evolution in *Homo erectus* and early *Homo sapiens*. *Am. J. Phys. Anthropol.* **87,** 1–13.

Lovejoy, C. O. (1978) A biochemical view of the locomotor diversity of early hominids. In *Early Hominids of Africa*, ed. Jolly, C. J. (St. Martins, New York), pp. 403–429.

Lovejoy, C. O. (1988) Evolution of human walking. *Sci. Am.* **259,** 118–126.

MacFadden, B. J. (1992) *Fossil Horses: Systematics, Paleobiology, and Evolution of the Family Equidae* (Cambridge Univ. Press, Cambridge, U.K.).

Martin, R. D. (1981) Relative brain size and basal metabolic rate in terrestrial vertebrates. *Nature (London)* **293,** 57–60.

Martin, R. D. (1983) *Human Brain Evolution in an Ecological Context* (American Museum of Natural History, New York).

Mayr, E. (1969) *Principles of Systematic Zoology* (McGraw–Hill, New York).

McHenry, H. M. (1974) How large were the Australopithecines? *Am. J. Phys. Anthropol.* **40,** 329–340.

McHenry, H. M. (1975a) Biomechanical interpretation of the early hominid hip. *J. Hum. Evol.* **4,** 343–356.

McHenry, H. M. (1975b) Fossils and the mosaic nature of human evolution. *Science* **190,** 425–431.

McHenry, H. M. (1982) The pattern of human evolution: studies on bipedalism, mastication and encephalization. *Annu. Rev. Anthropol.* **11,** 151–173.

McHenry, H. M. (1983) The capitate of *Australopithecus afarensis* and *Australopithecus afrcanus*. *Am. J. Phys. Anthropol.* **62,** 187–198.

McHenry, H. M. (1984) Relative cheek-tooth size in *Australopithecus*. *Am. J. Phys. Anthropol.* **64,** 297–306.

McHenry, H. M. (1986) The first bipeds: a comparison of the *Australopithecus afarensis* and *Australopithecus africanus* postcranium and implications for the evolution of bipedalism. *J. Hum. Evol.* **15,** 177–191.

McHenry, H. M. (1988) New estimates of body weights in early hominids and their significance to encephalization and megadontia in "robust" australopithecines. In *Evolutionary History of the "Robust" Australopithecines*. ed. Grine, F. E. (Aldine de Gruyter, New York), pp. 133–148.

McHenry, H. M. (1991) First Steps? Analyses of the postcranium of early hominids. In *Origine(s) de la Bipédie chez les Hominidés*, eds. Coppens, Y. & Senut, B. (Presses du CNRS, Paris), pp. 133–142.

McHenry, H. M. (1992) Body size and proportions in early hominids. *Am. J. Phys. Anthropol.* **87,** 407–431.

McHenry, H. M. (1994a) Early hominid postcrania: Phylogeny and function. In

Integrative Pathways to the Past: Paleoanthropological Papers in Honor of F. Clark Howell, eds. Corruccini, R. S. & Ciochon, R. L. (Prentice–Hall, Englewood Cliffs, NJ), pp. 251–268.

McHenry, H. M. (1994b) Behavioral ecological implications of early hominid body size. *J. Hum. Evol.* **27,** pp. 77–87.

McHenry, H. M. (n.d.) Sexual dimorphism in fossil hominids and its socioecological implications. In *Power, Sex, and Tradition: The Archaeology of Human Ancestry*, eds. Shennan, S. & Steele, J. (Routledge & Kegan Paul, London), in press.

Parker, S. T. (1990) Why big brains are so rare: Energy costs of intelligence and brain size in anthropoid primates. In *"Language" and Intelligence in Monkeys and Apes*, eds. Parker, S. T. & Gibson, K. R. (Cambridge Univ. Press, Cambridge, U.K.), pp. 129–154.

Passingham, R. E. (1985) Rates of brain development in mammals including man. *Brain Behav. Evol.* **26,** 167–175.

Pilbeam, D. R. & Gould, S. J. (1974) Size and scaling in human evolution. *Science* **186,** 892–901.

Rak, Y. (1983) *The Australopithecine Face* (Academic, New York).

Ricklan, D. E. (1987) Functional anatomy of the hand of *Australopithecus africanus*. *J. Hum. Evol.* **16,** 643–664.

Ricklan, D. E. (1990) The precision grip in *Australopithecus africanus*: Anatomical and behavioral correlates. In *From Apes to Angels: Essays in Honor of Phillip V. Tobias*, ed. Sperber, G. H. (Wiley-Liss, New York), pp. 171–183.

Rightmire, G. P. (1990) *The Evolution of Homo erectus* (Cambridge Univ. Press, Cambridge, U.K.).

Ruff, C. B. & Walker, A. (1993) Body size and body shape. In *The Nariokotome Homo erectus Skeleton*, eds. Walker, A. & Leakey, R. (Harvard Univ. Press, Cambridge, MA), pp. 234–265.

Schmid, P. (1983) Eine Rekonstruktion des skelettes von A.L. 288-1 (Hadar) und deren Konsequenzen. *Folia Primatol.* **40,** 283–306.

Schmid, P. (1991) The trunk of the Australopithecines. In *Origine(s) de la Bipédie chez les Hominidés*, eds. Coppens, Y. & Senut, B. (Presses du CNRS, Paris), pp. 225–234.

Senut, B. (1981) *L'Humérus et Ses Articulations chez les Hominidés Plio-Pléistocene* (Presses du CNRS, Paris).

Senut, B. (1991) Origine(s) de la bipedie humaine: une approche paleontologique. In *Origine(s) de la Bipédie chez les Hominidés*, eds. Coppens, Y. & Senut, B. (Presses du CNRS, Paris), pp. 245–258.

Senut, B. & Tardieu, C. (1985) Functional aspects of Plio-Pleistocene hominoid limb bones: implications for taxonomy and phylogeny/ In *Ancestors: The Hard Evidence*, ed. Delson, E. (Liss, New York), pp. 193–201.

Shea, B. T. (1987) Reproductive strategies, body size, and encephalization in primate evolution. *Int. J. Primatol.* **8,** 139–156.

Simpson, G. G. (1944) *Tempo and Mode in Evolution* (Columbia Univ. Press, New York).

Skelton, R. R. & McHenry, H. M. (1992) Evolutionary relationships among early hominids. *J. Hum. Evol.* **23,** 309–349.

Stern, J. T. & Susman, R. L. (1983) The locomotor anatomy of *Australopithecus afarensis*. *Am. J. Phys. Anthropol.* **60,** 279–318.

Stringer, C. B. (1994) in *Issues in Hominid Evolution*, ed. Howell, F. C. (California Acad. Sci., San Francisco).

Susman, R. L., Stern, J. T. & Jungers, W. L. (1984) Arboreality and bipedality in the Hadar hominids. *Folia Primatol.* **43,** 113–156.

Tague, R. G. & Lovejoy, C. O. (1986) The obstetric pelvis of A.L. 288-1 (Lucy). *J. Hum. Evol.* **15,** 237–255.

Tardieu, C. (1983) *L'articulation du Genou.* Analyse morpho-fonctionelle chez les primates et les hominides fossiles. (Presses du CNRS, Paris).

Tobias, P. V. (1971) *The Brain in Hominid Evolution* (Columbia Univ. Press, New York).

Tobias, P. V. (1991a) *Olduvai Gorge Volume 4: The Skulls, Endocasts and Teeth of Homo habilis* (Cambridge Univ. Press, Cambridge, U.K.).

Tobias, P. V. (1991b) Man, culture, and environment. In *Evolution of Life Fossils, Molecules, and Culture,* eds. Osawa, S. & Honjo, T. (Springer, Tokyo), pp. 363–377.

Turner, A. & Wood, B. (1993) Comparative palaeontological context for the evolution of the early hominid masticatory system. *J. Hum. Evol.* **24,** 301–318.

Tuttle, R. H. (1981) Evolution of hominid bipedalism and prehensile capabilities. *Philos. Trans. R. Soc. London B* **292,** 89–94.

Tuttle, R. H. (1987) Kinesiological inferences and evolutionary implications from Laetoli bipedal trails G-1, G-2/3 and A. *Laetoli: A Pliocene Site in Northern Tanzania,* eds. Leakey, M. D. & Harris, J. M. (Clarendon, Oxford), pp. 503–523.

Walker, A. (1993) Perspectives on the Nariokotome discovery. In *The Nariokotome Homo erectus Skeleton,* eds. Walker, A. & Leakey, R. (Harvard Univ. Press, Cambridge, MA), pp. 411–430.

Walker, A. & Leakey, R. (1993) *The Nariokotome Homo erectus Skeleton,* eds. Walker, A. & Leakey, R. (Harvard Univ. Press, Cambridge, MA).

Walker, A. & Ruff, C. B. (1993) The reconstruction of the pelvis. In *The Nariokotome Homo erectus Skeleton,* eds. Walker, A. & Leakey, R. (Harvard Univ. Press, Cambridge, MA), pp. 221–233.

Walker, A., Leakey, R. E. F., Harris, J. M. & Brown, F. H. (1986) 25 MYR *Australopithecus boisei* from west of Lake Turkana, Kenya. *Nature (London)* **322,** 517–522.

White, T. D. (1980) Evolutionary implications of pliocene hominid footprints. *Science* **208,** 175–176.

White, T. D., Johanson, D. C. & Kimbel, W. H. (1981) *Australopithecus africanus*: its phylogenetic position reconsidered. *S. Afr. J. Sci.* **77,** 445–470.

Wood, B. A. (1993) Early *Homo*: How many species? In *Species, Species Concepts, and Primate Evolution,* eds. Kimbel, W. H. & Martin, L. B. (Plenum, New York), pp. 485–522.

Wood, B. A. (1988) Are robust australopithecines a monophyletic group? In *Evolutionary History of the "Robust" Australopithecines,* ed. Grine, E. F. (de Gruyter, New York), pp. 269–284.

Wood, B. A. (1991) *Koobi Fora Research Project IV: Hominid Cranial Remains from Koobi Fora* (Clarendon, Oxford).

Wood, B. A. (1992) Origin and evolution of the genus *Homo. Nature (London)* **355,** 783–790.

Wood, B. A. & Chamberlain, A. T. (1987) The nature and affinities of the "robust" australopithecines: a review. *J. Hum. Evol.* **16,** 625–642.

10

Molecular Genetics of Speciation and Human Origins

FRANCISCO J. AYALA, ANANÍAS ESCALANTE, COLM O'HUIGIN, AND JAN KLEIN

Species are populations of organisms reproductively isolated from other organisms. Speciation is the process by which two gene pools, say X and Y, derived from an ancestral pool, say Z, acquire species-specific genes that keep them from interbreeding. The ancestral species may persist, so that speciation may simply involve the divergence of a new species, say X, from the ancestral one, Z.

Species continuously evolve; after a time, a descendant gene pool, say Z′, may be sufficiently different from the ancestral gene pool, Z, as to be considered a different species. Whether or not Z and Z′ would exhibit reproductive isolation from each other, and thus meet the criterion for speciation, is an empirically meaningless question since the matter cannot be tested. But ancestral and descendant populations are given different species names if they are as different from each other as other contemporary species—for example, if Z′ is as different from Z as it is from X or Y.

The process of speciation is notoriously refractory to investigation because it is complex and usually takes a long time, and also because speciation can happen in diverse ways (Ayala, 1982; Barigozzi, 1982; Giddings *et al.*, 1989; Otte and Endler, 1989). Moreover, genetic differ-

Francisco J. Ayala is Donald Bren Professor of Biological Sciences and Ananías Escalante is a Ph.D. candidate in ecology and evolutionary biology at the University of California, Irvine. Colm O'hUigin and Jan Klein are scientists at the Max-Planck-Institut für Biologie, Tübingen, Germany.

entiation between species cannot be investigated with the classical methods of genetics. With Mendelian methods, the presence of a gene is established by observing segregation in the progenies of crosses; but crosses between individuals from different species are characteristically not possible. Fortunately, the advent of molecular biology has opened up the investigation of the genetic changes associated with the speciation process. Thus, the proportion of genes that change during speciation has been ascertained in a variety of organisms (Ayala *et al.*, 1974; Ayala, 1975; Bullini, 1982; Avise, 1994). Moreover, DNA and protein sequence information has facilitated the reconstruction of phylogenetic relationships. Yet the DNA in the gene pool of species holds a largely untapped wealth of genetic and evolutionary information. In this paper we explore DNA sequence polymorphism in current human populations in order to shed light on some aspects of human evolution, particularly on the size of human ancestral populations as they evolved from one to another species.

Speciation by Founder Effect

The theories of "founder effect" speciation propose that speciation often occurs after a founder event or bottleneck; that is, when a new population is established by a pair or very few individuals, as may happen in the colonization of an island, or when an established population declines severely so that extremely few individuals survive, from which a population expands again (Mayr, 1954, 1963, 1982; Carson, 1968, 1975; Templeton, 1980; Carson and Templeton, 1984). These theories claim that founder events and drastic bottlenecks are associated with random genetic drift, inbreeding, and selective changes that refashion the gene pool ("genetic revolution," Mayr, 1963) and thus increase the likelihood that a new species will arise.

The prevalence of founder events (or severe bottlenecks) in speciation is a matter of great interest and acrimonious debate. Some authors argue that founder effects prevail, whereas others reject on theoretical grounds the purported genetic consequences associated with drastically reduced population numbers (Barton and Charlesworth, 1984; Giddings *et al.*, 1989; Otte and Endler, 1989; Galiana *et al.*, 1993).

The evolution from hominoid ancestors that lived a few million years ago to modern humans involved several transitional species, characterized by important biological changes, such as the evolution of bipedalism and a large brain. The issue arises whether these changes may have been associated with, and perhaps a consequence of, extreme population constrictions. It has been claimed, in particular, that a severe population bottleneck of only one pair or very few individuals preceded the evolution of modern humans. This claim has been erroneously founded on the inference that the mitochondrial DNA polymorphisms of modern humans can be traced to a single woman who lived some 200,000 years ago (Cann *et al.*, 1987; Stoneking *et al.*, 1990; Vigilant *et al.*, 1991). As we shall show, DNA polymorphisms in the major histocompatibility complex (*MHC*) of humans and other primates manifest that no severe population bottleneck has occurred in human evolution.

The Major Histocompatibility Complex

The *MHC* is an array of genetic loci that specify molecules with a major role in tissue compatibility and defense against foreign substances. The *MHC* is present in all mammals, birds, amphibians, and fishes, and it may in fact exist in all vertebrates (Klein, 1986; Bjorkman and Parham, 1990). MHC molecules present on the surfaces of certain cells bind fragments of proteins (antigens) and present them to thymus-derived lymphocytes (T cells) expressing T-cell receptors on their surfaces. The clone of T lymphocytes bearing receptors that match a particular combination of protein fragment and MHC molecule is stimulated, by the contact with the antigen-presenting cells, to proliferate and to initiate the specific arm of the immune response, including the secretion of specific antibodies. The MHC molecules thus protect against pathogens and parasites in general.

The recognition of protein fragments is mediated by a specialized groove on the surface of the MHC molecule, the so-called peptide-binding region (PBR) composed of some 50 amino acid residues (Bjorkman *et al.*, 1987a, b; Brown *et al.*, 1993). The composition of the amino acids in the PBR varies from one MHC molecule to another, and it is primarily this variation that is responsible for the tremendous polymorphism characteristic of the MHC molecules and their encoding genes. In people, as well as in some mammalian species (e.g., the house mouse), scores of alleles may exist at any one of several MHC loci, and some of the allelic pairs may differ by more than 100 nucleotide substitutions (Klein and Figueroa, 1986; Marsh and Bodmer, 1991, 1993; Bontrop, 1994).

The human *MHC*, also referred to as the human leukocyte antigen (*HLA*) complex, consists of about 100 genes located on chromosome 6, extending over a DNA region that is four million nucleotide pairs in length. The MHC genes fall into two distinct classes, I and II, separated by a set of genes with functions mostly unrelated to the immune response (Figure 1).

The origins of the *MHC* polymorphisms have been studied extensively in rodents (Arden and Klein, 1982; Figueroa *et al.*, 1988; McCon-

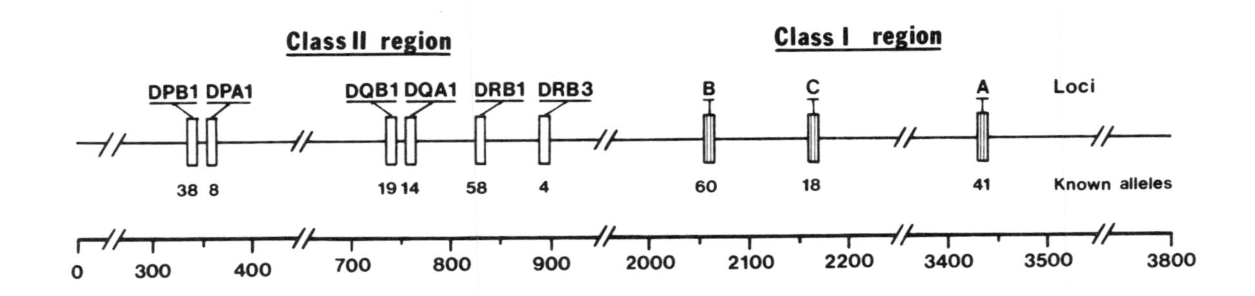

FIGURE 1 Location of some polymorphic genes within the HLA complex in human chromosome 6. There are two sets of genes, class I and class II, separated by a region with unrelated genes. The number of alleles known at a locus is written below the box that indicates the location of the gene.

nell *et al.*, 1988) and primates (Mayer *et al.*, 1988; Lawlor *et al.*, 1988; Fan *et al.*, 1989; Gyllenstein and Erlich, 1989; and Gyllenstein *et al.*, 1990). In both orders, convincing evidence has been obtained indicating that some of the allelic lineages at the *MHC* loci are several million years (Myr) old. We will review the allelic diversity found at one particular *HLA* class II locus, the *DRB1* gene. Length constrains us from reporting other *HLA* loci, such as *DPB1* and *DQB1* (see Figure 1), that lead to similar conclusions.

Trans-Specific Polymorphisms

We are interested in identifying trans-specific polymorphisms, that is, sets of allelic lineages that may have been passed from an ancestral species to its descendant species. Two ways are available for detecting trans-specific polymorphisms. The first method seeks to identify alleles in a species that are more different from each other than they are from alleles in a separate species. Consider a situation in which there are two alleles, *01* and *02*, at a locus in species Z, which are both passed on to descendant species X and Y (Figure 2). In the new species the two alleles accumulate mutations, turning into alleles *0101, 0102, 0103*, etc. and *0201, 0202, 0203*, etc. which are different from one another and from the original alleles *01* and *02*. If we compare these alleles, we find that some alleles of the same species, such as *0101* and *0201*, are more different from each other than some alleles from different species, such as *0101* and *0103*. If we now draw a tree depicting the descent of these alleles, it will not coincide with the phylogenetic tree of the species: the *01* and *02* alleles split from a common ancestor before the two species X and Y evolved from their common ancestral species Z.

An example of trans-species polymorphism is depicted in Figure 3. Here, two human alleles at the *DRB1* locus (*HLA DRB1*0302* and **0701*) differ in the exon encoding the PBR by 31 nucleotide substitutions, whereas one of them (*HLA DRB1*0302*) differs from the corresponding chimpanzee allele (*Patr DRB1*0305*) by 13 substitutions, and the other (*HLA DRB1*0701*) differs from the corresponding chimpanzee allele (*Patr DRB1*0702*) by 2 substitutions only. In terms of genetic distance (the number of nucleotide differences divided by the total number of sites compared), each of the two human alleles is more closely related to a chimpanzee allele than the two human alleles are to each other. This relationship is reflected in a family tree of the four alleles, which shows that the two human alleles diverged from a common ancestral gene before the ancestors of the human and chimpanzee species separated from each other around 6 Myr ago.

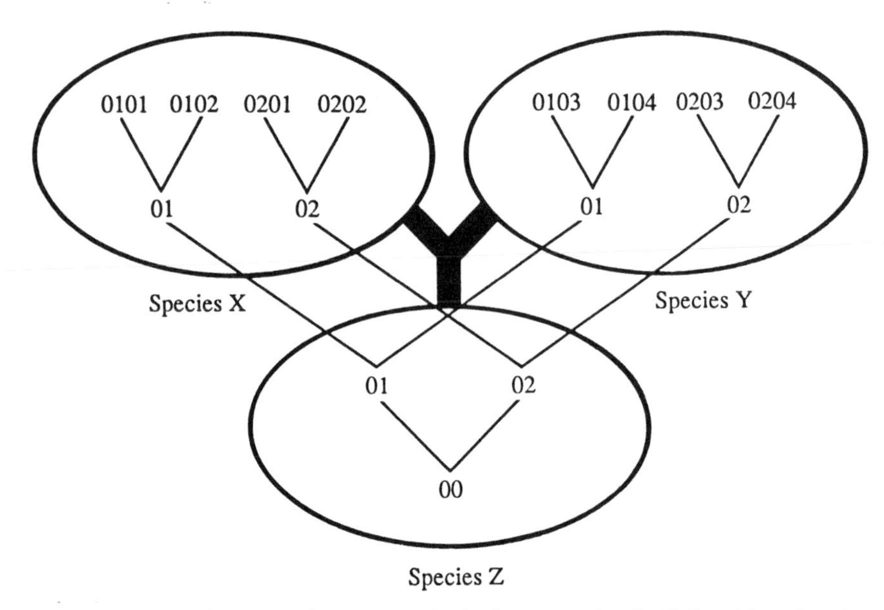

FIGURE 2 Phylogeny of species (thick lines) and of alleles (thin lines). Numbers indicate progressively diverging alleles at a given locus. The species and allele trees may not coincide whenever a polymorphism is passed from species Z on to species X and Y; i.e., the root of the alleles may be deeper than species Z. Some alleles, such as 0101 and 0201, of the same species are more different from each other than some alleles, such as 0101 and 0103, of different species.

The second method for detecting trans-species polymorphism depends on knowing the rate of evolution (i.e., rate of nucleotide substitution) for the particular gene. Genetic distances between alleles can

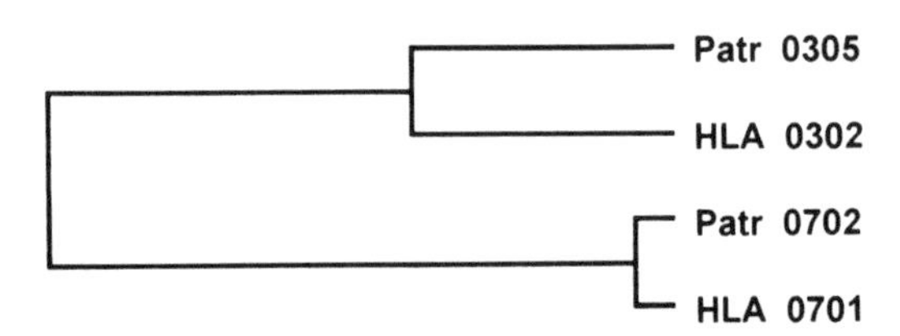

FIGURE 3 Trans-species polymorphism manifested by the genealogy of four alleles at the *MHC DRB1* locus in humans (HLA) and chimpanzee (*Pan troglodytes, Patr*). The two HLA allelic lineages predate the divergence of the human and chimpanzee lineages. The lengths of the horizontal branches are roughly proportional to the genetic distance, or amount of genetic change, separating the alleles.

then be transformed into times since divergence from the ancestral allele (Figure 4). (In reality, the two methods are equivalent. Thus, if 80 *DRB1* alleles sequenced in other primates were incorporated in Figure 4 with the 58 human alleles, we would see that many human alleles are more closely related to alleles from other primates than to one another. One practical reason not to represent all 138 alleles here is the enormous size of the required figure.)

Evolutionary History of the *DRB1* Gene Locus

Exon 2 of the *DRB1* gene consists of 270 nucleotides that specify all the β-chain amino acids involved in peptide binding. No fewer than 58 distinct human alleles have been identified that differ in their exon 2 nucleotide sequences. Figure 4 is a phylogenetic reconstruction of the 58 alleles.

The time scale has been determined by the "minimum-minimum" method (Satta *et al.*, 1991, 1993), which is based on the comparison between pairs of species that share the same divergence node, such as the three pairs orangutan–human, orangutan–gorilla, and orangutan–chimpanzee. The minimum distance observed in such a set of comparisons will correspond to alleles that diverge at, or close to, the time of the species divergence. (Alleles that were polymorphic at the time of the species divergence will show larger distances than the minimum-minimum value.) A plot of minimum-minimum distances versus the correspondent divergence times gives an estimate of rate of evolution.

As can be seen in Figure 4, all 58 *DRB1* alleles have persisted through the last 500,000 years, but coalesce into 44 ancestral lineages by 1.7 Myr before present (B.P.), 21 lineages by 6 Myr B.P., and 10 lineages by 13 Myr B.P. It is apparent that the *DRB1* polymorphism is ancient, so that numerous alleles have persisted through successive speciation events. In order to pass 58 alleles through the generations, no fewer than 29 individuals are required at any time. As we shall presently see, the size of the human populations needs to be much greater than that in order to retain the *DRB1* polymorphisms over the last few million years.

Persistence of Polymorphisms

What are the conditions under which we can expect a trans-specific passage of gene polymorphisms? There are several factors that determine the fate of gene polymorphisms during speciation. We will consider three of them: the time (t) elapsed since the divergence of the species, the selective value (s) of the alleles, and the size (N) of the gene pool.

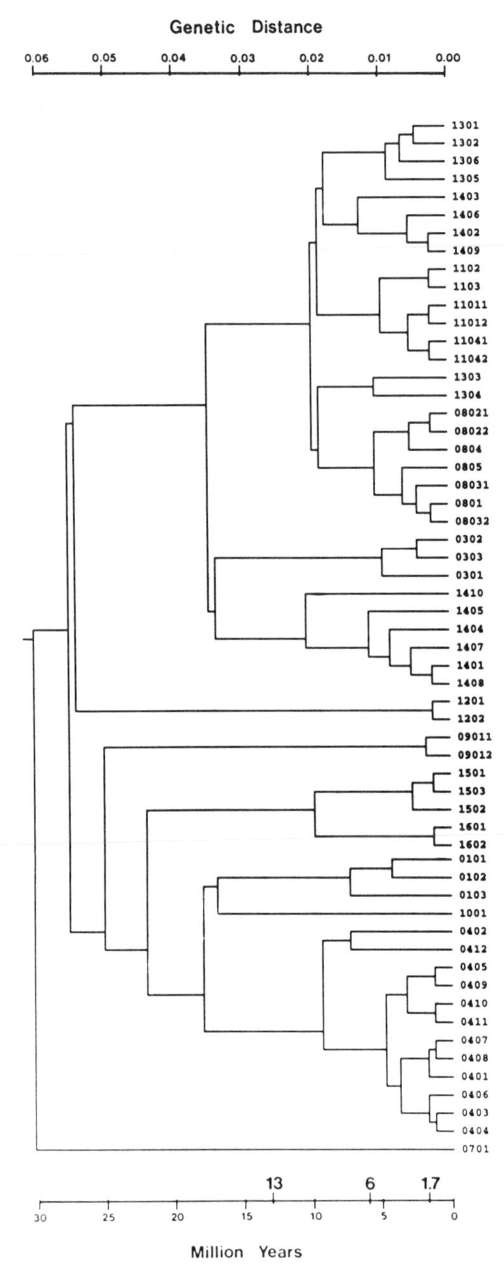

FIGURE 4 Evolutionary tree (genealogy) of 58 *HLA DRB1* alleles. The tree is based on the DNA sequence of exon 2, which consists of 270 nucleotides. Genetic distances are estimated by Kimura's (1983) two-parameter method (with

Alleles can have positive, negative, or no selective value relative to preexisting alleles. An allele with a negative selective value is usually eliminated from the gene pool soon after its appearance and does not contribute to polymorphism.

The fate of an allele with no selective value (a neutral allele) depends on chance only: the allele can drift and eventually become lost or fixed (i.e., replace all other alleles at that locus). Theory says that a neutral allele that eventually becomes fixed takes on average $4N$ generations (with a very large standard error) from its appearance by mutation to its fixation (Kimura, 1983; Nei, 1987), where N is the effective population size (which for our purposes is roughly the number of gene pairs at a given locus that are passed on from one generation to the next). In a population with $N = 10^6$ individuals and generation time = 15 years, it will take on the average 60 Myr for a new neutral allele to become fixed; a polymorphism will persist through that time. If the generation time is 1 year, the polymorphism will persist for 4 Myr.

If a new allele *01* has a positive selective value (it has a selective advantage, s, relative to another allele, *02*), the expected time to fixation is less than $4N$ generations and is, approximately, $(2/s)\log_e(2N)$ generations (Kimura and Ohta, 1969; Kimura, 1983; Nei, 1987). In the case of balancing selection due to overdominance (i.e., when heterozygotes have higher fitness than either homozygote) or frequency-dependent selection (i.e., if an allele is at an advantage when it is rare but becomes disadvantageous when it gets common), a polymorphism can persist without fixation for as long as the selection pressure persists (Takahata, 1990). Moreover, a balanced polymorphism will create a barrier that slows down or completely hinders the fixation of neutral mutations that are so closely linked to the positively selected site that recombination occurs between the two only rarely (Klein and Takahata, 1990).

The average duration of a species in the paleontological record is 4 Myr (Raup, 1994). Balanced as well as neutral polymorphisms may well

---

FIGURE 4 (Continued) pairwise elimination of any unsequenced sites). The tree was obtained with a standard UPGMA algorithm. The rate of nucleotide substitution, estimated by the minimum-minimum method (Satta *et al.*, 1991, 1993) for exon 2, is 2×10^{-9} per site, per year. As divergence times from the human lineage we use: chimpanzee and gorilla, 6 My; orangutan, 13 My; Old-World monkeys, 25 My; New-World monkeys, 40 My; prosimians, 65 My. Satta *et al.* (1993) have estimated rates of 1.84×10^{-9} for class I genes and 1.18×10^{-9} for *DRB1*, estimates that would increase the range of the time scale at bottom to 33 My and 51 My, respectively. Sequence data are mostly from Marsh and Bodmer (1993).

persist longer than it takes one species to evolve into a different species, or to diverge into two or more species.

Evolutionary Substitutions and Coalescence Theory

Now that we have defined the conditions for the passage of trans-species polymorphisms, we shall turn the argument around and attempt to extrapolate from the existing polymorphisms to the events converting an ancestral gene pool to a new pool.

The coalescence theory examines the genealogical relationships between genes (see Griffiths, 1980; Hudson, 1990). According to this theory, all alleles present in an extant pool must have descended from a single allele (to which they coalesce). For neutral alleles in a random mating population, the mean coalescence time is given by $4N[1 - (1/i)]$ generations, where i is the number of sampled genes. For any two genes ($i = 2$), the mean coalescence time reduces to $2N$ generations; for a large number of genes, the mean coalescence time approaches $4N$ generations. These relationships are important because, if we know the time when alleles coalesce, we can estimate N, the mean number of individuals in the species.

The coalescence theory was originally developed for neutral genes (Kingman, 1982a, b; Tajima, 1983; Tavaré, 1984; and Takahata and Nei, 1985) but has been recently extended to allelic genealogies under balancing selection (Takahata, 1990, 1993a, b; Takahata and Nei, 1990). In the latter situation, the theory has the same mathematical structure as in neutral gene genealogy except for a time-scaling factor, f_s. This factor is a function of population size N, selected mutation rate u, and selection coefficient s, and is given approximately by the formula

$$f_s = \frac{\sqrt{2Ns}}{2Nu} \cdot \left[\ln \frac{Ns}{8\pi (Nu)^2} \right]^{-3/2}.$$

Given i sampled genes, the coalescence theory permits one to estimate the number of distinct genes, j, that existed at a certain time, with time measured in $2Nf_s$ generations. The allelic phylogeny of the *DRB1* locus (Figure 4) shows 58 *HLA* distinct alleles, of which 10 lineages were already in existence 13 Myr ago. The complete set coalesces 30 Myr ago, that is, before the emergence of the hominoids.

If we assume an average generation time of 15 years, the coalescence would occur two million generations ago. In the case of neutral genes, the expected time to coalescence, given by $4N$ generations, yields a value of N = 500,000 individuals as the long-term mean effective size of the species. This estimate has a large variance (see below).

In the case of balanced selection, the population size required for the maintenance of a polymorphism is smaller. On the assumption of overdominance, with a heterozygote advantage on the order of 0.01, and 10^{-8} selected mutation rate per site per generation, the HLA polymorphism at the *DRB1* locus requires a mean effective size of 100,000 individuals over the last 30 million years (Takahata *et al.*, 1992). Computer simulations lead to estimates also on the order of 10^5 individuals as the long-term size of the ancestral populations leading to modern humans (Klein *et al.*, 1990; Takahata, 1991, 1993a).

If we assume a mean population size of 10^5 individuals and a long-term generation time of 15 years, the expected coalescence for neutral alleles is 6 Myr, which is much less than the 30 Myr coalescence of the *DRB1* alleles. Although the coalescence estimate has a large variance, it seems that either our ancestral population was even larger than 10^5 or, as assumed, balancing selection accounts for the long-term persistence of the MHC polymorphisms. The presence of balancing selection is supported by the analysis of the DNA sequences of HLA alleles. In codons specifying amino acids of the PBR, variation at the first and second positions is significantly higher than at the third position, and this observation is taken as evidence that positive selection acts on the first two positions (Hughes and Nei, 1988, 1992). Moreover, Hill *et al.* (1991, 1992a, b; see Miller, 1994) have shown that MHC polymorphism may increase resistance to *Plasmodium falciparum*, the parasite responsible for malignant malaria.

Estimates of the magnitude of the selection coefficient, *s*, that maintains the MHC polymorphisms vary from locus to locus, but range from 0.0007 to 0.019 (Satta *et al.*, 1994). It seems unlikely that the selection coefficient would be in any case much larger than 0.01–0.03, but even large selection coefficients do not allow for the long-term persistence of polymorphisms except in the presence of large populations. For example, only 7 alleles can be maintained in a population of $N = 1000$, even with overdominant selection as unreasonably large as 0.3 (Figure 5; Klein *et al.*, 1990).

Population Bottlenecks and the Origin of Modern Humans

We have concluded, assuming an overdominant selective advantage of 0.01, that the long-term effective population size of the human lineage for the last 30 Myr is 100,000 individuals or larger. When population size oscillates, smaller numbers have a large effect on the value of *N*, since

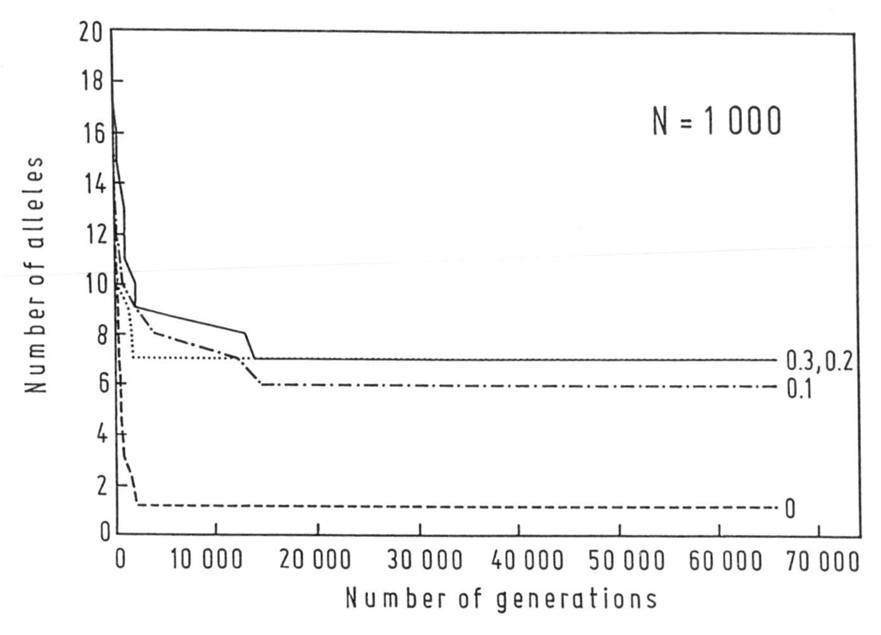

FIGURE 5 Computer simulation of the loss of alleles over 65,000 generations in a population consisting of 1000 individuals. Initially 20 alleles are present, each at 0.05 frequency. Mating is random and there is no mutation or migration. Heterozygotes reproduce with a higher probability s than homozygotes; $s = 0.1$, 0.2, and 0.3. Without selection, all alleles but one are soon eliminated. When strong overdominant selection is assumed, 6–7 alleles persist; but the magnitude of the selection is of little consequence within the range of s values used. (From Klein *et al.*, 1990.)

the effective population size is the harmonic mean of the population size over time. Thus, at nearly all times the population must have consisted of 100,000 or more individuals. The question, nevertheless, arises whether an occasional population bottleneck may have occurred, and how small could a bottleneck be. It has been suggested that a very narrow population bottleneck occurred at the transition from archaic to modern *Homo sapiens*, some 200,000 years ago (see Cann *et al.*, 1987; Stoneking *et al.*, 1990; Vigilant *et al.*, 1991).

The consequences of a bottleneck depend not only on the size, N_b, of the bottleneck, but also on the number, t_b, of bottleneck generations. The role of t_b is particularly significant in species with limited capacity for population increase. The rate of growth of human populations throughout the Pleistocene is estimated at 0.02% per generation (Spuhler, 1993, p. 279).

A useful measure for evaluating the effects of a bottleneck is the ratio

TABLE 1 Minimum population size at a bottleneck

			Alleles through bottleneck	
s	*R*	*N*	40/50	60/70
0	—	—	270–300	458–490
0.01	—	—	292–302	454–462
0	1.1	10,000	460–470	
0	1.1	5,000		750–790
0	1.05	5,000	980–1010	
0	1.01	10,000	1540–1560	
0	1.01	5,000		2120–2180
0.01	1.01	5,000		2140–2180

The two columns on the right give the minimum number of individuals required for passing either 40 (out of 50) or 60 (out of 70) alleles, with a 95% probability. The initial bottleneck is always 10 generations. In some cases it is assumed that the population grows at a rate *R* per generation before reaching the equilibrium size *N*. The selective value due to overdominance is *s*. Each value is based on 300 computer simulations.

N_b/t_b, which if smaller than 10 will have drastic effects in reducing genetic variation (Takahata, 1993a). Thus a bottleneck of 100 individuals would substantially reduce genetic variation if it would last 10 or more generations. Balancing selection facilitates the persistence of polymorphisms through a bottleneck. But because alleles behave as neutral whenever $Ns < 1$, if the selection is weak, such as $s = 0.01$, N has to be correspondingly large, at least 100, for selection to play a role. The persistence of HLA polymorphisms over millions of years requires that the size of human ancestral populations be at least $Ns = 10$ at all times (Takahata, 1990). If $s = 0.01$, the minimum population size possible at any time would be $N_b = 1000$. The minimum number must have been in fact much larger, because human population bottlenecks cannot last just a few generations, since many are required for a human population to grow from 1000 to the long-term mean of 100,000 individuals.

We have explored how small the population bottlenecks would be by computer simulations (Table 1). If a bottleneck lasts 10 generations and we ignore the time required for a population to grow back to its mean size, the smallest bottleneck that allows the persistence of 40 allelic lineages, out of 50 present before the bottleneck, with a probability of 95%, is 270–300 individuals. If the number of alleles passing through the bottleneck is 60 as in *DRB1* (and assuming 70 alleles before the bottleneck), the minimum population size is 458–490. When we take into account the time required for the population to recover to its average size, the minimum population size at the bottleneck becomes substantially larger. For example, if we assume a rate of population increase of 1% per generation (which is 50 times greater

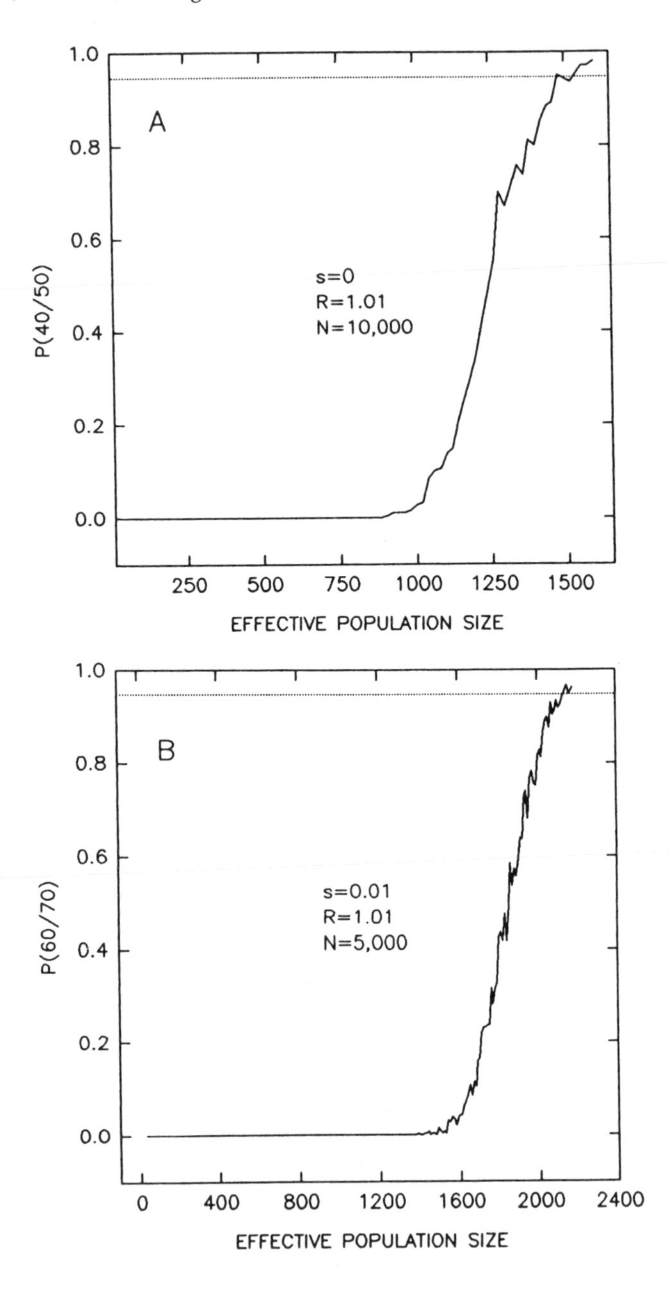

FIGURE 6 Probability that a number of alleles will persist after a bottleneck lasting 10 generations, as a function of population size at the bottleneck. The graphs represent averages of 300 computer simulations. (*A*) Probability that 40

than the average rate of growth of human populations throughout the Pleistocene; Spuhler, 1993, p. 279), a minimum effective population size of 2120–2180 individuals is required for passing 60 out of 70 alleles. (Recovery is simulated only to a population size of 5000 individuals in this case; recovery to 100,000 individuals will further increase the minimum population size.) Selection has very little effect under these conditions (compare the two top rows or the two bottom rows in Table 1). Reducing the probability of passing the polymorphisms to some value smaller than 95% is also of little consequence (Figure 6).

The conclusion follows that human ancestral populations could never have been smaller than two or three thousand individuals at any time over the last several million years. This conclusion might be strengthened by taking into account that extensive polymorphisms exist at other HLA loci (and at other genes as well). For example, 63 primate alleles are known of the *DQB1* gene, 17 of them in humans. As many as 14 *DQB1* human alleles predate the origin of *Homo erectus*, 9 alleles predate the divergence of the human and chimpanzee lineages, and 7 alleles predate the divergence of the human and orangutan lineages. Another instance is the occurrence of multiallelic polymorphisms in the β-globin family that yield at least 17 haplotypes, the coalescence of which goes back to 450,000 years B.P. or earlier (Fullerton *et al.*, 1994) and would be consistent with an effective population of 10,000 individuals through that time span.

Theories of Human Origins

The origin of anatomically modern humans, *Homo sapiens sapiens*, occurred around 200,000 years B.P. The transition from *H. erectus* to *H. sapiens* happened around 500,000 years B.P., and the emergence of *H. erectus* occurred in Africa somewhat earlier than 1.7 Myr B.P. (Swisher *et al.*, 1994, Gibbons, 1994). The hominid lineage diverged from the chimpanzee lineage about 6 Myr B.P., and the orangutan lineage

FIGURE 6 (Continued) alleles will persist out of 50 present at the start of the bottleneck; alleles are neutral ($s = 0$) and the population grows at a rate $R = 1.01$ per generation until it reaches a size $N = 10,000$ individuals. (*B*) Probability that 60 alleles will persist out of 70 present at the start of the bottleneck; heterozygotes have 1% fitness advantage over homozygotes ($s = 0.01$); $R = 1.01$; $N = 5000$. A probability of 95% requires minimum population sizes of 1540–1600 (*A*) and 2140–2180 (*B*) individuals. The probability increases sharply with population size and, thus, the minimum size is only moderately reduced for probabilities less than 0.95.

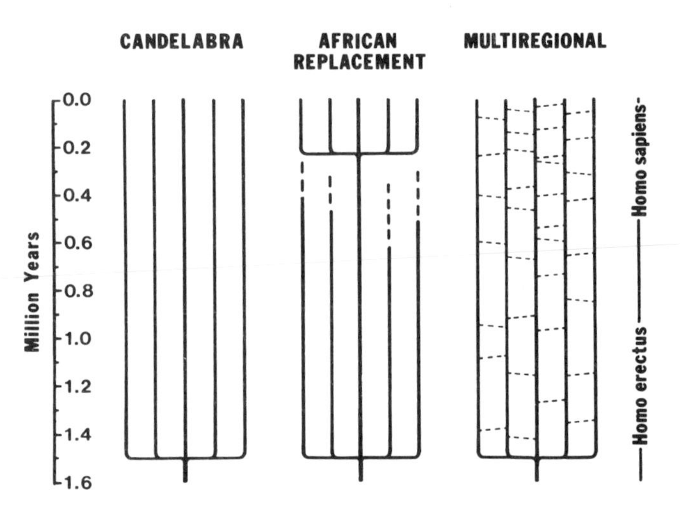

FIGURE 7 Three models of human evolution during the Pleistocene. The *candelabra* model proposes that the transition from *H. erectus* to *H. sapiens* occurred independently in different parts of the World. The *African replacement* model proposes that populations of *H. erectus* and archaic *H. sapiens* became extinct in Asia and Europe and were replaced by anatomically modern humans immigrating from Africa. The *multiregional* model proposes regional continuity and local selection pressures in different parts of the world, but with gene flow (indicated by the dashes connecting the vertical lines that represent different regional populations) due to occasional migrations between populations.

diverged from the lineage of African apes and humans about 13 Myr B.P.

H. erectus spread out of Africa shortly after its emergence from *Homo habilis*. Fossil remains of *H. erectus* are known from Indonesia (Java), China, and the Middle East, as well as Africa (Gibbons, 1994). *H. erectus* fossils from Java have been dated 1.81 ± 0.04 and 1.66 ± 0.04 Myr B.P. (Swisher *et al.*, 1994). There are three models concerning the geographic origin of modern humans: candelabra, African replacement, and multiregional (Figure 7).

The *candelabra model* was proposed by Carleton S. Coon (1962), who argued that "*Homo erectus* evolved into *Homo sapiens* not once but five times, as each subspecies, living in its own territory, passed a critical threshold from a more brutal to a more *sapient* state." Moreover, the threshold was crossed at different times, the Caucasoid race becoming *sapiens* first, whereas the Congoloids reached the *sapiens* condition some 200,000 years later. Coon's proposal was effectively criticized by several authors who pointed out the impossibility that the same species would evolve independently twice, let alone five times, in different regions and at different times (Dobzhansky, 1963). The candelabra model is, moreover, contradicted by fossil (Aiello, 1993; Spuhler, 1993) and molecular (Takahata, 1993a, pp. 13, 19) evidence.

The *African replacement model* proposes that modern humans first arose in Africa somewhat prior to 100,000 years B.P. and from there spread throughout the world, replacing elsewhere the preexisting populations of *H. erectus* or archaic *H. sapiens* (Stringer, 1990, 1992; Stringer and Andrews, 1988). One extreme version of this hypothesis (the *Noah's Ark model*) proposes that the transition of archaic to modern *H. sapiens* was associated with a very narrow bottleneck, consisting of only two or very few individuals who are the ancestors of all modern mankind.

The *Noah's Ark model* is supported by an interpretation of mitochondrial DNA analysis showing that the diverse mitochondrial DNA sequences found in modern humans coalesce to one ancestral sequence, the "mitochondrial Eve" or "mother of us all," that existed in Africa about 200,000 years ago (Cann *et al.*, 1987; Stoneking *et al.*, 1990; Vigilant *et al.*, 1991). This conclusion has been challenged on grounds concerning (1) whether the coalescence is to Africa, (2) the time of the coalescence, and (3) the inference of a population bottleneck (e.g., Templeton, 1992). The actual date of coalescence depends on assumptions about evolutionary rates. Based on a time of divergence between humans and chimpanzees of 6 Myr, the time to coalescence for mitochondrial DNA polymorphism has been recently estimated at 298,000 years B.P., with a 95% confidence interval of 129,000–536,000 years (Ruvolo *et al.*, 1993).

The inference that a narrow bottleneck occurred at the time of the coalescence is based on a confusion between gene genealogies and individual genealogies. Gene genealogies gradually coalesce towards a unique DNA ancestral sequence, whereas individual genealogies increase by a factor of 2 per generation: an individual has two parents in the previous generation, four ancestors in the generation before that, and so on. (The theoretical number of ancestors for any one individual becomes enormous after some tens of generations, but "inbreeding" occurs: after some generations, ancestors appear more than once in the genealogy.)

As we pointed out above, assuming an effective population of N individuals, mean coalescence is $4N$ generations for nuclear polymorphisms. Mitochondrial DNA is haploid and maternally inherited; hence, the mean coalescence is $2N_f$, where N_f is the number of mothers. If we assume 20 years for a human generation throughout the Pleistocene, 298,000 years to coalescence implies an effective mean population size of 7450 mothers or an effective population size of 14,900 humans. The 95% confidence time estimate of 129,000–536,000 years yields 6,450–26,800 generations, corresponding to a mean effective population size of 6,450–26,800 individuals.

There is one more factor to take into account in the calculations just made. We have used *mean* estimates to coalescence time, but these estimates have large variances. When the sample of genes is large, the standard deviation of the mean for nuclear genes is larger than $2N$ (Nei, 1987, eq. 13.74); for mitochondrial DNA it is larger than $N/2$. The 95% confidence interval for the number of generations to coalescence will correspondingly extend at the upper end to more than $2N$ (53,600) generations for mitochondrial DNA.

Thus the mitochondrial DNA sequence data are consistent with a mean effective population size between somewhat less than 10,000 and more than 50,000 individuals throughout the Pleistocene. This population size is, in turn, consistent with the estimate based on the *HLA DRB1* polymorphism of a mean population size of 100,000 individuals for human ancestors over the last 30–40 Myr. The mitochondrial DNA data are also consistent with the result that no bottleneck smaller than several thousand individuals could have happened in hominid history.

The *multiregional model* emphasizes regional continuity in the evolution from *H. erectus* to archaic *H. sapiens,* and later to anatomically modern humans. Several variations of this model have been formulated (Aiello, 1993) that differ in the relative role given to two factors: (*i*) the prevalence of Africa in the gradual emergence of modern features, and (*ii*) the amount of migration between regions. At one extreme, the "African hybridization and replacement model" proposes that modern humans first evolved in Africa, but their dispersal throughout the world was accompanied by certain amount of hybridization with indigenous premodern populations (Braüer, 1984, 1992). A variant that might be called the "Levantine hybridization and replacement model" is statistically supported by a correlation analysis of morphological traits that favors the Middle East over Africa as the origin of modern human traits (Waddle, 1994). At the other extreme, some authors deny a recent African origin for modern humans while emphasizing regional continuity and gene exchange throughout Africa, Europe, and Asia (Wolpoff *et al.*, 1988; Wolpoff, 1989). Other authors have proposed an intermediate "assimilation model" that gives preponderance to Africa in the emergence of anatomically modern traits but attributes major importance to gene flow and differential selection pressures resulting in morphological differentiation among regions (Smith *et al.*, 1989; Smith, 1992; Spuhler, 1993).

The paleontological data favor some form of the multiregional model, but one that gives preponderance to Africa (or the Levant, according to Waddle, 1994) as the locus where modern human traits first appear

(Aiello, 1993). There is evidence of regional continuity, and also of the incorporation of traits from one into another population (Aiello, 1993; Spuhler, 1993).

Molecular Evidence and the Multiregional Model

Does the molecular evidence favor any particular model of the origin of modern humans? The *HLA* evidence contradicts the Noah's Ark model and evinces that the ancestral population of modern humans was at no time smaller than several thousand individuals, a result also consistent with the mitochondrial DNA data.

Two recent molecular studies favor some degree of regional continuity over complete African replacement. The first study concerns polymorphisms in the genes for green and red visual pigments. Color vision in animals is mediated by light-sensitive pigments consisting of a chromophore covalently linked to a protein moiety (opsin). The genes coding for opsins in the red and green pigments are located on the long arm of chromosome X, whereas the one for the blue pigment is on chromosome 7 (Deeb *et al.*, 1994). In humans, the red and green opsin genes are highly homologous and consist of six exons. The duplication of these two genes has been dated to 30–40 Myr B.P., shortly after the divergence of the Old- and New-World primates.

The green and red opsin genes have now been sequenced in a sample of 16 chimpanzees, 7 gorillas, and 4 orangutans, yielding a total of 14 biallelic polymorphic sites (all in either exon 2 or 4) (Deeb *et al.*, 1994). Six of these polymorphisms are also found in humans, which indicates that they are of ancient origin predating the divergence of humans and apes.

One of these trans-specific polymorphisms involves the amino acid residue 65, which in the green opsin gene of orangutans and humans can be either threonine or isoleucine (Deeb *et al.*, 1994). The relevant result is that this polymorphism has been found in Caucasians (the Ile-65 allele in 4 out of 120 individuals) but not in a sample of 56 individuals of African ancestry and 49 of Asian ancestry (Deeb *et al.*, 1994). It is possible that the Ile-65 allele may eventually be found in African populations. It is also possible that it may have been lost from African and Asian populations in recent times, that is, after the emergence of modern humans. But since this polymorphism is millions of years old, loss of the allele over the long period since the migration of *H. erectus* out of Africa is more likely than a recent loss. In the replacement model, migrants from Africa colonize other parts of the world and replace any preexisting populations within the last 200,000 years. It would seem unlikely that the polymorphism would have been passed on to Caucasian populations and become thereafter lost in the larger African population. Thus, the opsin polymorphism argues (mildly) against a complete replacement of the Caucasian gene pool by African populations.

The second example concerns an autosomal recessive disorder in lipid metabolism due to the absence of apolipoprotein C-II, the physiological activator of lipoprotein lipase, a key enzyme in very low density lipoprotein metabolism. Two deleterious alleles, one from a Venezuelan Caucasian family and one from a Japanese family, share a frameshift mutation suggesting common ancestry (Li *et al.*, 1993). These two mutants diverged from the normal allele at least 2 Myr B.P. (Li *et al.*, 1993). The persistence of two defective alleles over such a long time is a puzzle, perhaps a consequence of small heterozygote advantage. But this persistence (*i*) argues against extremely small population bottlenecks throughout Pleistocene human history, and (*ii*) favors the conclusion that European and Asian *H. erectus* have contributed to the gene pool of modern *H. sapiens* (Li *et al.*, 1993).

SUMMARY

The major histocompatibility complex (*MHC*) plays a cardinal role in the defense of vertebrates against parasites and other pathogens. In some genes there are extensive and ancient polymorphisms that have passed from ancestral to descendant species and are shared among contemporary species. The polymorphism at the *DRB1* locus, represented by 58 known alleles in humans, has existed for at least 30 million years and is shared by humans, apes, and other primates. The coalescence theory of population genetics leads to the conclusion that the *DRB1* polymorphism requires that the population ancestral to modern humans has maintained a mean effective size of 100,000 individuals over the 30-million-year persistence of this polymorphism. We explore the possibility of occasional population bottlenecks and conclude that the ancestral population could not have at any time consisted of fewer than several thousand individuals.

The *MHC* polymorphisms exclude the theory claiming, on the basis of mitochondrial DNA polymorphisms, that a constriction down to one or few women occurred in Africa, at the transition from archaic to anatomically modern humans, some 200,000 years ago. The data are consistent with, but do not provide specific support for, the claim that human populations throughout the World were at that time replaced by populations migrating from Africa. The *MHC* and other molecular polymorphisms are consistent with a "multiregional" theory of Pleistocene human evolution that proposes regional continuity of human populations since the time of migrations of *Homo erectus* to the present, with distinctive regional selective pressures and occasional migrations between populations.

We are grateful to Drs. W. M. Fitch, R. R. Hudson, and Y. Satta for reading the manuscript and offering suggestions, and to Dr. E. Barrio for help with the figures. Research in molecular evolution by the senior author is supported by National Institutes of Health Grant GM42397. A.E. is supported by a fellowship from the Banco Interamericano de Desarrollo–Consejo Nacional de Investigaciones Científicas y Tecnológicas Program, Venezuela.

REFERENCES

Aiello, L. C. (1993) The fossil evidence for modern human origins in Africa: A revised view. *Am. Anthropol.* **95,** 73–96.

Arden, B. & Klein, J. (1982) Biochemical comparison of major histocompatibility complex molecules from different subspecies of *Mus musculus*: Evidence for *trans*-specific evolution of alleles. *Proc. Natl. Acad. Sci. USA* **79,** 2342–2346.

Avise, J. C. (1994) *Molecular Markers, Natural History and Evolution* (Chapman & Hall, New York).

Ayala, F. J. (1975) Genetic differentiation during the speciation process. *Evol. Biol.* **8,** 1–78.

Ayala, F. J. (1982) *Population and Evolutionary Genetics* (Benjamin/Cummings, Menlo Park, CA).

Ayala, F. J., Tracey, M. L., Barr, L. G. & Ehrenfeld, J. G. (1974) Genetic and reproductive differentiation of *Drosophila equinoxialis caribbensis*. *Evolution* **28,** 24–41.

Barigozzi, C. (1982) *Mechanisms of Speciation* (Liss, New York).

Barton, N. H. & Charlesworth, B. (1984) Founder effects, genetic revolutions, and speciation. *Annu. Rev. Ecol. Syst.* **15,** 133–164.

Bjorkman, P. J., Saper, M. A., Samraoui, B., Bennett, W. S., Strominger, J. L. & Wiley, D. C. (1987a) Structure of the human class I histocompatibility antigen, HLA-A2. *Nature (London)* **329,** 506–512.

Bjorkman, P. J., Saper, M. A., Samraoui, B., Bennett, W. S., Strominger, J. L. & Wiley, D. C. (1987b) The foreign antigen binding site and T cell recognition regions of class I histocompatibility antigens. *Nature (London)* **329,** 512–521.

Bjorkman, P. J. & Parham, P. (1990) Structure, function, and diversity of class-I major histocompatibility complex molecules *Annu. Rev. Biochem.* **59,** 253–288.

Bontrop, R. E. (1994) Nonhuman primate *Mhc-DQA* and *-DQB* second exon nucleotide sequences: a compilation. *Immunogenetics* **39,** 81–92.

Braüer, G. (1984) A craniological approach to the origin of anatomically modern *Homo sapiens* in Africa and implications for the appearance of modern Europeans. In *The Origins of Modern Humans: A World Survey of the Fossil Evidence*, eds. Smith, F. H. & Spencer, F. (Liss, New York), pp. 327–410.

Braüer, G. (1992) Africa's place in the evolution of *Home sapiens*. In *Continuity or Replacement? Controversies in Homo sapiens Evolution*, eds. Braüer, G. & Smith, F. H. (Balkema, Rotterdam, The Netherlands), pp. 83–98.

Brown, J. H., Jardetzky, T. S., Gorga, J. C., Stern, L. J., Urban, R. G., Strominger, J. L. & Wiley, D. C. (1993) Three-dimensional structure of the human class II histocompatibility antigen HLA-DR1. *Nature (London)* **364,** 33–39.

Bullini, L. (1982) Ecological and ethological aspects of the speciation process. In *Mechanisms of Speciation*, ed. Barigozzi, C. (Liss, New York), pp. 241–264.

Cann, R. L., Stoneking, M. & Wilson, A. C. (1987) Mitochondrial DNA and human evolution. *Nature (London)* **325,** 31–36.

Carson, H. L. (1968) The population flush and its genetic consequences. In *Population Biology and Evolution*, ed. Lewontin, R. C. (Syracuse Univ. Press, New York), pp. 123–127.

Carson, H. L. (1975) The genetics of speciation at the diploid level. *Am. Nat.* **109,** 83–92.

Carson, H. L. & Templeton, A. R. (1984) Genetic revolution in relation to speciation phenomena: The founding of new populations. *Annu. Rev. Ecol. Syst.* **15,** 97–131.

Coon, C. S. (1962) *The Origin of Races* (Knopf, New York).

Deeb, S. S., Jørgensen, A. L., Battisti, L., Iwasaki, L. & Motulsky, A. G. (1994) Sequence divergence of the red and green visual pigments in the great apes and man. *Proc. Natl. Acad. Sci. USA* **91,** 7262–7266.

Dobzhansky, T. (1963) A debatable account of the origin of races. *Sci. Am.* **208 (2),** 169–172.

Fan, W., Kasahara, M., Gutknecht, J., Klein, D., Mayer, W. E., Jonker, M. & Klein, J. (1989) Shared class II MHC polymorphism between humans and chimpanzees. *Hum. Immunol.* **26,** 107–121.

Figueroa, F., Günther, E. & Klein, J. (1988) MHC polymorphism predating speciation. *Nature (London)* **335,** 265–267.

Fullerton, S. M., Harding, R. M., Boyce, A. J. & Clegg, J. B. (1994) Molecular and population genetic analysis of allelic sequence diversity at the human β-globin locus. *Proc. Natl. Acad. Sci. USA* **91,** 1805–1809.

Galiana, A., Moya, A. & Ayala, F. J. (1993) Founder-flush speciation in *Drosophila pseudoobscura*: A large-scale experiment. *Evolution* **47,** 432–444.

Gibbons, A. (1994) Rewriting—and redating—prehistory. *Science* **263,** 1087–1088.

Giddings, L. V., Kaneshiro, K. Y. & Anderson, W. W. (1989) *Genetics, Speciation, and the Founder Principle* (Oxford Univ. Press, New York).

Griffiths, R. C. (1980) Lines of decent in the diffusion approximation of neutral Wright-Fisher models. *Theor. Popul. Biol.* **17,** 37–50.

Gyllenstein, U. B. & Erlich, H. A. (1989) Ancient roots for polymorphism at the HLA-DQα locus in primates. *Proc. Natl. Acad. Sci. USA* **86,** 9986–9990.

Gyllenstein, U. B., Lashkari, D. & Erlich, H. A. (1990) Allelic diversification at the class II *DQB* locus of the mammalian major histocompatibility complex. *Proc. Natl. Acad. Sci. USA* **87,** 1835–1839.

Hill, A. V. S., Allsopp, C. E. M., Kwiatowski, D., Anstey, N. M., Twumasi, P., Rowe, P. A., Bennett, S., Brewster, D., McMichael, A. J. & Greenwood, B. M. (1991) Common West African HLA antigens are associated with protection from severe malaria. *Nature (London)* **352,** 595–600.

Hill, A. V. S., Kwiatkowski, D., McMichael, A. J., Greenwood, B. M. & Bennett, S. (1992a) Maintenance of MHC polymorphism. *Nature (London)* **355,** 402–403.

Hill, A. V. S., Elvin, J., Willis, A. C., Aidoo, M., Alsopp, E. E. M., Gotch, F. M., Gao, X. M., Takiguchi, M., Greenwood, B. M., Townsend, A. R. M., McMichael, A. J. & Whittle, H. C. (1992b) Molecular analysis of the association of HLA-B53 and resistance to severe malaria. *Nature (London)* **360,** 434–439.

Hudson, R. R. (1990) Gene genealogies and the coalescent process. *Oxford Surv. Evol. Biol.* **7,** 1–44.

Hughes, A. L. & Nei, M. (1988) Pattern of nucleotide substitution at major histocompatibility complex class I loci reveals overdominant selection. *Nature (London)* **335,** 167–170.

Hughes, A. L. & Nei, M. (1992) Maintenance of *MHC* polymorphism. *Nature (London)* **355,** 402–403.

Kimura, M. (1983) *The Neutral Theory of Molecular Evolution* (Cambridge Univ. Press, Cambridge, U.K.).

Kimura, M. & Ohta, T. (1969) The average number of generations until fixation of a mutant gene in a finite population. *Genetics* **61,** 763–771.

Kingman, J. F. C. (1982a) The coalescent. *Stochast. Processes Appl.* **13,** 235–248.

Kingman, J. F. C. (1982b) On the genealogy of large populations. *J. Appl. Probab.* **19,** 27–43.

Klein, J. (1986) *The Natural History of the Major Histocompatibility Complex* (Wiley, New York).

Klein, J. & Figueroa, F. (1986) Evolution of the Major Histocompatibility Complex. *CRC Crit. Rev. Immunol.* **6,** 295–386.

Klein, J., Gutknecht, J. & Fischer, N. (1990) The major histocompatibility complex and human evolution. *Trends Genet.* **6,** 7–11.

Klein, J. & Takahata, N. (1990) The major histocompatibility complex and the quest for origins. *Immunol. Rev.* **113,** 5–25.

Lawlor, D. A., Ward, F. E., Ennis, P. D., Jackson, A. P. & Parham, P. (1988) *HLA-A* and *B* polymorphisms predate the divergence of humans and chimpanzees. *Nature (London)* **355,** 268–271.

Li, W.-H., Xiong, W., Liu, S. A.-W. & Chan, L. (1993) Nucleotide diversity in humans and evidence for the absence of a severe bottleneck during human evolution. In *Genetics of Cellular, Individual, Family, and Population Variability*, eds. Sing, C. F. & Hanis, C. L. (Oxford Univ. Press, New York), pp. 253–261.

Marsh, S. G. E. & Bodmer, J. G. (1991) HLA class II nucleotide sequences, 1991. *Immunogenetics* **33,** 321–334.

Marsh, S. G. E. & Bodmer, J. G. (1993) HLA class II nucleotide sequences, 1992. *Immunogenetics* **37,** 79–94.

Mayer, W. E., Jonker, M., Klein, D., Ivanyi, P., van Seventer, G. & Klein, J. (1988) Nucleotide sequences of chimpanzee MHC class I alleles: evidence for *trans*-species mode of evolution. *EMBO J.* **7,** 2765–2774.

Mayr, E. (1954) Change of genetic environment and evolution. In *Evolution as a Process*, eds. Huxley, J., Hardy, A. C. & Ford, E. B. (Allen & Unwin, London), pp. 157–180.

Mayr, E. (1963) *Animal Species and Evolution* (Harvard Univ. Press, Cambridge, MA).

Mayr, E. (1982) Processes of speciation in animals. In *Mechanisms of Speciation*, ed. Barigozzi, C. (Liss, New York), pp. 1–19.

McConnell, T. J., Talbot, W. S., McIndoe, R. A. & Wakeland, E. K. (1988) The origin of MHC class II gene polymorphism within the genus *Mus. Nature (London)* **332,** 651–654.

Miller, L. H. (1994) Impact of malaria on genetic polymorphism and genetic diseases in Africans and African Americans. *Proc. Natl. Acad. Sci. USA* **91,** 2415–2419.

Nei, M. (1987) *Molecular Evolutionary Genetics* (Columbia Univ. Press, New York).

Otte, D. & Endler, J. A. (1989) *Speciation and Its Consequences* (Sinauer, Sunderland, MA).

Raup, D. M. (1994) The Role of Extinction in Evolution. *Proc. Natl. Acad. Sci. USA* **91,** 6758–6763.

Ruvolo, M., Zehr, S., von Dornum, M., Pan, D., Chang, B. & Lin, J. (1993)

Mitochondrial COII sequences and modern human origins. *Mol. Biol. Evol.* **10,** 1115–1135.

Satta, Y., Takahata, N., Schönbach, C., Gutknecht, J. & Klein, J. (1991) Calibrating evolutionary rates at major histocompatibility complex loci. In *Molecular Evolution of the Major Histocompatibility Complex,* eds. Klein, J. & Klein, D. (Springer, Heidelberg), pp. 51–62.

Satta, Y., O'hUigin, C., Takahata, N. & Klein, J. (1993) The synonymous substitution rate of the major histocompatibility complex loci in primates. *Proc. Natl. Acad. Sci. USA* **90,** 7480–7484.

Satta, Y., O'hUigin, C., Takahata, N. & Klein, J. (1994) Intensity of natural selection at the major histocompatibility complex loci. *Proc. Natl. Acad. Sci. USA* **91,** 7184–7188.

Smith, F. H. (1992) The role of continuity in modern human origins. In *Continuity or Replacement? Controversies in Homo sapiens Evolution,* eds. Braüer, G. & Smith, F. H. (Balkema, Rotterdam, The Netherlands), pp. 145–156.

Smith, F. H., Falsetti, A. B. & Donnelly, S. M. (1989) Modern human origins. *Yearb. Phys. Anthropol.* **32,** 35–68.

Spuhler, J. N. (1993) Population genetics and evolution in the genus *Homo* in the last two million years. In *Genetics of Cellular, Individual, Family, and Population Variability,* eds. Sing, C. F. & Hanis, C. L. (Oxford Univ. Press, New York), pp. 262–297.

Stoneking, M., Jorde, L. B., Bhatia, K. & Wilson, A. C. (1990) Geographic variation of human mitochondrial DNA from Papua New Guinea. *Genetics* **124,** 717–733.

Stringer, C. B. (1990) The emergence of modern humans. *Sci. Am.* **263 (6),** 98–104.

Stringer, C. B. (1992) Reconstructing recent human evolution. *Philos. Trans. R. Soc. London B* **337,** 217–224.

Stringer, C. B. & Andrews, P. (1988) Genetic and fossil evidence for the origin of modern humans. *Science* **239,** 1263–1268.

Swisher, C. C., III, Curtis, G. H., Jacob, T., Getty, A. G., Suprijo, A. & Widiasmoro (1994) Age of the earliest known hominids in Java, Indonesia. *Science* **263,** 1118–1121.

Tajima, F. (1983) Evolutionary relationship of DNA sequences in finite populations. *Genetics* **105,** 437–460.

Takahata, N. (1990) A simple genealogical structure of strongly balanced allelic lines and trans-species evolution of polymorphism. *Proc. Natl. Acad. Sci. USA* **87,** 2419–2423.

Takahata, N. (1991) in *Molecular Evolution of the Major Histocompatibility Complex,* eds. Klein, J. & Klein, D. (Springer, Heidelberg), pp. 29–49.

Takahata, N. (1993a) Allelic genealogy and human evolution. *Mol. Biol. Evol.* **10,** 2–22.

Takahata, N. (1993b) Evolutionary genetics of human paleopopulations. In: *Mechanisms of Molecular Evolution,* eds. Takahata, N. & Clark, A. G. (Japan Scientific Societies Press, Tokyo and Sinauer Associates, Inc., Sunderland, MA), pp. 1–21.

Takahata, N. & Nei, M. (1985) Gene genealogy and variance of interpopulational nucleotide differences. *Genetics* **110,** 325–344.

Takahata, N. & Nei, M. (1990) Allelic genealogy under overdominant and frequency-dependent selection and polymorphism of major histocompatibility complex loci. *Genetics* **124,** 967–978.

Takahata, N., Satta, Y. & Klein, J. (1992) Polymorphism and balancing selection at major histocompatibility complex loci. *Genetics* **130,** 925–938.

Tavaré, S. (1984) Line-of-descent and genealogical processes, and their applications in population genetics models. *Theor. Popul. Biol.* **26,** 119–164.

Templeton, A. R. (1980) The theory of speciation via the founder principle. *Genetics* **94,** 1011–1038.

Templeton, A. R. (1992) Human origins and analysis of mitochondrial DNA sequences. *Science* **255,** 737.

Vigilant, L., Stoneking, M., Harpending, H., Hawkes, K. & Wilson, A. C. (1991) African populations and the evolution of human mitochondrial DNA. *Science* **253,** 1503–1507.

Waddle, D. M. (1994) Matrix correlation tests support a single origin for modern humans. *Nature (London)* **368,** 452–454.

Wolpoff, M. H. (1989) Multiregional evolution: The fossil alternative to Eden. In *The Human Revolution,* eds. Mellars, P. & Stringer, C. (Edinburgh Univ. Press, Edinburgh), pp. 62–108.

Wolpoff, M. H., Spuhler, J. N., Smith, F. H., Radovčič, J., Pope, G., Frayer, D. W., Eckhardt, R. & Clark, G. (1988) Modern human origins. *Science* **241,** 772–773.

Part IV

RATES

The chloroplast is an essential organelle derived from a cyanobacteria-like organism that was acquired as an endosymbiont by a remote ancestor of modern plants. The chloroplast's genome is a DNA molecule consisting of 150 kilobase pairs that encode 100 gene products. It has been completely sequenced in six very diverse plants and investigated for various purposes in several score species, yielding a tremendous wealth of information available for comparative evolutionary investigations. Clegg and his colleagues, in Chapter 11, uncover a complex evolutionary pattern. Some noncoding regions include hot spots for insertions and deletions and exhibit complex recombinational features. Selective drives in codon utilization have changed over evolutionary time. Patterns of amino acid replacements reflect functional constraints imposed by natural selection on protein configuration. Rates of evolution are quite variable from one order to another, although much of the variation can be accounted for by differences in generation time.

The constancy of evolutionary rates is the subject of Chapter 12. The Cu,Zn superoxide dismutase (SOD) seems to behave like a very erratic clock: the rate of amino acid replacement is 5 times faster among mammals than between fungi and animals. Walter M. Fitch and Francisco J. Ayala (19) analyze the amino acid sequences of several score species and show that SOD behaves like a fairly accurate clock by assuming a complex pattern in which different sets of amino acids have different probabilities of change that are nevertheless constant through time. The model for constancy requires that a set of only 28 amino acids be replaceable at any one time. Although the elements of the set vary from time to time and from lineage to lineage, a total of 44 amino acids are permanently unreplaceable. Moreover, the number of different amino acids that can occur at any particular variable site is very small, limited to 2–4 alternatives. The conclusion is that molecular clocks have complex features that must be ascertained before drawing out inferences about the topology and timing of historical relationships.

11

Rates and Patterns of Chloroplast DNA Evolution

MICHAEL T. CLEGG, BRANDON S. GAUT, GERALD H. LEARN, Jr., AND BRIAN R. MORTON

The chloroplast genome was almost certainly contributed to the eukaryotic cell through an endosymbiotic association with a cyanobacteria-like prokaryote. Moreover, present evidence suggests that the association that led to land plants occurred only once in evolution (Gray, 1993). The derivative plastid genome (cpDNA) is a relictual molecule of about 150 kbp that encodes roughly 100 genetic functions (reviews in Palmer, 1991; Clegg *et al.*, 1991). This genome is the most widely studied plant genome with regard to both molecular organization and evolution. Complete sequences of six chloroplast genomes are now available, and these represent virtually the full range of plant diversity [an alga, *Euglena* (Hallick *et al.*, 1993); a bryophyte, *Marchantia* (Ohyama *et al.*, 1986); a conifer, black pine (Wakasugi *et al.*, 1993); a dicot, tobacco (Shinozaki *et al.*, 1986); a monocot, rice (Hiratsuka *et al.*, 1989); and a parasitic dicot, *Epifagus* (Wolfe *et al.*, 1992)]. With the possible exception of *Euglena* and *Epifagus*, the picture that has emerged is one of a relatively stable genome with marked conservation of gene content and a substantial conservation of structural organization. Mapping studies that span land plants and algae confirm the impression of strong conservation in gene content (Palmer, 1985).

Michael T. Clegg is professor of genetics and Gerald H. Learn, Jr. and Brian R. Morton are postdoctoral research associates in the Department of Botany and Plant Science at the University of California, Riverside. Brandon S. Gaut is assistant professor of genetics at Rutgers University, New Brunswick, New Jersey.

Because the photosynthetic machinery requires many more gene functions than are specified on the cpDNA of plants and algae, it is assumed that many gene functions were transferred to the eukaryotic nuclear genome. The strong conservation of cpDNA gene content cited above indicates that most transfers of gene function from the chloroplast to the nuclear genome occurred early following the primordial endosymbiotic event. Among land plants, gene content is almost perfectly conserved, although a few recent transfers of function from the chloroplast to the nuclear genome have been demonstrated (Downie and Palmer 1991; Gantt *et al.*, 1991).

Conservation of gene content and a relatively slow rate of nucleotide substitution in protein-coding genes has made the chloroplast genome an ideal focus for studies of plant evolutionary history (reviewed in Clegg, 1993). These efforts have culminated in the past year with the publication of a volume of papers that presents a detailed molecular phylogeny for seed plants (Chase *et al.*, 1993). This effort has involved many laboratories that have together determined the DNA sequence for more than 750 copies (by latest count) of the cpDNA gene *rbcL* encoding the large subunit of ribulose-1,5-bisphosphate carboxylase (RuBisCo). The sequence data span the full range of plant diversity from algae to flowering plants. This very large collection of gene sequence data has allowed the reconstruction of plant evolutionary history at a level of detail that is unprecedented in molecular systematics.

Accurate phylogenies are of more than academic interest. They provide an organizing framework for addressing a host of other important questions about biological change. For example, questions about the origins of major morphological adaptions must be placed in a phylogenetic context to reconstruct the precise sequence of genetic (and molecular) changes that give rise to novel structures. The mutational processes that subsume biological change can be revealed in great detail through a phylogenetic analysis. And, questions about the frequency and mode of horizontal transfer of mobile genetic elements can only be addressed in a phylogenetic context. One can make a very long list of biological problems that are illuminated by phylogenetic analyses.

Despite their obvious utility, many important questions remain about the accuracy of molecular phylogenies. All methods of phylogenetic reconstruction assume a fairly high degree of statistical regularity in the underlying mutational dynamics. Our goal in this article is to analyze the tempo and mode of cpDNA evolution in greater depth. We will show that below the surface impression of conservation there are a variety of mutational processes that often violate assumptions of rate constancy and site independence of mutational change. To facilitate an analysis of patterns of cpDNA mutational change, we divide the genome into functional categories and we then study the pattern of mutational change within each functional category. The functional categories are (*i*) DNA regions that do not code for tRNA, ribosomal RNA (rRNA), or protein (referred to as "noncoding DNA"); (*ii*) protein-coding genes; and (*iii*) chloroplast introns. The methodological approach is that of comparative sequence analysis, where complete DNA sequences from phylogenetically structured samples are analyzed to reveal the pattern of mutational change in evolution.

Evolution of Noncoding DNA Regions

The chloroplast genome is highly condensed compared with eukaryotic genomes; for example, only 32% of the rice genome is noncoding. Most of this noncoding DNA is found in very short segments separating functional genes. A number of recent studies have revealed complex patterns of mutational change in noncoding regions. The most widely studied noncoding region of the chloroplast genome is the region downstream of the *rbcL* gene in the grass family (Poaceae). This noncoding sequence is flanked by the genes *rbcL* and *psaI* (the gene encoding photosystem I polypeptide I) and is 1694 bases long in the rice genome, making it one of the longest noncoding regions of the genome and the longest when introns are excluded (Figure 1). A pseudogene for the chloroplast gene *rpl23* is also located within this noncoding segment (Hiratsuka *et al.*, 1989). Hiratsuka *et al.* (1989) argued that a large inversion, unique to the grass family, arose through a recombinational interaction between nonhomologous tRNA genes, and the same process of recombination between short repeats has been invoked as the mechanism responsible for the origin of the *rpl23* pseudogene *ψrpl23* (Ogihara *et al.*, 1992).

The functional *rpl23* gene of the rice genome is located in the inverted repeat about 27 kb away. Bowman *et al.* (1988) suggested that the pseudogene was being converted by the functional *rpl23* gene because the genetic divergence among pseudogenes was lower than the divergence observed for the surrounding noncoding regions. Subsequent work, based on a phylogenetic analysis (Morton and Clegg, 1993), has provided additional support for a model of gene conversion between the *rpl23* pseudogene and its functional counterpart in at least two lineages of the grass family.

Four independent deletion events of at least 850 bases in length have been observed spanning almost identical stretches of the noncoding region between *rbcL* and *psaI* (Morton and Clegg, 1993; Ogihara *et al.*, 1988). Based on flanking sequence data, it has been suggested that recombination between short direct repeats was responsible for these

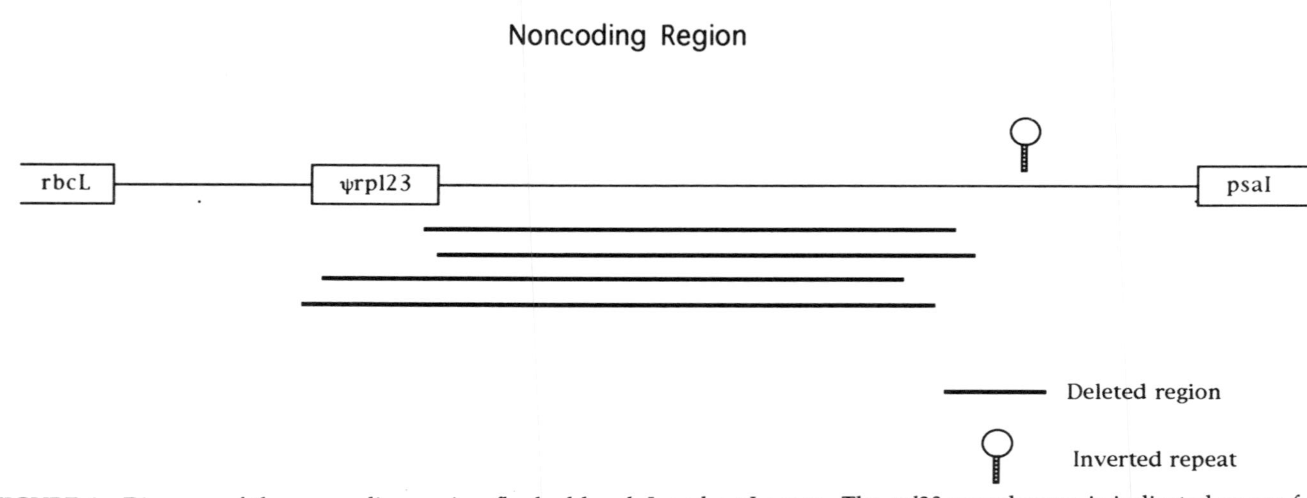

FIGURE 1 Diagram of the noncoding region flanked by *rbcL* and *psaI* genes. The *rpl23* pseudogene is indicated as are four independent deletions. The hypervariable inverted repeat region referred to in the text is marked by a hairpin structure.

deletions (Ogihara *et al.*, 1988). The four large deletions observed in the grasses all extend from within, or very close to, *ψrpl23* to an area about 400 bases upstream of *psaI*. The region spanned by the deletions has a low A+T content relative to other chloroplast noncoding sequences (including the surrounding noncoding sequences) and may have been a coding sequence inserted by recombination, as was *ψrpl23*, prior to the divergence of the grass family (Morton and Clegg, 1993). The rate of nucleotide substitution in this noncoding region is roughly equal to the rate of synonymous substitution (with the exception of *ψrpl23*) for the neighboring *rbcL* gene (Morton and Clegg, 1993), but a large number of short insertion/deletions (indels) have occurred. In addition to both the high rate of indel mutation and the large deletions, an inverted repeat in the noncoding region about 300 bases upstream of the *psal* gene appears to be labile to complex rearrangement events in the grass family (Morton and Clegg, 1993).

A study of the noncoding region upstream of *rbcL* in the grass family revealed similar patterns of complex change (Golenberg *et al.*, 1993). Indels were found to occur at a greater frequency than nucleotide substitutions, and more complex changes, including multiple tandem duplication events, were found to be confined to highly labile sites, resulting in similar although independent indels at identical sites. Taken together, both noncoding regions reveal a marked incidence of parallel mutations at labile sites that may be positively misleading for phylogenetic studies based on restriction fragment length polymorphism (RFLP) data.

Such complex patterns of mutation are not confined to noncoding sequences of the chloroplast. The chloroplast gene for the RNA polymerase subunit B'', *rpoC2*, has an insertion within the coding sequence in rice relative to tobacco. This insertion has been shown to be confined to the grass family. A comparison of the insertion among members of the grass family revealed widespread diversity in terms of indel mutations as well as an increased rate of nucleotide substitution (Cummings *et al.*, 1994).

The low noncoding content of the cpDNA generally and the high rate of large deletion events observed in the largest contiguous noncoding segment in the rice genome suggest that nonessential sequences are rapidly removed from the chloroplast genome as a result of both intra- and intermolecular recombination. This is supported by the rapid loss of all photosynthetic genes from the chloroplast genome of the nonphotosynthetic plant *Epifagus* (Wolfe *et al.*, 1992). Given the low similarity of the noncoding sequence between the chloroplast genomes of rice and tobacco and the observed degree of variation in noncoding sequences within the grasses, the conserved chloroplast genes appear to exist in a very fluid medium of surrounding noncoding sequence.

Evolution of Protein Coding Genes

Codon Bias of Chloroplast Genes. The codon utilization pattern of chloroplast genes of the green algae *Chlamydomonas reinhardtii* and *Chlamydomonas moewusii* appears to be closely adapted to the chloroplast tRNA population. In contrast, land plant chloroplast genes are dominated by a genomic bias towards a high A+T content, which is reflected in a high third-codon position A+T content. An interesting exception is the gene coding for the central protein of the photosystem II reaction center (denoted *psbA*) (Umesono *et al.*, 1988). The *psbA* protein turns over at a very high rate and, consequently, is the most abundant translation product of the plant chloroplast (Mullet and Klein, 1987). The *psbA* gene of flowering plants has a codon use very similar to *Chlamydomonas* chloroplast genes (Morton, 1993). Given the tRNAs encoded by the sequenced chloroplast genomes, the pattern of codon bias of the *Chlamydomonas* genes appears to be the result of selection for an intermediate codon–anticodon interaction strength (Morton, unpublished data). Further, based on the fact that *psbA* is the sole flowering plant chloroplast gene following this codon utilization pattern, selection most likely acted on *psbA* codon use to increase translation efficiency (Morton, 1993).

Despite the difference in codon use by *psbA* relative to other plant chloroplast genes, this gene has a much lower bias in codon use than do *Chlamydomonas* chloroplast genes. In fact, there is good evidence that selection no longer acts on *psbA* codon use in flowering plant lineages; it is merely a remnant of the ancestral codon use. Further, the highly expressed gene *rbcL* displays a codon use more like *psbA* than does any other flowering plant chloroplast gene (Morton, 1994). These two factors, the apparently recent loss of selection on flowering plant *psbA* codon use and the similarity of *rbcL* to *psbA*, indicate that loss of selection for codon adaptation on plant chloroplast genes may have occurred at different times for each plant chloroplast gene, most likely as genome number increased over time (Morton, unpublished). Such a scenario may have important implications for studies of chloroplast origins because standard phylogenetic estimation methods are likely to be biased and, therefore, unreliable at very deep evolutionary levels. The codon bias results also suggest that nucleotide composition cannot be assumed to be at equilibrium, contrary to the assumptions incorporated in most distance estimators.

Relative Rates of Nucleotide Substitution. Early work on chloroplast sequences suggested that substitution rates do not follow a constant molecular clock. Rodermel and Bogorad (1987) first claimed that substitution rates in chloroplast loci can vary among evolutionary lineages. They found that the *atpE* gene had slower missense rates, and the *atpH* gene had slower synonymous rates, in grass lineages relative to dicot lineages. Similar comparisons of substitution rates in the *rbcL* locus indicated that both missense and synonymous rates were faster in grass species relative to palm species (Wilson *et al.*, 1990).

These studies relied on fossil-based divergence times for substitution rate estimates. Fossil-based divergence times can have large errors that introduce large (and unmeasurable) uncertainty into rate estimates. Relative rate tests allow comparison of substitution rates between evolutionary lineages without dependence on knowledge of the time dimension. First utilized by Sarich and Wilson (1973), Wu and Li (1985) later extended relative rate tests to nucleotide sequences using a distance measure formulation. Subsequent refinements include a maximum likelihood relative test (Muse and Weir, 1992) and a simple counting method (Tajima, 1993). The power of these methods to detect deviations from a molecular clock depends on a number of factors (e.g., the number of substitution events in the lineages and the transition/transversion bias), but in many situations the three methods have comparable power (Muse and Weir, 1992; Tajima, 1993). Relative rate tests cannot detect changes in evolutionary rates if rates change proportionally among lineages (Fitch, 1976), although this precise condition seems unlikely to occur often. Relative rate tests may also fail to detect stochastic changes in rate within a lineage because the test compares average substitution rates. More importantly, relative rates contain less information on variation than do absolute rates, but absolute rates are difficult to estimate because of imperfect knowledge of the time dimension.

The large *rbcL* data base has facilitated the characterization of substitution rate variation in a number of evolutionary lineages. A few studies have applied relative rate tests to *rbcL* sequences from intrafamilial taxa (Soltis *et al.*, 1990; Doebley *et al.*, 1990; Bousquet *et al.*, 1992a); on the whole, these studies have uncovered limited rate variation. Two studies have characterized *rbcL* rate variation over a wider sampling of plant taxa (Bousquet *et al.*, 1992b; Gaut *et al.*, 1992). Gaut *et al.* (1992) examined *rbcL* sequences from 35 monocotyledonous taxa and found substantial synonymous rate variation among evolutionary lineages, but little missense rate variation. The analyses revealed rate homogeneity for *rbcL* sequences within well-defined families, but substantial rate heterogeneity in interfamilial contrasts. The pattern of interfamilial rate variation revealed the most rapid nucleotide substitution rate in the grass family, followed by families in the orders Orchidales, Liliales, Bromeliales, and Arecales (represented by the palms). Rates of synonymous nucleotide substitution in grass *rbcL* sequences exceed rates in other monocot sequences by as little as 130% and by as much as 800%.

Bousquet *et al.* (1992b) reported extensive rate variation among 15 *rbcL* sequences representing monocot, dicot, and gymnosperm taxa. They concluded that missense rate variation is more extensive than synonymous rate variation. This result differs with that reported by Gaut *et al.* (1992). The differences between these studies can probably be ascribed to the wide phylogenetic range of taxa surveyed. The relatively narrow monocot comparisons included few missense substitutions; the paucity of nonsynonymous substitutions may have made detection of missense rate heterogeneity difficult. On the other hand, the study of Bousquet *et al.* (1992b) may have underestimated synonymous rate variation because some of their sequence comparisons had probably been saturated for synonymous substitutions.

Clearly a simple stochastically constant molecular clock does not hold for the *rbcL* locus; however, there may be a generation-time effect (Li, 1993). While it is difficult to estimate generation times in most plant taxa, the monocot sequences show a clear negative correlation between the minimum generation time and substitution rates (Gaut *et al.*, 1992). Given that rate variation in monocot sequences is largely synonymous (and therefore presumably close to neutral) and that rate variation is correlated with minimum generation time, rate variation at the *rbcL* locus of monocot taxa is consistent with neutral predictions. The conclusions of Bousquet *et al.* (1992b) are not as straightforward, but their results do indicate clear differences in substitution rates among annual and perennial taxa, a result not inconsistent with a generation-time effect. Bousquet *et al.* (1992b) also speculate that speciation rates influence substitution rates. Other factors hypothesized to contribute to rate variation among evolutionary lineages include polymerase fidelity (Wu and Li, 1985; Li *et al.*, 1985), selection (Gillespie, 1986), and G+C content (Bulmer *et al.*, 1991).

What is the pattern of rate variation at other chloroplast loci? If rate variation is predominantly a function of an evolutionary factor that affects the whole genome (like, presumably, the generation-time effect), then one would expect to find similar patterns of rate variation in most chloroplast loci. Conversely, widely divergent patterns of rate variation among chloroplast loci would argue for locus-specific factors (like selection) as the motive force behind substitution rate variation. Ideally, one should sample chloroplast loci from the taxa used in the *rbcL* studies to directly compare patterns of rate variation among loci. Unfortunately, data for these kinds of studies are not yet available. As a first step in the analysis of rate variation among chloroplast loci, Gaut *et al.* (1993) examined rate heterogeneity among a number of chloroplast loci from three taxa (maize, rice, and tobacco, using *Marchantia* as an outgroup). Comparison of sequence data from the maize and rice chloroplast genomes revealed little rate heterogeneity (using tobacco as an outgroup). However, comparisons of sequence data from rice and tobacco (using *Marchantia* as the outgroup) revealed much heterogeneity: significant deviation from a molecular clock was detected at 14 of 40 loci. All 14 loci have accelerated substitution rates in rice relative to tobacco, suggesting concerted rate increases in the rice lineage. In addition, 17 loci had nonsignificantly accelerated rates in rice, while the remaining 9 loci had nonsignificantly slower rates in rice, relative to tobacco.

Interestingly, many of the 14 loci that exhibit significant rate heterogeneity between rice and tobacco lineages encode protein products of related function. For example, three of the four loci that encode RNA polymerase subunits demonstrate rate heterogeneity between rice and tobacco lineages. Further analysis of RNA polymerase genes suggests rate acceleration with subsequent rate deceleration (B. S. Gaut, unpublished data), perhaps indicating the episodic rate pattern thought to result from selective pressures (Gillespie, 1986).

Patterns of Amino Acid Replacement in the RuBisCo Protein

The question of site-dependent probabilities of amino acid replacement can be addressed in considerable detail by using the very large *rbcL* data base. This large data base may be used to ask whether models of nucleotide substitution provide an acceptable fit to the data, and it is even more important to ask whether the pattern of accepted amino acid change provides useful information on protein adaptation. The three-dimensional structure of the RuBisCo protein has been determined for a wide phylogenetic range of species (Chapman *et al.*, 1988; Knight *et al.*, 1989; Schneider *et al.*, 1990). The pattern of amino acid replacement can be mapped onto the physical structure to identify major constraints in molecular change. Such an analysis, when placed in a phylogenetic context, may also help to identify amino acid replacement of functional importance.

The large subunit of RuBisCo contains two domains: (*i*) a carboxyl-terminal C domain that includes (*a*) an α/β-barrel consisting of alternating α-helices and β-strands, with the parallel β-strands forming an interior barrel; (*b*) an interior extension with an α-helix followed by a pair of antiparallel β-strands; and (*c*) a terminal extension of two to four α-helices; and (*ii*) an amino-terminal N domain consisting of a five-stranded antiparallel β-sheet and four or five α-helices that form a lid-like structure covering the barrel of an adjacent large subunit. Furthermore, the active site is well known. It consists of charged and polar residues at the carboxyl-terminal ends of the β-strands in the α/β-barrel of the C domain of one subunit and asparagine and glutamic acid residues from the lid of the adjacent N domain.

One approach to examining patterns of amino acid replacements in a structural context is to map amino acid replacements on a fully resolved, unambiguous tree. Since the evolutionary relationships among all of the >750 taxa are not known to any degree of certainty, a subset of the taxa were chosen for this analysis. We used 105 taxa, including three prokaryotes, four algae (including Charophytes), five bryophytes and ferns, eight gymnosperms, and 85 angiosperms (for details, contact the authors). In general, conventional phylogenetic relationships that were supported by the analysis of the *rbcL* nucleotide sequence data (Chase *et al.*, 1993; Duvall *et al.*, 1993) were used to construct this tree. The amino acid sequences for the 105 taxa were translated from the nucleotide sequences and aligned (along with the >700 amino acid sequences in the large data set) by using the following method. A preliminary alignment obtained using CLUSTAL V (Higgins *et al.*, 1992) was refined by aligning similar, presumably homologous features of the solved three-dimensional crystal structures for the RuBisCo large subunit. Nine gaps were required to align the sequences. There were 494 sites in the aligned data set. The computer program MACCLADE version 3.04 (Maddison and Maddison, 1992) was then used to locate amino acid replacement on branches of the tree and to count the total number of amino acid changes required through the phylogeny.

Of the 1350 amino acid replacements, 762 could be unambiguously inferred; 568 and 182 were in α-helices and β-strands, respectively. Only 488 (36%) of the replacements were in the α/β-barrel structure, which constitutes 46% of the sites. For the complete sequence, the most common unambiguous replacements were Glu → Asp, and Ala → Ser (40 and 35 changes, respectively). When the number and types of replacements were examined for the various structures, interesting patterns emerged. Figure 2 shows the fractions of sites with replacements and the number of changes for the complete sequences, α-helices, β-strands, and other structures. The distributions are highly skewed, indicating that some sites may accept as many as 26 replacement events while other sites do not accept amino acid replacements. [Character state distributions (amino acids at a given site) for the 105 taxon dendrogram does not allow unequivocal reconstruction of the particular residues at all nodes in the dendrogram. For these residues the number

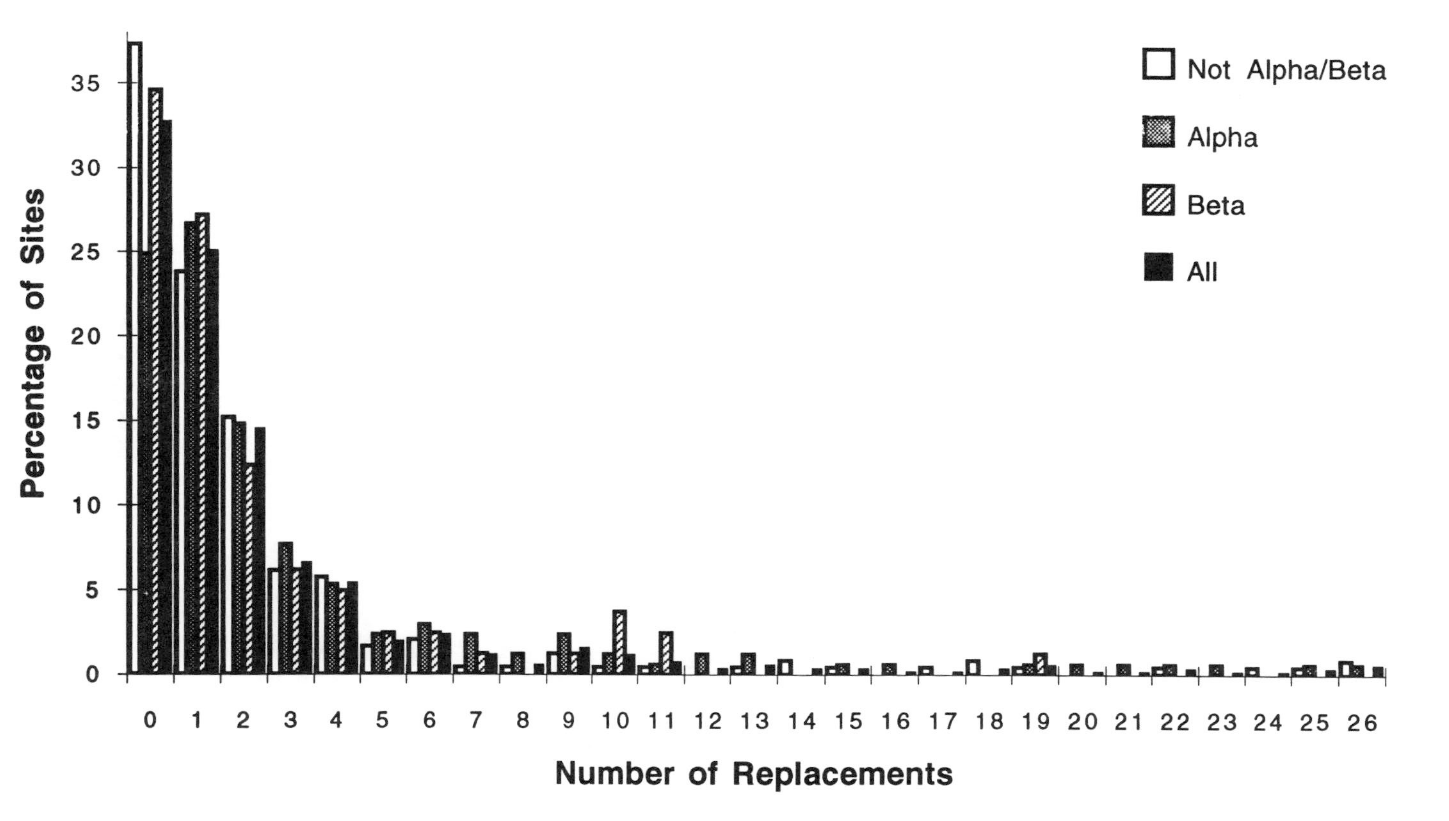

FIGURE 2 The percentage of sites with replacements versus the number of changes at a given site for the complete sequences, α-helix, β-strand, and other structures. The replacements were inferred for a dendrogram of 105 taxa (see text) by using the computer program MACCLADE version 3.04. Of the 494 aligned positions, 169 lie in α-helices, and 81 lie in β-strands.

of unambiguous replacements underestimates the total number of replacements.]

For the complete sequence, 33% of all sites are unvaried, showing no changes. For α-helices, β-strands, and non-α/β structures, the percentages of unvaried sites are 25%, 35%, and 37%, respectively. The length of the tails of the distributions also differ among structural classes; there are as many as 26 unambiguous changes for the α and non-α/β structural classes, while the most changes allowed for sites in β-strands is 19. The unambiguous amino acid replacements that were most common among all sites in the α-helical regions were Ala → Ser (31 changes), Glu → Asp (15), and Ser → Thr (15). For the β regions among all sites, the most common replacements were Ile → Met (9) and Ile → Val (8). The most common replacements for the non-α/β regions were Glu → Asp (21) and Tyr → Phe (14). These are all fairly conservative replacements. More noteworthy is the fact that the predominant changes for the β-strands involve replacements among nonpolar residues.

Tables 1A and 1B show the sites that are most variable for α and non-α/β (>19 changes) and β (>8 changes) structures. Many of the highly variable sites for all of the structural classes involve replacements among nonpolar residues. Also interesting is the replacement of Ala-103 by Cys. This change apparently occurs in six different lineages. In none of the lineages is the change reversed.

What conclusions may we draw from these analyses of the patterns of amino acid variation? The analyses suggest that in angiosperms, amino acid replacements among hydrophobic (nonpolar) residues occur more frequently at some sites in the large subunit of RuBisCo. While the frequency of the highly variable sites and the degree of variability differ for β-strand sites as compared to other structural regions, these variable sites may be found throughout the large subunit sequence. This sort of variability may have an impact on the use of nucleotide sequence data for phylogeny reconstruction because methods of phylogenetic inference assume the probability of replacement is independent of site. The fact that third-nucleotide positions in a codon differ in substitution rate compared to first and second position sites is widely appreciated, but this is a simple and easily corrected source of variation. It is much more difficult to correct for complex site-dependent probabilities that may also change between major evolutionary lineages.

Some of the more variable sites show high levels of replacement for amino acids that are coded for by codons with third-site differences [Asp (GAY) ↔ Glu (GAR), Ile (ATH) ↔ Met (ATG)], where Y = T or C, R = G or A, and H = A, C, or T. The fact that these replacements are fairly frequent and readily tolerated is concordant with the higher rate of

TABLE 1 Highly variable sites in RuBisCo large subunits in the phylogenetic tree for 105 taxa for α and non-α/β sites ($\geq$20 replacements per site) and for β-strand sites ($\geq$9 replacements per site)

a. α and non-α/β sites

Site[a]	Replacements	States[b]	Consistency index	Predominant replacement[c]
32	22	2	0.05	Asp → Glu (6)
95	26	5	0.15	
99	25	7	0.24	
146 α	23	6	0.22	
149 α	20	7	0.30	
229 α	21	4	0.14	Ile → Leu (13)
255 α	26	5	0.15	Ile → Met (9)
259 α	22	7	0.27	
332	24	5	0.17	
453	26	6	0.19	

b. β stand sites

Site[a]	Replacements	States[b]	Consistency index	Predominant replacement[c]
90	9	7	0.67	
103	10	3	0.20	Ala → Cys (6)
313	19	3	0.11	Ile → Met (8)
330	10	4	0.30	Ile → Val (3).
357	10	4	0.30	Phe → Tyr (6)
358	11	5	0.36	Ile → Val (3)

[a]Site is the position number in the 494 aligned positions. An α is appended to the number of the residue for those in α-helices to distinguish them from non-α/β sites.
[b]States is the number of different kinds of amino acids found at the indicated site.
[c]Predominant replacement indicates the most frequent unambiguously inferred amino acid replacement in the character state reconstruction for the dendrogram at the indicated site; the number of unambiguously inferred independent occurrences is in parentheses. The number of unambiguous replacements is an underestimate of the true number of replacements for some sites (see text).

substitution for third-position sites. Other third-position substitutions [e.g., Gln (CAR) ↔ His (CAY), Lys (AAR) ↔ Asn (AAY), Arg (AGR) ↔ Ser (AGY)] are not among the replacements seen for highly variable sites; furthermore, replacements such as Ser (TCN) → Ala (GCN), Ile (ATH) → Val (GTN), and Phe (TTY) → Tyr (TAY) are similarly favorable even though they require first- or second-position substitutions. The rate heterogeneities that follow from such patterns make methods of phylogenetic inference using a simple weighting scheme for different codon positions problematical.

Patterns of Intron Evolution

Data from the plastid genomes that have been completely sequenced reveal that introns are a general feature of these genomes. Each of the three classes of introns (groups I, II, and III) have a characteristic secondary structure that is related to the mechanisms by which the intervening sequence is excised from RNA precursors. The secondary structure requirements, along with other constraints, appear to limit the acceptance of mutational changes in intron sequences through evolutionary time. Group I introns, which were the first self-splicing introns identified (Zaug and Cech, 1980), are found in a single tRNA gene, *trnL*(UAA), from a number of plastid genomes. Group III introns have thus far only been identified in the euglenophytes *Euglena gracilis* and *Astasia longa* (Christopher and Hallick, 1989) and appear to be truncated (or streamlined) group II introns. It is noteworthy that group I introns are absent from the newly sequenced plastid genomes of *Euglena gracilis* (Hallick *et al.*, 1993) and beechdrops, *Epifagus virginiana* (Wolfe *et al.*, 1992). Both genomes are atypical, however: the *Euglena* genome is markedly different in structural organization and may have had a separate origin from the plastid genomes of land plants; beechdrops is a nonphotosynthetic parasitic plant, and its genome is highly reduced, lacking *trnL*(UAA) among other genes. Although comparative sequence analyses of both group I and group III introns could provide information about rates and mechanisms of chloroplast DNA evolution, there are few comparative data, and consequently our discussion will be limited to group II introns.

Group II introns form the most numerous and best characterized class of plastid intervening sequences. They are found in both protein-coding genes and tRNA genes. The secondary structure of group II introns is characterized by six domains (I–VI) (reviewed in Michel *et al.*, 1989). Domain I has a complicated structure and contains sites that probably form base pairs with the 5′ exon and are important for intron processing (Jacquier and Michel, 1987). Domains II–VI are typically simple stem–loop structures. Domains V and VI have also been shown to be required for proper processing of the transcript (Schmelzer and Müller, 1987; Jarrell *et al.*, 1988). Learn *et al.* (1992) examined the evolutionary constraints on the various domains of group II introns in a comparative study of the intron found in a tRNA gene, *trnV*(UAC), from seven land plants. They found that domain II evolves most rapidly, comparable to the synonymous substitution rate of protein-coding genes, consistent with the finding that domain II may be dispensable in self-splicing introns (Kwakman *et al.*, 1989). Portions of domain I (that are important in binding to the 5′ exon) and domain V evolve at the slowest rates, comparable to missense rates for a number of protein-coding genes. These results illustrate that evolutionary rates within group II introns vary and appear to be related to the functional importance of intron structural features.

Conclusions

The uniformitarian assumption plays a fundamental role in the science of molecular evolution, just as it did in the evolutionary paleontology of G. G. Simpson. We must assume that the kinds of mutational changes that we can demonstrate in "microscopic detail" from comparative sequence analyses are the substance of molecular evolutionary change. When viewed in detail, the patterns of mutational change in the chloroplast genome are complex, and they belie the notion that the cpDNA is a staid and conservative molecule.

The use of cpDNA RFLPs and cpDNA gene sequence data for phylogenetic inference has had an enormous impact on studies of plant phylogenetics and systematics. This has been facilitated by the belief that cpDNA-based mutational change is regular. By and large, the assumption of statistical regularity is adequate, and cpDNA data are an important addition to the previous evidence available to students of plant evolution. Nevertheless, it is important to investigate the ways that mutational change in this genome departs from the kinds of regularity assumed by most methods of phylogenetic inference. When this question is asked, we find that noncoding regions exhibit a number of mutational mechanisms. Some sites are clearly labile to small indel mutations, and in at least one case, large deletions also appear to be site dependent. Complex recombinational processes are also found to influence the evolution of at least some noncoding regions of cpDNA. Group II introns show a strong relationship between structure and probability of mutational change. Analyses of protein-coding genes also reveal complex patterns of mutational change. Finally, rates of nucleotide substitution are quite variable at the level of order and above. Current data suggest that much of the variation in evolutionary rate can be accounted for by the generation-time effect hypothesis, although this requires further investigation. There are also interesting patterns of codon bias. When land plant and algal genes are compared, it is evident that patterns of selection for codon utilization have changed over evolutionary time and that models that assume equilibrium nucleotide frequencies are likely to be violated. Patterns of amino acid replacement in the *rbcL* gene also reveal substantial variation in site-dependent probabilities of substitution. Taken in toto, these complexities in mutational change should motivate the development of more realistic algorithms for phylogenetic inference when based on molecular data. In the interim, one must view molecular phylogenetic reconstructions as approximate, especially at very deep levels of evolution.

Comparative sequence analyses have utility beyond the study of phylogeny. The pattern of amino acid replacements observed over evolutionary time reflects the pattern of functional constraints that are imposed by natural selection on the protein molecule. Natural selection is a sensitive filter because it is capable of detecting subtle changes associated with small fitness effects. Regions that do not accept change are clearly strongly constrained by functional requirements, but those regions that do accept change may be simply unconstrained, or in a few cases they may reflect responses to adaptive change. It is difficult to distinguish these latter two possibilities, but unusual patterns, such as the repeated replacement of Ala-103 by Cys, may be suggestive of adaptive change.

Comparative molecular sequence analyses also reveal aspects of the evolutionary process that would otherwise be opaque. For example, the recombinational processes that acted to convert *ψrpl23* to the functional gene are only resolvable at the sequence level. Similarly, the identification of labile sites for indel mutation, and more importantly, the identification of local sequence features that may promote indel mutation, depend on a substantial comparative sequence base. When viewed broadly, comparative sequence analyses reveal the rich variety of mutational mechanisms that subsume the processes of molecular evolution.

SUMMARY

The chloroplast genome (cpDNA) of plants has been a focus of research in plant molecular evolution and systematics. Several features of this genome have facilitated molecular evolutionary analyses. First, the genome is small and constitutes an abundant component of cellular DNA. Second, the chloroplast genome has been extensively characterized at the molecular level providing the basic information to support comparative evolutionary research. And third, rates of nucleotide substitution are relatively slow and therefore provide the appropriate window of resolution to study plant phylogeny at deep levels of evolution. Despite a conservative rate of evolution and a relatively stable gene content, comparative molecular analyses reveal complex patterns of mutational change. Noncoding regions of cpDNA diverge through insertion/deletion changes that are sometimes site dependent. Coding genes exhibit different patterns of codon bias that appear to violate the equilibrium assumptions of some evolutionary models. Rates of molecular change often vary among plant families and orders in a manner that violates the assumption of a simple molecular clock. Finally, protein-coding genes exhibit patterns of amino acid change that appear to depend on protein structure, and these patterns may reveal subtle aspects of structure/function relationships. Only comparative studies of molecular sequences have the resolution to reveal this underlying complexity. A complete description of the complexity of molecular change is essential to a full understanding of the mechanisms of evolutionary change and in the formulation of realistic models of mutational processes.

Support from National Institutes of Health Grants GM 45144 (to M.T.C.), GM 15528 (to B.S.G.) and GM 45344 (to North Carolina State University) is gratefully acknowledged. The order of authorship is listed alphabetically. All authors made an equal contribution to this article.

REFERENCES

Bousquet, J., Strauss, S. H. & Li, P. (1992a) Complete congruence between morphological and rbcL-based molecular phylogenies in birches and related species (Betulaceae). *Mol. Biol. Evol.* **9,** 1076–1088.

Bousquet, J., Strauss, S. H., Doerksen, A. H. & Price, R. A. (1992b) Extensive variation in evolutionary rate of rbcL gene sequences among seed plants. *Proc. Natl. Acad. Sci. USA* **89,** 7844–7848.

Bowman, C. M., Barker, R. F. & Dyer, T. (1988) In wheat ctDNA, segments of ribosomal protein genes are dispersed repeats, probably maintained by nonreciprocal recombination. *Curr. Genet.* **14,** 127–136.

Bulmer, R., Wolfe, K. H. & Sharp, P. M. (1991) Synonymous nucleotide substitution rates in mammalian genes: implication for the molecular clock and the relationship of mammalian orders. *Proc. Natl. Acad. Sci. USA* **88,** 5974–5978.

Chapman, M. S., Suh, S. W., Curmi, P. M. G., Cascio, D., Smith, W. W. & Eisenberg, D. S. (1988) Tertiary structure of plant RuBisCO: Domains and their contacts. *Science* **241,** 71–74.

Chase, M. W., Soltis, D. E., Olmstead, R. G., Morgan, D., Les, D. H., Duvall, M. R., Price, R. A., Hills, H. G., Qiu, Y.-L., Kron, K. A., Rettig, J. H., Conti, E., Palmer, J. D., Manhart, J. R., Sytsma, K. J., Michaels, H. J., Kress, W. J., Karol, K. G., Clark, W. D., Hedren, M., Gaut, B. S., Jansen, R. K., Kim, K.-J., Wimpee, C. F., Smith, J. F., Furnier, G. R., Straus, S. H., Xiang, Q.-Y., Plunkett, G. M., Soltis, P. S., Swensen, S. M., Williams, S. E., Gadek, P. A., Quinn, C. J., Equiarte, L. E., Golenberg, E., Learn, G. H., Jr., Graham, S. W., Barrett, S. C. H., Dayanandan, S. & Albert, V. A. (1993) Phylogenetics of seed plants: an analysis of nucleotide sequences from the plastid gene rbcL. *Ann. Mo. Bot. Gard.* **80,** 528–580.

Christopher, D. A. & Hallick, R. B. (1989) *Euglena gracilis* chloroplast ribosomal protein operon: a new chloroplast gene for ribosomal protein L5 and a description of a novel organelle intron category designated group III. *Nucleic Acids Res.* **17,** 7591–7608.

Clegg, M. T. (1993) Chloroplast gene sequences and the study of plant evolution. *Proc. Natl. Acad. Sci. USA* **90,** 363–367.

Clegg, M. T., Learn, G. H. & Golenberg, E. M. (1991) Molecular evolution of Chloroplast DNA. Chapter 7 In *Evolution at the Molecular Level*, eds. Selander, R. K., Clark, A. G. & Whittam, T. S. (Sinauer, Sunderland, MA), pp. 135–149.

Cummings, M. P., Mertens King, L. & Kellogg, E. A. (1994) Slipped-strand mispairing in a plastid gene: rpoC2 in grasses (Poaceae) *Mol. Biol. Evol.* **11,** 1–8.

Doebley, J., Durbin, M., Golenberg, E. M., Clegg, M. T. & Ma, D. P. (1990) Evolutionary analysis of the large subunit of carboxylase (*rbc*L) nucleotide sequence among the grasses (Gramineae). *Evolution* **44,** 1097–1108.

Downie, S. R. & Palmer, J. D. (1991) Use of chloroplast DNA rearrangements in reconstructing plant phylogeny. In *Molecular Systematics of Plants*, eds. Soltis, P. S., Soltis, D. E. & Doyle, J. J. (Chapman and Hall, New York), pp. 14–35.

Duvall, M. R., Clegg, M. T., Chase, M. W., Clark, W. D., Kress, W. J., Hills, H. G., Eguiarte, L. E., Smith, J. F., Gaut, B. S., Zimmer, E. A. & Learn, G. H., Jr. (1993) Phylogenetic hypotheses for the monocotyledons constructed from *rbc*L sequence data. *Ann. Mo. Bot. Gard.* **80,** 607–619.

Fitch, W. M. (1976) Molecular evolutionary clocks. In *Molecular Evolution*, ed., Ayala, F. J. (Sinauer, Sunderland, MA), pp. 160–178.

Gantt, J. S., Baldauf, S. L., Calie, P. J., Weeden, N. F. & Palmer, J. D. (1991) Transfer of *rpl*22 to the nucleus greatly preceded its loss from the chloroplast and involved the gain of an intron. *EMBO J.* **10,** 3073–3078.

Gaut, B. S., Muse, S. V. & Clegg, M. T. (1993) Relative rates of nucleotide substitution in the chloroplast genome. *Mol. Phylogenet. Evol.* **2,** 89–96.

Gaut, B. S., Muse, S. V., Clark, W. D. & Clegg, M. T. (1992) Relative rates of nucleotide substitution at the rbcL locus of monocotyledonous plants. *J. Mol. Evol.* **35,** 292–303.

Gillespie, J. H. (1986) Natural selection and the molecular clock. *Mol. Biol. Evol.* **3,** 138–155.

Golenberg, E. M., Clegg, M. T., Durbin, M. L., Doebley, J. & Ma, D. P. (1993) Evolution of a noncoding region of the chloroplast genome. *Mol. Phylogenet. Evol.* **2,** 52–64.

Gray, M. W. (1993) Origin and evolution of organelle genomes. *Curr. Opin. Genet. Dev.* **3,** 884–890.

Hallick, R. B., Hong, L., Drager, R. G., Favreau, M. R., Monfort, A., Orsat, B., Spielmann, A. & Stutz, E. (1993) Complete sequence of *Euglena gracilis* chloroplast DNA. *Nucleic Acids Res.* **21,** 3537–3544.

Higgins, D. G., Bleasby, A. J. & Fuchs, R. (1992) Phylogenetic hypotheses for the monocotyledons constructed from *rbc*L sequence data. *Comp. Appl. Biosci.* **8,** 189–191.

Hiratsuka, J., Shamida, H., Whittier, R., Ishibashi, T., Sakamoto, M., Mori, M., Knodo, C., Honji, Y., Sun, C.-R., Meng, B.-Y., Li, Y.-Q., Kano, A., Nishizawa, Y., Hirai, A., Shinozaki, K. & Sugiura, M. (1989) The complete sequence of the rice (Oryza sativa) chloroplast genome: Intermolecular recombination between distinct tRNA genes accounts for a major plastid DNA inversion during the evolution of the cereals. *Mol. Gen. Genet.* **217,** 185–194.

Jacquier, A. & Michel, F. (1987) Multiple exon-binding sites in class II self-splicing introns. *Cell* **50,** 17–29.

Jarrell, K. A., Dietrich, R. C. & Perlman, P. (1988) Group II intron domain 5 facilitates a trans-splicing reaction. *Mol. Cell. Biol.* **8,** 2361–2366.

Knight, S., Andersson, I. & Branden, C.-I. (1989) Reexamination of the three-dimensional structure of RuBisCo from higher plants. *Science* **244,** 702–705.

Kwakman, J. H., Konings, D., Pel, H. J. & Grivell, L. A. (1989) Structure-function relationships in a self-splicing group II intron: a large part of domain II of the mitochondrial intron aI5 is not essential for self-splicing. *Nucleic Acids Res.* **17,** 4205–4216.

Learn, G. H., Jr., Shore, J. S., Furnier, G. R., Zurawski, G. & Clegg, M. T. (1992) Constraints on the evolution of chloroplast introns: the intron in the gene encoding tRNA-Val(UAC). *Mol. Biol. Evol.* **9,** 856–871.

Li, W.-H. (1993) So, what about the molecular clock hypothesis? *Curr. Opin. Genet. Dev.* **3,** 896–901.

Li, W.-H., Luo, C.-C., & Wu, C.-I. (1985) Evolution of DNA sequences. In *Molecular Evolutionary Genetics*, ed. MacIntyre, R. J. (Plenum, New York), pp. 1–94.

Maddison, W. P. and D. R. Maddison (1992) MacClade: Analysis of phylogeny and character evolution. Version 3.0. Sinauer Associates, Sunderland, Massachusetts.

Michel, F., Umesono, K. & Ozeki, H. (1989) Comparative and functional anatomy of group II catalytic introns—a review. *Gene* **82,** 5–30.

Morton, B. R. (1993) Chloroplast DNA codon use: evidence for selection at the psbA locus based on tRNA availability. *J. Mol. Evol.* **37,** 273–280.

Morton, B. R. (1994) Codon use and the rate of divergence of land plant chloroplast genes. *Mol. Biol. Evol.* **11,** 231–238.

Morton, B. R. & Clegg, M. T. (1993) A chloroplast DNa mutational hotspot and gene conversion in a noncoding region near rbcL in the grass family (Poaceae). *Curr. Genet.* **24,** 357–365.

Mullet, J. E. & Klein, R. R. (1987) Transcription and RNA stability are important determinants of higher plant chloroplast RNA levels. *EMBO J.* **6,** 1571–1579.

Muse, S. V. & Weir, B. S. (1992) Testing for equality of evolutionary rates. *Genetics* **132,** 269–276.

Ogihara, Y., Terachi, T. & Sasakuma, T. (1988) Intramolecular recombination of chloroplast genome mediated by short direct-repeat sequences in wheat species. *Proc. Natl. Acad. Sci. USA* **85,** 8573–8577.

Ogihara, Y., Terachi, T. & Sasakuma, T. (1992) Structural analysis of length mutations in a hot-spot region of wheat chloroplast DNAs. *Curr. Genet.* **22,** 251–258.

Ohyama, K., Fukuzawa, H., Kohchi, T., Shirai, H., Sano, T., Sano, S., Umesono, K., Shiki, Y., Takeuchi, M., Chang, Z., Aota, S.-I., Inokuchi, H. & Ozeki, H. (1986) Chloroplast gene organization deduced from complete sequence of liverwort *Marchantia polymorpha* chloroplast DNA. *Nature (London)* **322,** 572–574.

Palmer, J. D. (1985) Comparative organization of chloroplast genomes. *Annu. Rev. Genet.* **19,** 325–354.

Palmer, J. D. (1991) Plastid chromosomes: structure and evolution. In *Cell Culture and Somatic Cell Genetics of Plants*, eds. Bogorad, L. & Vasil, K. (Academic, New York), Vol. 7A, pp. 5–53.

Rodermel, S. R. & Bogorad, L. (1987) Molecular evolution and nucleotide sequences of the maize plasmid genes for the subunit of CF1 (atpA) and the proteolipid subunit of CF0 (atpH). *Genetics* **116,** 127–139.

Sarich, V. M. & Wilson, A. C. (1973) Generation time and genomic evolution in primates. *Science* **179,** 1144–1147.

Schmelzer, C. & Mueller, M. W. (1987) Self-splicing of group II introns in vitro: lariat formation and 3′ splice site selection in mutant RNAs. *Cell* **51,** 753–767.

Schneider, G., Lindqvist, Y. & Lundqvist, T. (1990) Crystallographic refinement and structure of ribulose-1,5-biphosphate carboxylase from Rhodospirillum rubrum at 1.7 A resolution. *J. Mol. Biol.* **211,** 989–1008.

Shinozaki, K., Ohme, M., Tanaka, M., Wakasugi, T., Hayashida, N., Matsubayashi, T., Zaita, N., Chunwongse, J., Obokata, J., Yamaguchi-Shinozaki, K., Ohto, C., Torazawa, K., Meng, B. Y., Sugita, M., Deno, H., Kamogashira, T., Yamada, K., Kusuda, J., Takaiwa, F., Kato, A., Tohdoh, N., Shimada, H. & Sugiura, M. (1986) The complete nucleotide sequence of the tobacco chloroplast genome: its gene organization and expression. *EMBO J.* **5,** 2043–2049.

Soltis, D. E., Soltis, P. S., Clegg, M. T. & Durbin, M. (1990) *rbc*L sequence divergence and phylogenetic relationships in Saxifragaceae sensu lato. *Proc. Natl. Acad. Sci. USA* **87,** 4640–4644.

Tajima, F. (1993) Simple methods for testing the molecular evolutionary clock hypothesis. *Genetics* **135,** 599–607.

Umesono, K., Inokuchi, H., Shiki, Y., Takeuchi, M., Chang, Z., Fukuzawa, H., Kohchi, T., Shirai, H., Ohyama, K. & Ozeki, H. (1988) Structure and organization of Marchantia polymorpha chloroplast genome II. Gene organization of the large single copy region from rps'12 to atpB. *J. Mol. Biol.* **203,** 299–331.

Wakasugi, T., Tsudzuki, J., Itoh, S., Nakashima, K., Tsudzuki, T., Shibata, M. and M. Sugiura (1993) The entire structure of the chloroplast genome from black pine (*Pinus Thubergii*). XV International Botanical Congress Abstract 7.2.1-6. p. 161.

Wilson, M., Gaut, B. & Clegg, M. T. (1990) Chloroplast DNA evolves slowly in the palm family (Arecaceae). *Mol. Biol. Evol.* **7,** 303–314.

Wolfe, K. H., Morden, C. W. & Palmer, J. D. (1992) Function and evolution of a minimal plastid genome from a nonphotosynthetic parasitic plant. *Proc. Natl. Acad. Sci. USA* **89,** 10648–10652.

Wu, C.-I. & Li, W.-H. (1985) Evidence for higher rates of nucleotide substitution in rodents than in man. *Proc. Natl. Acad. Sci. USA* **82,** 1741–1745.

Zaug, A. J. & Cech, T. R. (1980) In vitro splicing of the ribosomal RNA precursor in nuclei of Tetrahymena. *Cell* **19,** 331–338.

12

The Superoxide Dismutase Molecular Clock Revisited

WALTER M. FITCH AND FRANCISCO J. AYALA

In 1985, Ayala, in his Wilhelmine E. Key lecture, spoke on the enormous difference between the way one might expect a molecular clock to work and the way Cu,Zn superoxide dismutase (SOD) seems to have evolved since the last common ancestor of the metazoans and fungi (Ayala, 1986). Even after an accepted point mutation (PAM) correction for multiple amino acid replacements at a single position in the sequence (Dayhoff, 1978), it was clear that the rates of change were discrepant; there were 10 replacements in the last 75 million years (Myr) but only 21 replacements in the first 600 Myr. This represents a nearly 4-fold difference in rates that could not be ascribed to an incorrect dating of the time of the common ancestor [75 million years ago (Mya) for cow–human; 1200 Mya for fungi–metazoans].

It seemed to one of us (W.M.F.) that the following could be true (or close to the truth): (*i*) the dates are correct, (*ii*) the observed differences are correct, and (*iii*) the clock is working. But these three assertions could all be true only if the correction for multiple replacements had been in error. Fitch and Markowitz introduced the concept of concomitantly variable codons (covarions), which asserts that at any one point in time and in any one lineage, there is a limited number of amino acid sequence positions that can tolerate an amino acid replacement and

Walter M. Fitch is professor and chairman of the Department of Ecology and Evolutionary Biology and Francisco J. Ayala is Donald Bren Professor of Biological Sciences at the University of California, Irvine.

perhaps some positions can never tolerate a replacement (Fitch and Markowitz, 1970). Thus multiple replacements at a site may occur more frequently than would be the estimate when the whole sequence is considered to be variable. However, as replacements are accepted, those very changes may affect and change the sites that are covariable. Thus the sites that have changed across very different taxa may be considerably greater than the number that are covariable in any given species. It seemed possible that the unrecognized presence of covarions could hide the extent of multiple replacements, thus causing the deeper (older) portions of the tree to appear as if evolving too slowly. As the other of us (F.J.A.) has embarked on a program of extensive SOD sequencing, it seems a propitious time to collaborate on a deeper evaluation of this problem.

From this collaboration, we have determined the number of covarions in SOD (28 in *Drosophila*) and shown that the molecular clock could be working fairly well. This is done by showing, by simulation, that the observed amino acid differences could have arisen if, among the fungi and metazoans, SOD had 44 of its codons permanently invariable and there was a high rate of exchange between codons that are covarions and those that are not variable at any particular time.

METHODS

The data analyzed were 67 SOD sequences, aligned by eye. The sequences are from GenBank (see Figure 1 for sequence numbers), except for most of the flies, which came from the lab of F.J.A. The tree was obtained using the ANCESTOR program (Fitch, 1971; Fitch and Farris, 1974). The number of covarions was determined by the method of Fitch and Markowitz (1970).

Simulations were performed (Fitch and Ye, 1991) using a program for which the user prescribes five parameters: (*i*), the length of the sequence; (*ii*), the number of covarions; (*iii*), the persistence of the covarion set (the probability that no covarion will be exchanged for a presently invariable codon; if an exchange does occur, only one of the covarions is exchanged; the possibility of an exchange exists after each replacement); (*iv*), the number of alternative amino acids allowed at a site, and (*v*) the times at which the number of amino acid differences are to be determined. The times are in units of amino acid replacements and hence the clock is perfect in the simulations.

The divergence times can have any value but, to test the possibility that the SOD data can arise by the simulated clock, the replacements must be in the same proportion as the paleontological time estimates. The time estimates we used are given in Table 1.

The parameters used for the simulations were (*i*) length of 118 potentially variable amino acids (162 aligned positions minus 44 permanently invariable positions), (*ii*) number of covarions = 28 (determined as shown in *Results*), (*iii*) persistence = 0.01, (*iv*) half the variable sites had two alternative amino acids at each site; the other half had three alternatives, and (*v*) clock times are set by the paleontological dates in Table 1 with a rate of six replacements per 10 Myr.

RESULTS

The SODs are abundant enzymes in aerobic organisms, with highly specific activity that protects the cell against the harmfulness of free oxygen radicals (Fridovich, 1986). The SODs have active centers that contain either iron, manganese, or both copper and zinc (Fridovich, 1986). The Cu,Zn SOD is a well-studied protein found in eukaryotes but also in some bacteria (Steinman, 1988). The amino acid sequence is known in many organisms, plants, animals, fungi, and bacteria. The three-dimensional structure for the bovine SOD has been determined at a 2-Å resolution (Tainer *et al.*, 1982); it is conserved in humans (Getzoff *et al.*, 1989) and presumably in *Drosophila* (Kwiatowski *et al.*, 1992) and bacteria (Bannister and Parker, 1985). The amino acids essential for catalytic action (Tainer *et al.*, 1983), as well as those for protein structures, are strongly conserved (Getzoff *et al.*, 1989; Kwiatowski *et al.*, 1992).

The Phylogeny. The tree used for this study is shown in Figure 1. It is not the most parsimonious. Rather, we required the tree to conform to what is believed to be correct on *a priori* grounds (i.e., based on knowledge that is independent of the SOD data). We only let parsimony dictate regions of the tree where other evidence does not seem to us to be determinative. The reason for this is that we wish to optimize the correctness of the tree, rather than its being most parsimonious, because estimates of divergence, against which clock measures are to be tested, will be more valid the closer the tree is to reality. The accuracy of the covarion estimate is similarly constrained. The most parsimonious tree we have found requires 1940 nucleotide substitutions; the tree in Figure 1 requires 1984. The sequences used were amino acid sequences, back translated into ambiguous codons so that the changes are in substitutions rather than replacements, although nearly all substitutions are replacement substitutions. The number of differences between pairs of sequences used for the clock test does not depend upon the topology of the tree. The average differences are shown in Table 1 for those contrasts for which there are reasonable, nonmolecular, paleontological dates.

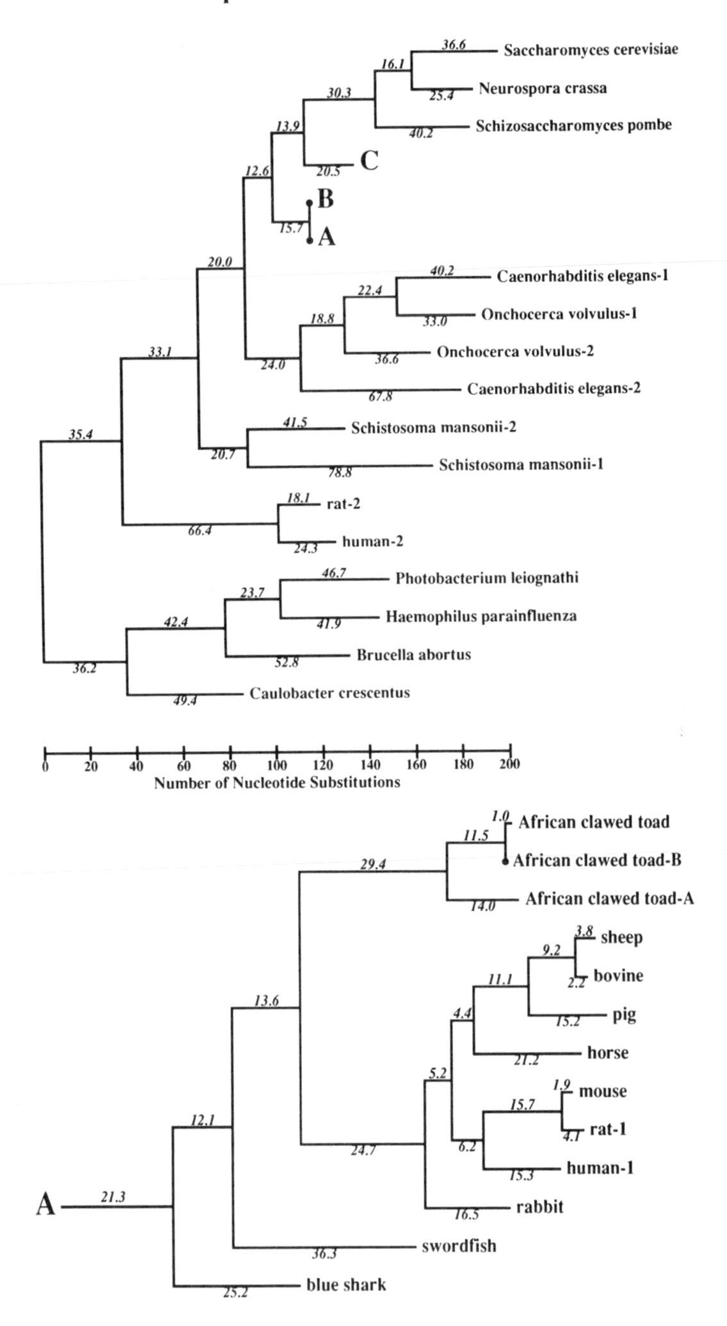

FIGURE 1 Superoxide dismutase tree. The amino acid sequences (back translated to ambiguous codons) were fit most parsimoniously to this tree, which is 1984 nucleotide substitutions long by the method of Fitch (1971) and Fitch and Farris (1974). It is not the most parsimonious tree, which was not used for

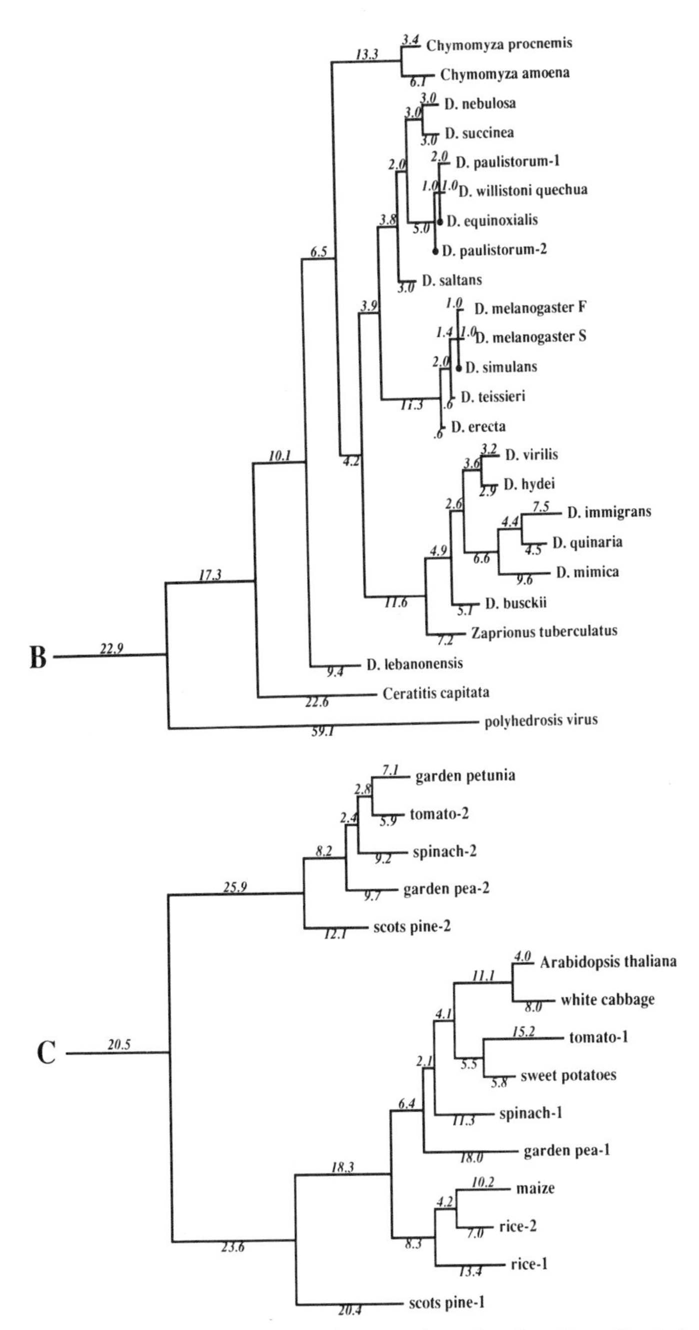

FIGURE 1 (Continued) reasons discussed in *Results*. Part C of the figure shows that the Cu,Zn SOD gene became duplicated early in plant evolution, at least before the divergence of gymnosperms and angiosperms. The GenBank numbers for these sequences are as follows: A00512, human-1 (*Homo sapiens*); S01134, rabbit (*Oryctolagus cuniculus*); P07632, rat-1 (*Rattus norvegicus*); JQ0915,

Covarions. Figure 2 shows a plot that estimates the fraction of the codons that are invariable, based on a two-Poisson fit. The plotted value for each clade is the average number (±SD) of substitutions from the root of the clade to the descendents at the tips. The average is obtained by weighing equally each bifurcating branch irrespective of the number of subsequent branchings. The y intercept yields the estimate of the fraction of invariable codons in a single species and is 0.826. Thus, 0.174 is the fraction of codons that is covarying and, therefore, the number of covarions is $0.174 \times 162 = 28$. Strictly speaking, this number is for the *Drosophila* species. Other work (e.g., on cytochrome *c* where plants and mammals have the same number of covarions; see figure 5 of Fitch, 1976) suggests that the number of covarions is not highly variable across phyla, although this has not been widely tested.

DISCUSSION

Interest in dating molecular evolutionary events extends back at least to Zuckerkandl and Pauling (1962) who observed 6, 36, and 78 differences between the human β subunit of hemoglobin and each of three other human globin subunits δ, γ, and α, respectively. Assuming a rate calculated from the divergence of the α subunit of human and horse hemoglobin, they transformed the above differences into the dates when the gene duplications occurred—namely, 44, 260, and 565 Mya. This calculation assumes that the number of replacements equals the

FIGURE 1 (Continued) mouse (*Mus musculus*); P08294, human-2; X68041, rat-2; A00514, pig (*Sus scrofa*); A00513, bovine (*Bos primigenius taurus*); P09670, sheep (*Ovis aries*); A00515, horse (*Equus caballus*); S05021, African clawed toad-A (*Xenopus laevis*); S05022, African clawed toad-B; S09568, African clawed toad; P03946, swordfish (*Xiphias gladius*); S04623, blue shark (*Prionace glauca*); S84896, scots pine-1 (*Pinus sylvestris*); S00999, rice-1 (*Oryza sativa*); D01000, rice-2; A29077, maize (*Zea mays*); A25569, white cabbage (*Brassica oleracea*); X60935, *Arabidopsis thaliana*; X73139, sweet potatoes (*Ipomoea batatas*); S08350, tomato-1 (*Lycopersicon esculentum*); P22233, spinach-1 (*Spinacia oleracea*); M63003, garden pea-1 (*Pisum sativum*); S84902, scots pine-2; JS0011, spinach-2; S12313, garden pea-2; S08497, tomato-2; P10792, garden petunia (*Petunia hybrida*); M58687, *Neurospora crassa*; A36171, *Saccharomyces cerevisiae* (yeast); X66722, *Schizosaccharomyces pombe* (yeast); X57105, *Onchocerca volvulus*-1 (nematode); Z27080, *Caenorhabditis elegans*-1 (nematode); L20135, *Caenorhabditis elegans*-2; L13778, *Onchocerca volvulus*-2; M68862, polyhedrosis virus; A37019, *Schistosoma mansonii*-1 (liver fluke); M86867, *Schistosoma mansonii*-2; M84013 and M84012, *Haemophilus parainfluenza*; M55259, *Caulobacter crescentus* (bacterium); A00519, *Photobacterium leiognathi* (bacterium); A33893, *Brucella abortus* (bacterium). In part B, D. stands for *Drosophila*.

TABLE 1 Paleontological dates and numbers of replacements

Sister groups	Mya	Replacements	Amino acid differences: Observed SOD	Amino acid differences: Simulated
D. nebulosa–D. melanogaster	55 ± 5	33	18 ± 2	18 ± 2
D. hydei–D. melanogaster	60 ± ?	36	19 ± 3	20 ± 4
Chymomyza–D. melanogaster	65 ± ?	39	23 ± 2	20 ± 4
Homo sapiens–Bos taurus	70 ± 10	42	27 ± 2	22 ± 4
Ceratitis–D. melanogaster	100 ± ?	60	31 ± 2	28 ± 3
Monocot–dicot	125 ± ?	74	28 ± 3	31 ± 5
Angiosperm–gymnosperm	220 ± ?	132	29 ± 7	42 ± 5
Frog–mammals	350 ± 10	210	49 ± 2	53 ± 6
Fish–tetrapods	400 ± 20	240	44 ± 4	56 ± 7
Yeast–*Neurospora*	?		46 ± 1	
Insect–vertebrate	580 ± 20	348	59 ± 3	60 ± 6
Fungi–metazoans	1000 ± ?	600	67 ± 4	66 ± 7

Replacements refer to the number of replacements used in the simulation, which is equivalent to six replacements every 10 Myr. Observed is the average number of amino acid differences observed between members of the two sister groups shown. The sister group names (e.g., *Drosophila nebulosa–Drosophila melanogaster*) should be understood as indicating the groups to which these species belong and not just two individual species. Simulated is the observed number of amino acid differences obtained after the number of replacements shown had been incorporated. The plus/minus values are crude estimates of error for Mya but are SDs for observed and simulated differences. The simulated values are based on 40 simulated instances for each entry. See *Methods* for details of the simulation. D. stands for *Drosophila.*

number of differences. It was soon recognized that changes could occur at positions where changes had already occurred. Margoliash and Smith (1965) introduced a correction as follows

$$r = -n \ln(1 - d/n), \qquad \text{(equation 1[1])}$$

where r is the number of replacements, n is the number of *variable* positions in the sequence, and d is the number of divergent positions (differences). It is interesting that Margoliash and Smith contemplated the possibility of invariable positions so early, even though the concern that this might be an important consideration was not present in the field for nearly 20 years, except for Fitch (Fitch and Markowitz, 1970; Shoemaker and Fitch, 1989).

[1] The equation they said they used was $r = n \ln(n/d)$, which would be incorrect, but their results are those obtained by the correct equation shown.

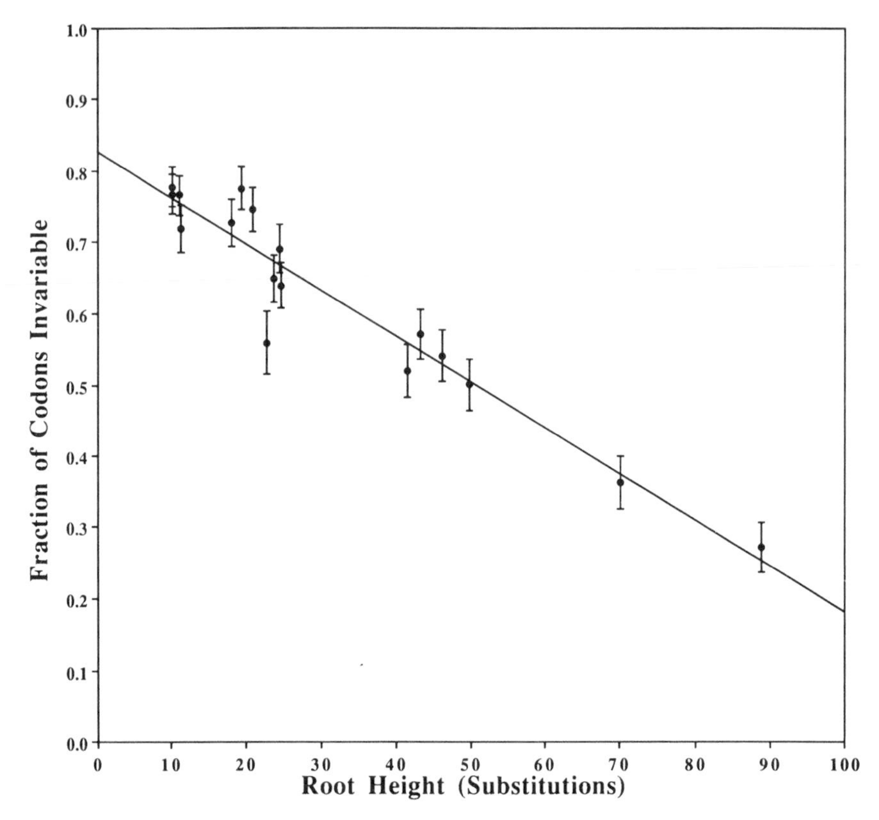

FIGURE 2 Estimation of the number of covarions. For each clade in the tree in Figure 1, we counted the number of codons suffering zero, one, two, . . . substitutions and fitted these data to two-Poissons by the method of Fitch and Markowitz (1970). One Poisson fits the varied codons and thus estimates a number of variable but unvaried codons. This number, subtracted from the number of codons with zero changes, is the number of invariable codons that, as a fraction of the total, is plotted on the y axis. On the x axis is plotted the weighted average of the number of substitutions from the root of the clade to the descendents of that root. Weighting is described in *Results*. The y intercept is the estimated fraction of codons that are not covarions. The vertical bars represent 1 SD of the estimate of the fraction of invariable codons.

In 1969, Jukes and Cantor (1969) introduced a correction that has become commonly used:

$$r/n = -(3/4)\ln(1 - 4\delta/3), \qquad \text{(equation 2)}$$

where n is the length of the sequence and $\delta = d/n$ is the fraction of the variable sites that differ. This correction is for nucleotide substitutions, not amino acid replacements, and, moreover, assumes that all sites are variable. For the first time, this equation recognizes that even two unrelated sequences will still have positions that match. The 3/4 and 4/3 terms reflect that there are four kinds of nucleotides and thus there are three ways in which a second nucleotide may not match the first. To make equation **2** suitable for replacements, the 3/4 and 4/3 should be 19/20 and 20/19, respectively. Note that equation **2**, to be technically correct, requires n to be the number of variable sites, although generally n is taken as the length of the sequence, the implicit assumption being that all sites are variable.

Equation **2** assumes that every nucleotide (or every amino acid when modified as stated) is equally likely to be present at a position. The equation can be made more accurate if one knows the frequencies of the elements. If $b = 1 - \Sigma p_i^2$, where p_i is the frequency of the ith element, and $1 \leq i \leq k$, where $k = 4$ for nucleotides and $k = 20$ for amino acids, then

$$r/n = -b \ln(1 - \delta/b), \qquad \text{(equation 3)}$$

where b is the probability that two randomly drawn elements do not match.

The above modifications (equations **2** and **3**) improve the estimate of r by recognizing and accounting for additional biological facts. However, there are other biological features that may be important. In particular, it may be important to know whether n is all sites or only a fraction thereof. Moreover, we may need to consider not only that n is less than the length of the sequence but also that the sites that make up n may not be the same (e.g., in fungi as in mammals). This is the concept incorporated in the notion of covarions, which is supported by reasonable evidence, derived not only from the observed fitting of Poisson distributions to data such as in Fitch and Markowitz (1970) but based also on a test that showed that the invariable positions of cytochrome *c* are, in fact, not the same in the fungi as in the metazoans (Fitch, 1971).

The first parameter determined was the number of covarions in SOD, estimated to be 28. The fixation of this number reduces the number of parameters that are free to change in order to get a good fit to the SOD observations. In a complementary vein, the 11 clades for which paleontological dates were estimated constitute a large extent of variant data, all of which must be fit to demonstrate that the SOD observations could arise via a perfect clock.

The second parameter fixed was the number of potentially variable codons. This was set rather arbitrarily at 118 by the following logic. Of the 162 codon positions, 44 were unvaried in our data set or were positions present only in the liverfluke, nematode, and/or bacteria but not in the other sequences. If all 44 unvaried sites were permanently invariable sites within the plant–animal–fungal sequences, then 162 − 44 = 118 were potentially variable.

The third parameter set was the number of alternatives that, on average, are allowed in a variable site. A site that must have a negative charge (aspartate or glutamate) has only two alternatives. Other possible pairs include serine–threonine, phenylalanine–tyrosine, asparagine–glutamine, and lysine–arginine. At the other extreme, there may be sites at which any amino acid can be present, in which case there are 20 alternatives. If all possibly variable sites do vary at some point, then one would expect that $(\alpha - 1)/\alpha$ of them will differ in distant pairwise comparisons (α is the average number of alternatives at variable sites; it is like the 3/4 or 19/20, in Eq. **2** above). If $\alpha = 2.5$, then one expects 1.5/2.5 = 0.6 of the potentially variable sites to differ when the number of replacements per site is large. This number may be estimated as 0.6 × 118 = 70.8 differences, a number slightly greater than the average number of differences observed between fungi and metazoans. Thus, we have set α at 2.5, although it may be somewhat low in view of the fact that the number is 3.1 in the Ayala (1986) data.

The other parameters are obtained by trying, in a hit-or-miss fashion, various combinations of the persistence and the number of replacements per million years. The results shown in Figure 3 were obtained by setting the persistence at 0.01 and the replacements per million years at 0.6. The persistence needs to be low for the vast majority of potentially variable sites to have been variable and experienced a replacement while variable—that is, to get an average of 67 differences in 118 codons after 600 replacements. A larger α plus a larger persistence could yield a comparable result but would make the short-duration times yield simulated differences that are too small (compared to observed differences) because too many replacements would occur in sites with prior replacements. A larger α and a smaller number of potentially variable codons would also improve the fit, but it does not seem correct to reduce the number of potentially variable codons below the number that had in fact varied at least once among the 67 sequences in the group. Data similar to that shown in Figure 3 were presented by Kwiatowski *et al.* (1991) who tried to fit them with both a double Poisson and a rectangular parabola. They found, as we do here, that it was difficult to fit well all the data at once.

We present the best fit that we have obtained, but there is no reason to believe that it is optimal. Therefore, we do not wish to assert that we have determined the real values of the parameters. What we do wish to assert, however, is that a reasonable model of the biological processes

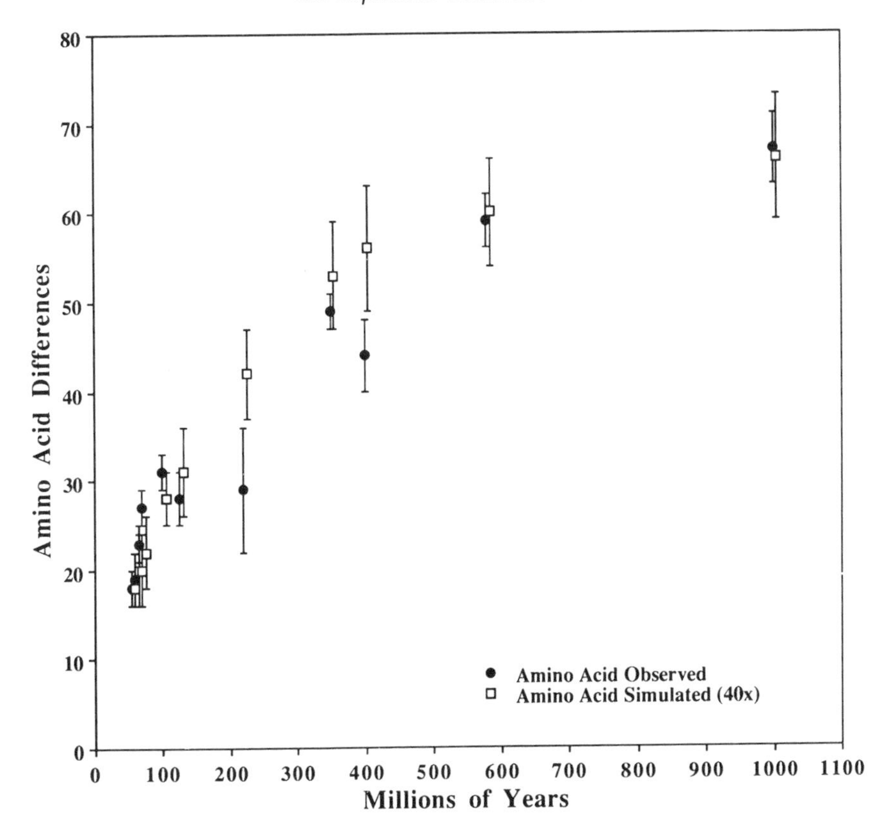

FIGURE 3 Comparison of observed SOD differences with those from a simulated perfect clock. The closed circles are observed differences in SOD; The open squares are simulated with a perfect clock (see text for clock parameters). The vertical bars show 1 SD about the mean. Observed and simulated values are for the same time despite their being horizontally offset slightly for clarity.

involved permits one to conclude that what at first appeared to be a very inaccurate clock (1) may have been inaccurate simply because the necessary corrections had not been made. In the same spirit as it was appropriate to correct Zuckerkandl–Pauling differences (1962) by accounting for multiple hits (Margoliash and Smith, 1965) and then correcting that for a finite number of alternative character states (Jukes and Cantor, 1969), it may then be necessary to take into account that the variable positions are not the same in different lineages.

Any estimate of divergence time, where the covarion process has not been taken into account, is in danger of significant error. One might naively believe that, if the differences between species are less than 10%, no Poisson or Jukes–Cantor correction is needed. That is true only if all sites are variable. For example, if only 12 of 100 sites are variable, observing 10 differences implies 23.9 replacements [$r = -12(19/20)\ln[1 - (20/19)(10/12)]$, which is more than double the 10.6 replacements [$r = -100(19/20)\ln[1 - (20/19)(10/100)]$ calculated on the assumption that all 100 sites were variable. The example may well be somewhat extreme, but any minor problem gets amplified as the ancestor becomes increasingly distant.

We do not assert that the parameters used to get the fit in Figure 3 and Table 1 are necessarily correct, but we want to make five observations.

The first observation is that the implications of the parameter values are likely to be correct. These implications are (*i*) there are about 28 covarions that (*ii*) tend to turn over fairly often, although (*iii*) the number of allowable alternatives at a variable position is, on average, limited (two to four alternatives), and (*iv*) that there is a sizeable number of positions ($\approx$40) that cannot fix a replacement even though the covarions fix them (*v*) with a reasonable rate of approximately six replacements per 10 Myr.

The second observation follows from the pattern seen in Figure 3 (and Table 1) that the simulated values are consistently lower than the observed ones during the first 100 Myr but consistently higher in the intermediate years (200–600 Mya). This pattern is expected if a sizeable fraction of the potentially variable codons has a significantly lower probability of becoming variable than the rest. This is biologically reasonable, although an alternative possibility is that the paleontological dates are either systematically underestimated for the early dates or overestimated for the intermediate dates. This would make it easier to obtain a better fit for both regions, simply by assuming a somewhat larger or smaller replacement rate respectively.

The third observation is that while we know of no good estimates of the divergence time of *S. cerevisiae* and *N. crassa*, our data, based simply on differences and not on covarions, place their divergence around 380 $\pm$ 40 Mya.

Fourth, we have tried a correction for superimposed replacements of the Jukes–Cantor type where $b = 19/20$ for the fungi–metazoan divergence. The value obtained is $r = -162 \times (0.95)\ln[1 - 67/(0.95 \times 162)] =$ 88 replacements, a 31% increase over the 67 observed differences, whereas the number of replacements required to obtain 67 differences in the simulations is 600, which is a 796% increase. Thus, corrections of the Jukes–Cantor type may still yield divergences that are gross underestimates if a covarion model is operating.

Fifth, the rate of amino acid replacement that fits our data is 6×10^{-7} replacements per year for the entire gene. This is the same as

$(6 \times 10^7)/[(3 \text{ nucleotides/covarion}) \times 28 \text{ covarions}] = 7 \times 10^{-9}$ replacement substitutions per variable nucleotide per year. This may be compared to the neutral rate, which, for cow and goat β hemoglobin pseudogenes (Li and Graur, 1991; p. 72) is 3.6×10^{-9} substitutions per nucleotide per year. Since only 3/4 of those substitutions lead to replacements, the neutral replacement substitution rate is 2.7×10^{-9}. As the fitted replacement rate is 2.6 times the pseudogene replacement rate, it suggests that there may be positive selection occurring at the SOD locus.

In conclusion, the following observations may be derived from our analysis. First, a molecular clock (i.e., a particular gene or protein) may appear to be very unreliable, yet be fairly accurate. The apparent distortion may emerge because relevant components of the clock (such as covarions, persistence, and the number of alternatives allowed per site) have not been taken into account. As we have shown, the apparently erratic SOD becomes a fairly accurate clock when the appropriate components are taken into account.

The flip side of the observation just made is that inferences of divergence time between lineages derived from a particular clock cannot be assumed to be correct unless the relevant components of the clock have been ascertained. The apparent rate of SOD divergence observed among mammals or flies would yield grossly erroneous time estimates when simply extrapolated to the differences observed between the vertebrate classes (fish versus tetrapod or amphibian versus mammal) or between metazoans and fungi. The SOD "clock" is a complex of several component parts subject to different constraints and that interact with each other.

Finally, divergence times inferred from a particular molecular clock are subject to the possibility of variations caused by natural selection or other extraneous factors. Thus, for example, some chloroplast genes evolve at rates that are similar in several grass lineages [such as maize and rice (Gaut *et al.*, 1993)] but different from those observed in the tobacco lineage (Kwiatowski *et al.*, 1991). Another example is the acceleration of the rate of lysozyme *c* evolution in ruminant lineages as lysozyme was recruited for a distinctive stomach function (Stewart and Wilson, 1987; Jolles *et al.*, 1990). A conspicuous anomaly in the SOD data is that there are fewer amino acid differences between fish and tetrapods than between amphibians and mammals. Whether or not the differences are statistically significant, they would seem to support the wrong branching order among the corresponding lineages. The conclusion derived from this anomaly, as well as from the chloroplast, lysozyme, and other examples of uneven rates, is the obvious one that inferences based on a particular clock must be taken with caution, but they become stronger as they become supported by several independent molecular clocks.

SUMMARY

The Cu,Zn superoxide dismutase (SOD) was examined earlier by Ayala (1986) and found to behave in a very unclocklike manner despite (accepted point mutation, or PAM) corrections for multiple replacements per site. Depending upon the time span involved, rates could differ 5-fold. We have sought to determine whether the data might be clocklike if a covarion model were used. We first determined that the number of concomitantly variable codons (covarions) in SOD is 28. With that value fixed we found that the observations for SOD could fit reasonably well a molecular clock if, given 28 covarions, (*i*) there are approximately six replacements every 10 million years, (*ii*) the total number of codons is 162, (*iii*) the number of codons that are permanently invariable across the range of taxa from fungi to mammals is 44, and (*iv*) the persistence of variability is quite low (0.01). Thus, the inconsistent number of amino acid differences between various pairs of descendent sequences could well be the result of a fairly accurate molecular clock. The general conclusion has two sides: (*i*) the inference that a given gene is a bad clock may sometimes arise through a failure to take all the relevant biology into account and (*ii*) one should examine the possibility that different subsets of amino acids are evolving at different rates, because otherwise the assumption of a clock may yield erroneous estimates of divergence times on the basis of the observed number of amino acid differences.

We thank Ms. Helene Van for her valuable technical assistance. This work was supported by National Institutes of Health Grant GM42397 to F.J.A. (principal investigator) and W.M.F. (co-principal investigator).

REFERENCES

Ayala, F. J. (1986) On the virtues and pitfalls of the molecular evolutionary clock. *J. Hered.* **77,** 226–235.

Bannister, J. V. and Parker, M. W. (1985) The presence of a copper/zinc superoxide dismutase in the bacterium *Photobacterium leiognathi*: A likely case of gene transfer from eukaryotes to prokaryotes. *Proc. Natl. Acad. Sci. USA* **82,** 149–152.

Dayhoff, M. O. (1978) *Atlas of Protein Sequence and Structure* (Natl. Biomed. Res. Found., Washington, DC).

Fitch, W. M. (1971) Toward defining the course of evolution: Minimum change for a specific tree topology. *Syst. Zool.* **20,** 406–416.

Fitch, W. M. (1971) The non-identity of invariant positions in the Cytochrome c of different species. *Biochem. Genet.* **5,** 231–241.

Fitch, W. M. (1976) The molecular evolution of cytochrome c in eukaryotes. *J. Mol. Evol.* **8,** 13–40.
Fitch, W. M. and Farris, J. S. (1974) Evolutionary trees with minimum nucleotide replacements from amino acid sequences. *J. Mol. Evol.* **3,** 263–278.
Fitch, W. M. and Markowitz, E. (1970) An improved method for determining codon variability in a gene and its application to the rate of fixations of mutations in evolution. *Biochem. Genet.* **4,** 579–593.
Fitch, W. M. and Ye, J. (1991) Weighted parsimony: Does it work? In *Phylogenetic Analysis of DNA Sequences,* eds. Miyamoto, M. and Cracraft, J. (Oxford Univ. Press, New York), pp. 147–154.
Fridovich, I. (1986) Superoxide dismutase. *Adv. Enzymol.* **58,** 61–97.
Gaut, B. S., Muse, S. V. and Clegg, M. T. (1993) Relative rates of nucleotide substitution in the chloroplast genome. *Mol. Phylogenet. Evol.* **2,** 89–96.
Getzoff, E. D., Tainer, J. A., Stempien, M. M., Bell, G. I. and Hallewell, R. A. (1989) Evolution of CuZn superoxide dismutase and the Greek key beta-barrel structural motif. *Proteins* **5,** 322–336.
Jolles, J., Prager, E. M., Alnemri, E. S., Jolles, P., Ibrahimi, I. M. and Wilson, A. C. (1990) Amino acid sequences of stomach and nonstomach lysozymes of ruminants. *J. Mol. Evol.* **30,** 370–382.
Jukes, T. H. and Cantor, C. R. (1969) Evolution of protein molecules In *Mammalian Protein Metabolism,* ed. Munro, H. N. (Academic, New York), pp. 21–132.
Kwiatowski, J., Hudson, R. R. and Ayala, F. J. (1991) The rate of Cu,Zn Superoxide dismutase evolution. *Free Radical Res. Commun.* **13,** 363–370.
Kwiatowski, J., Skarecky, D. and Ayala, F. J. (1992) Structure and sequence of the *Cu,Zn Sod* gene in the Mediterranean fruit fly, *Ceratitis capitata*: Intron insertion/deletion and evolution of the gene. *Mol. Phylogenet. Evol.* **1,** 72–82.
Li, W. H. and Graur, D. (1991) *Fundamentals of Molecular Evolution* (Sinauer, Sunderland, MA).
Margoliash, E. and Smith, E. (1965) Structural and functional aspects of cytochrome *c* in relation to evolution. In *Evolving Genes and Proteins,* eds. Bryson, V. and Vogel, H. J. (Academic, New York), pp. 221–242.
Shoemaker, J. S. and Fitch, W. M. (1989) Evidence from nuclear sequences that invariable sites should be considered when calculating sequence divergence. *Mol. Biol. Evol.* **6,** 270–289.
Steinman, H. M. (1988) Bacterial superoxide dismutases. *Basic Life Sci.* **49,** 641–646.
Stewart, C. B. & Wilson, A. C. (1987) Sequence convergence and functional adaptation of stomach lysozymes from foregut fermenters. Cold Spring Harbor Symposium. *Quant. Biol.* **52,** 891–899.
Tainer, J. A., Getzoff, E. D., Beem, K. M., Richardson, J. S. and Richardson, D. C. (1982) Determination and analysis of the 2Å structure of copper, zinc superoxide dismutase. *J. Mol. Biol.* **160,** 181–217.
Tainer, J. A., Getzoff, E. D., Richardson, J. S. and Richardson, D. C. (1983) Structure and mechanism of copper, zinc superoxide dismutase. *Nature (London)* **306,** 284–287.
Zuckerkandl, E. and Pauling, L. (1962) Molecular disease, evolution and genetic heterogeneity. In *Horizons in Biochemistry* (Academic, New York), pp. 189–225.

Part V

PATTERNS

Ten thousand human generations take 250,000 years and 10,000 fruit fly generations take 500 years. Ten thousand generations are but an instant of evolutionary time, but in humans and flies demand too much time for direct experimentation. Not so in the case of bacteria; 10,000 generations of *Escherichia coli* require "only" 4 years. Richard E. Lenski and Michael Travisano, in Chapter 13, show that in new but constant environments evolution occurs rapidly during the first 2000 generations, slowly during the following 3000, and not at all over the last 5000. They have 12 separate populations derived from identical ancestors and evolving in identical environments, but their trajectories are different in both morphology and fitness. The conclusion is inescapable that chance events play an important role in adaptive evolution.

DNA polymorphisms along the chromosomes of *Drosophila* flies exhibit an unanticipated pattern. Where the incidence of genetic recombination is low, such as near the centromere and the tips, the level of polymorphism also is low. This is not a consequence of different mutation rates, concludes Richard R. Hudson in Chapter 14, because divergence between species is indifferent to incidence of recombination. The pattern can be explained by "hitchhiking"—that is, by selection of favorable mutations that carry along other mutations as they spread through a population; how much DNA will be carried along is determined by the incidence of recombination. The reciprocal of selection of favorable mutations is selection against unfavorable ones. This possibility, however, does not quite explain the observed pattern.

1944 was a propitious year for evolutionary studies. In addition to

Simpson's *Tempo and Mode* it saw the publication of a monograph by Theodosius Dobzhansky and Carl Epling (*Contributions to the Genetics, Taxonomy, and Ecology of Drosophila pseudoobscura and Its Relatives*) that would usher in an interest in reconstructing phylogenetic history on the basis of genetic information. The method relies on the sequential composition of chromosomes, a premonition of the molecular methods that rely on the sequential composition of the DNA. The third chromosome of *Drosophila pseudoobscura* exhibits a rich polymorphism with more than 40 alternatives. One vexing problem is rooting the topological relationships, that is, identifying the ancestral element. Aleksandar Popadić and Wyatt W. Anderson, in Chapter 15, examine the nucleotide sequence of a DNA fragment included within the chromosomal elements and conclude that only two of the elements are possible ancestors, one of them ("Santa Cruz") with higher probability.

The last chapter is a display of molecular biology virtuosity. Daniel L. Hartl and colleagues have produced a physical map of the chromosomes of the fruit fly *Drosophila melanogaster*, by ordering sequentially 2461 different DNA fragments, each about 80,000 nucleotides long. Eighty-five percent of all genes are included in these fragments. The methods are the same in use for mapping human chromosomes, and they can readily be extended to other *Drosophila* species, a possibility of genetic and evolutionary consequence.

13

Dynamics of Adaptation and Diversification: A 10,000-Generation Experiment with Bacterial Populations

RICHARD E. LENSKI AND MICHAEL TRAVISANO

Fifty years after the publication of Simpson's *Tempo and Mode in Evolution* (1944), evolutionary biologists are still fascinated by—and struggling to understand—the dynamics of adaptation and diversification, especially for those traits that affect the reproductive success of individual organisms. How quickly do populations change in these traits, and are their rates of change constant or variable? How rapidly do populations diverge from one another in these traits, and are rates of adaptation and diversification tightly or loosely coupled? How repeatable is evolution, and how sensitive are evolutionary outcomes to a population's initial genetic state? What are the relative roles of chance, phylogeny, and adaptation in evolution? How do the answers to these questions depend on the genetic system of an organism and on the traits examined?

We have embarked on an experimental program to investigate these questions. We believe that experiments complement historical and comparative studies and, when appropriately designed, may forge an important link between micro- and macroevolutionary studies. Before describing our experiments, however, we present an *imaginary* framework for such research. This imaginary framework illustrates the profound problems of inference inherent in purely observational ap-

Richard E. Lenski is the John Hannah Professor of Microbial Ecology at Michigan State University, East Lansing, Michigan. Michael Travisano is a postdoctoral researcher at the RIKEN Institute, Wako, Japan.

proaches to studying evolutionary dynamics, while also highlighting the power of our particular experimental system.

Imagine, then, that you have discovered a well-preserved and clearly stratified fossil bed that provides a record of evolution extending thousands of generations for the particular organism that you study. You could measure the size and shape of the organisms that were preserved and perhaps deduce the rate of change in these traits. But even from a near-perfect record, you would have great difficulty inferring the evolutionary processes—selection, drift, mutation, recombination, and migration—affecting these morphological traits. It might be difficult even to exclude the hypothesis that any phenotypic trends reflect nonheritable changes caused by the direct effects of a changing environment on the organism.

But imagine that you could infer that the environment had not changed for thousands of generations, so that any phenotypic trends must have resulted from underlying genetic changes. Moreover, you could be sure that there was no influx of genotypes from other populations and that the population was initially homogeneous, so that all of the genetic variation in the fossil population must have arisen *in situ*. You could then confidently assess the tempo and mode of morphological evolution.

Now imagine that you found many fossil beds, all in identical environments and having the same initial genetic state. You could evaluate the repeatability of evolution by examining the parallelism or divergence of the populations from one another. Any repeatability (or lack thereof) would also bear on the success of specific hypotheses that sought to address the adaptive significance of particular phenotypic trends.

And the fantasy continues. Imagine that you could resurrect these organisms (not merely bits of fossil DNA but the entire living organisms) and reconstruct their environment exactly as it was during the thousands of generations preserved in the fossil bed. You could measure not only the organism's morphology, but also its functional capacities and genetic composition. You could even place derived and ancestral forms in competition to determine their relative fitness in the "fossil" environment. You could assess which phenotypes promoted ecological success, and you could evaluate the similarity of the adaptive solutions achieved by the replicate populations, thereby disentangling the roles of "chance and necessity" (Monod, 1971) in evolutionary dynamics.

Still more opportunities exist in this fantastic world. You could travel back in time and manipulate populations by altering their evolutionary history or their environment, and then return to the present to examine the effect of these conditions on the dynamics of adaptation and diversification. With this power, you could gain insight into the potency of phylogenetic constraints, examine the effects of environmental constancy or complexity, and pursue a host of other evolutionary questions.

Yet this fantasy is not fiction; it is fact. We have many such "fossil beds" preserved, and we have "traveled in time" to manipulate populations with respect to their history and environment. The fossil beds are preserved in a freezer and contain populations of the bacterium *Escherichia coli*. Our time travel thus far extends over 5 years, representing >10,000 generations in this system, and we have manipulated many populations each comprising millions of individual organisms. In essence, our approach might be called experimental paleontology.

The following section gives a quick overview of our experimental system. We then analyze and interpret our experiments, which are organized around the analogy to an increasingly fantastic exploration of fossil beds. The discussion relates our findings on the dynamics of adaptation and diversification to theories of micro- and macroevolution. Finally, we briefly discuss tensions that inevitably encroach on any effort to forge an experimental link between micro- and macroevolution.

Experimental Overview

Twelve populations of *E. coli* B were propagated in replicate environments for 1500 days (10,000 generations). Each population was founded by a single cell from an asexual clone, and so there was initially no genetic variation either within or between replicate populations (except for a neutral marker used to identify populations). The experimental environment consisted of a serial transfer regime, in which populations were diluted (1:100) each day into 10 ml of a glucose-limited minimal salts medium that supports $\approx 5 \times 10^7$ cells per ml. Populations were maintained at 37°C with aeration. Every day, the bacteria underwent a lag phase prior to growth, followed by a period of sustained growth, eventual depletion of the limiting glucose, and starvation until the next serial transfer. The 1:100 dilution permits ≈ 6.6 ($\log_2 100$) cell generations per day. Samples from each population were periodically stored at −80°C, along with the common ancestor.

In this paper, we report the dynamics of two properties of the evolving bacterial populations, cell size and mean fitness. Size is a morphological trait commonly studied by paleontologists and influences many functional properties of organisms. Fitness is the most important property of any organism according to evolutionary theory. The mean fitness of a population was obtained by allowing it to compete against the common ancestor. Relative fitness was then calculated as the ratio of the competitors' realized rates of increase (Malthusian parameters). Cell sizes were obtained by using an electronic device that measures the volume displaced by a particle (Lenski and Bennett, 1993; Vasi *et al.*, 1994). Prior to assays of either fitness or cell size, bacteria were removed from the freezer and allowed to acclimate for 1 day (several generations) to the experimental conditions. This procedure eliminates confounding effects of different physiological states of organisms and thereby establishes that phenotypic differences between populations have an underlying genetic basis.

We emphasize that our experiments employ natural selection, and not artificial selection as practiced by breeders and many experimentalists. That is, we do not select individual organisms based on any particular trait, but rather impose an environment on the experimental populations. Any heritable properties that systematically enhance an individual's reproductive success in that environment can respond to natural selection.

Except as indicated otherwise, materials and methods are identical to those previously published. In particular, see Lenski *et al.* (1991) for details concerning the ancestral strain, culture conditions, methods for exclusion of contamination from external sources and cross-contamination between replicate populations, and procedures for estimating mean fitness of derived populations relative to their ancestor.

Analyses and Interpretation

It may seem peculiar that we begin with analyses of morphological data (given that we have data on fitness) and that we focus initially on a single population (given that we have replicate populations). We do so to emphasize the potential difficulties of drawing inferences from morphological data (without information on fitness) and from single populations (without independent replication). This organization also facilitates comparison of anagenetic (within a lineage) and cladogenetic (diversifying) trends. We conclude our analyses by examining the relationship between morphology and fitness.

The Record of Morphological Evolution in One Fossil Bed. Figure 1 shows the trajectory for average cell size in one bacterial population during 10,000 generations of evolution. For ≈2000 generations after its introduction into the experimental environment, cell size increased quite rapidly. But after the environment was unchanged for several thousand generations more, any further evolution of cell size was imperceptible. A hyperbolic model [$y = x_0 + ax/(b + x)$] accounts for ≈99% of the temporal variation in average cell size [$r = 0.995$, $n = 9$, $P < 0.001$; relative to linear model, partial $F = 162.9$, 1 and 6 df (df = degrees of freedom), $P < 0.001$]. These data therefore indicate a rapid bout of morphological evolution after the population was placed in the experimental environment, followed by evolutionary stasis (or near stasis).

From these data, one might speculate that selection had favored larger cell size *per se*, although one could not exclude the possibilities that size was a correlated response to selection on some other trait or, more remotely, that cell size was subject solely to random genetic drift. If cell size was indeed a target of natural selection, then the eventual stasis might be interpreted either as a genetic/developmental constraint (such that no new mutations appeared that could increase cell size further) or as stabilizing selection (such that both larger and smaller variants continued to appear but were purged by natural selection). And whether or not cell size was a target of selection, the population may have continued to adapt to the environment (after size was static) by changing other traits.

Replicating History. Inferences concerning the tempo of morphological evolution and the adaptive significance of cell size would be greatly strengthened if similar trends were seen in several independent fossil beds. This opportunity is not usually possible: even when several contemporaneous fossil beds exist, one cannot exclude the possibility

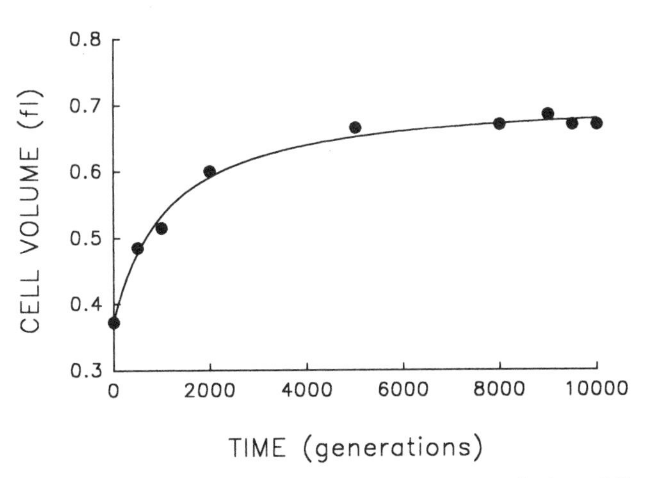

FIGURE 1 Trajectory for average cell volume in one population of *E. coli* during 10,000 generations of experimental evolution. Each point is the mean of two assays. Curve is the best fit of a hyperbolic model. Averages were calculated after removing particles of <0.25 fl, except for the ancestor ($t = 0$) where particles of <0.15 fl were removed; these criteria gave a clear separation between cells and background particles. Cell sizes were measured in stationary-phase populations, at the end of the 24-h serial transfer cycle.

that the several populations responded similarly because they shared genetic variation (i.e., alleles identical by descent that were either present in the common ancestor or introduced by migration), or that they diversified because they were in subtly different environments. But our experimental system affords the opportunity to examine rigorously the repeatability of evolutionary dynamics, including the origin as well as the fate of phenotypic modifications.

Figure 2 shows the estimated trajectories for average cell size in the replicate populations during 10,000 generations. All 12 independently evolved larger cells and, moreover, all 12 underwent much more rapid change soon after their introduction into the experimental environment than when their environment had been constant for several thousand generations.

Despite their superficial similarity, these trajectories also suggest that the replicate populations were approaching somewhat different plateaus for cell size. This inference must be made cautiously, however, because the parameters for each fitted curve have an associated statistical uncertainty. We performed analyses of variance on the raw data used to obtain these trajectories in order to estimate directly the among-population variance component for average cell size at each time point. We know that the populations were initially identical; if they are, in fact, approaching different plateaus for average cell size, then the among-population variance component must increase from zero to some

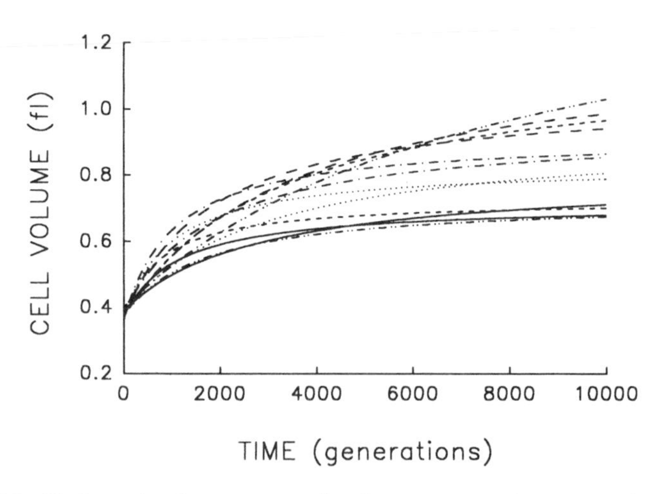

FIGURE 2 Trajectories for average cell volume in 12 replicate populations of *E. coli* during 10,000 generations. Each curve represents the best fit of a hyperbolic model to data obtained for one population at intervals indicated in Figure 1.

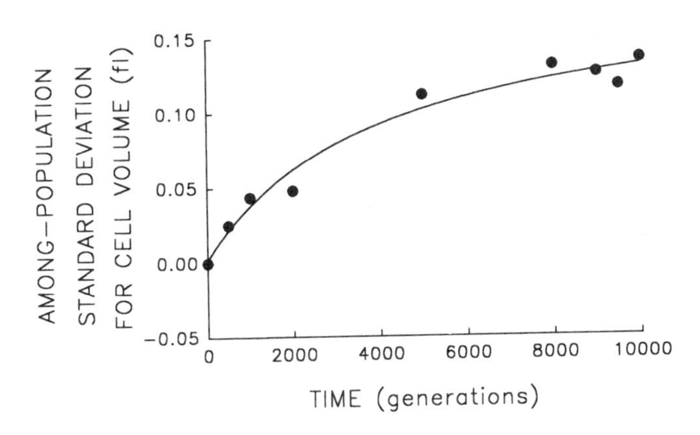

FIGURE 3 Trajectory for diversification of the 12 populations in their average cell volume. Analyses of variance were performed to partition the observed variation in cell size into its components. Each point represents the among-population SD for cell volume (i.e., the square root of the among-population variance component). Curve shows the best fit of a hyperbolic model.

plateau. Figure 3 shows the trajectory for the among-population standard deviation (i.e., square root of the variance component). The fit of the hyperbolic model to these data is very good (r = 0.987, n = 9, $P <$ 0.001; relative to linear model, partial F = 15.6, 1 and 6 df, P = 0.008). Thus, we conclude that the populations have diverged, not only from their common ancestor but from one another, in cell size. After 10,000 generations, the among-population standard deviation was ≈0.14 fl, as compared with the average change in cell size from the ancestral state of ≈0.44 fl.

The observation that all 12 independently evolving populations responded similarly, even if not identically, seems to rule out random genetic drift, unless we invoke some profound asymmetry in mutational effects on cell size. However, we still cannot determine whether cell size was the actual target of selection (or merely a correlated response to selection on other traits), nor can we discern whether adaptation to the environment continued apace for the entire 10,000 generations (but without producing further changes in cell size).

The Quick and the Dead. A remarkable feature of our experimental system is that we can measure the mean fitness of a derived population relative to its actual ancestor. That is, populations of cells can be "resurrected from the dead" (i.e., removed from the freezer) at any time and placed in direct competition. This ability to measure fitness permits investigation of a host of intriguing questions. Has mean fitness im-

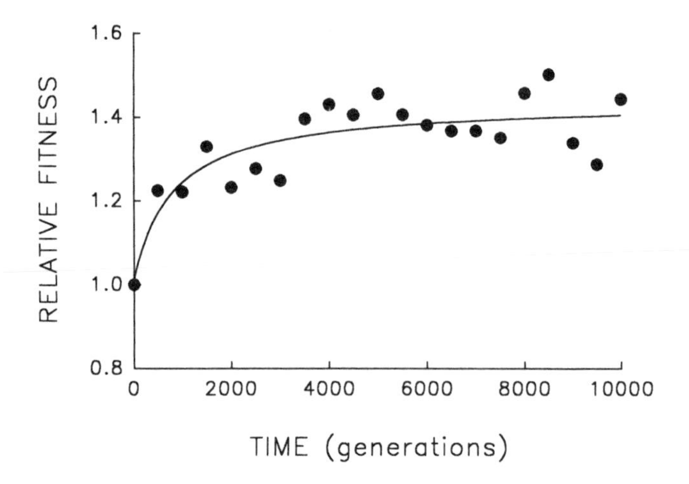

FIGURE 4 Trajectory for mean fitness relative to the ancestor in one population of *E. coli* during 10,000 generations of experimental evolution. Each point is the mean of three assays. Curve is the best fit of a hyperbolic model.

proved, thus demonstrating adaptation by natural selection? Is the evolutionary trajectory for fitness similar to that for cell size, or did fitness improve at a constant rate throughout the experimental evolution? Have the replicate populations also diverged from one another in mean fitness, suggesting that they scaled different adaptive peaks? Or have the populations converged on similar fitnesses, implying that their differences in cell size are inconsequential for fitness? How tightly correlated are morphology and fitness?

Figure 4 shows the dynamics of mean fitness relative to the ancestor for the same population whose evolutionary trajectory for average cell size is shown in Figure 1. Mean fitness evolved rapidly for ≈2000 generations in the experimental environment but was nearly constant for the final several thousand generations. Although the fitness data are subject to more "noise" than the data on cell size, the hyperbolic model explains ≈70% of the variation in mean fitness ($r = 0.843$, $n = 21$, $P < 0.001$; relative to linear model, partial $F = 17.9$, 1 and 18 df, $P < 0.001$). The fact that fitness shows the same decelerating trajectory belies the hypothesis that adaptation continued apace for the 10,000 generations but was not reflected in further changes in cell size.

Figure 5 provides finer resolution of the fitness trajectory for the first 2000 generations in the same population; fitness was estimated every 100 (rather than 500) generations and with 10-fold (rather than 3-fold) replication. A step model (Lenski *et al.*, 1991) [$y = c_0 + c_1$ (if $t > T_1$) +

c_2 (if $t > T_2$) + . . . + c_n (if $t > T_n$)] with three steps provides an excellent fit to the trajectory ($r = 0.978$; 3 and 17 df; $P < 0.001$). What accounts for the initial delay and the seemingly discontinuous jumps? Both features are expected from population genetic theory, given the uniformity of the founding population, its asexuality, and the resulting dependence of the selection response on new mutations (Muller, 1932; Lenski *et al.*, 1991). Any favorable allele must first appear and then increase from a very low frequency. Assuming constant selection, it takes as long for a favored new allele to increase from a frequency of 10^{-7} to 10^{-6} as it takes it to increase from 10% to 90%; and yet only after the allele has reached high frequency does it appreciably affect the mean properties of a population. Thus, the smoothness of the trajectories shown in Figures 1 and 4 is, to some extent, a product of relatively infrequent sampling; discontinuities revealed by more frequent sampling (Figure 5) indicate nothing more than the dynamics of selection when a population depends on new mutations (rather than abundant standing variation) for its response.

The Adaptive Landscape. Wright's (1932, 1982, 1988) concept of the adaptive landscape (or fitness surface) provides one of the most vivid images in all of evolutionary theory, but it is also one of the most difficult to firmly grasp and study. The essential idea is that natural selection tends to drive a population to a local optimum, which is not necessarily a global optimum. Thus, a population may be stuck with a suboptimal solution to its environment because natural selection (which is not goal

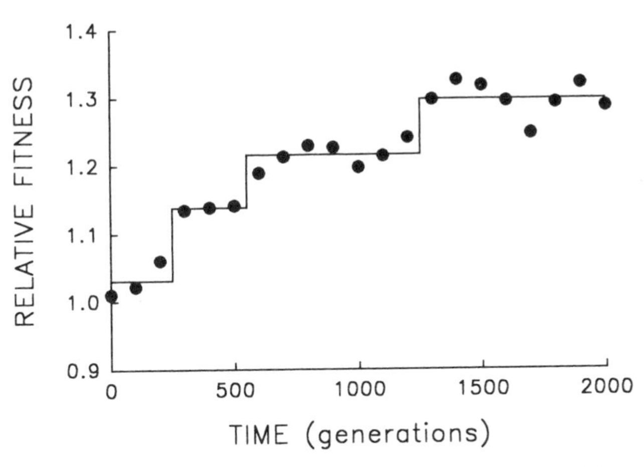

FIGURE 5 Finer scale analysis of the trajectory for mean fitness in one population of *E. coli* during its first 2000 generations of experimental evolution. Each point is the mean of 10 assays. Solid lines indicate the fit of a step model.

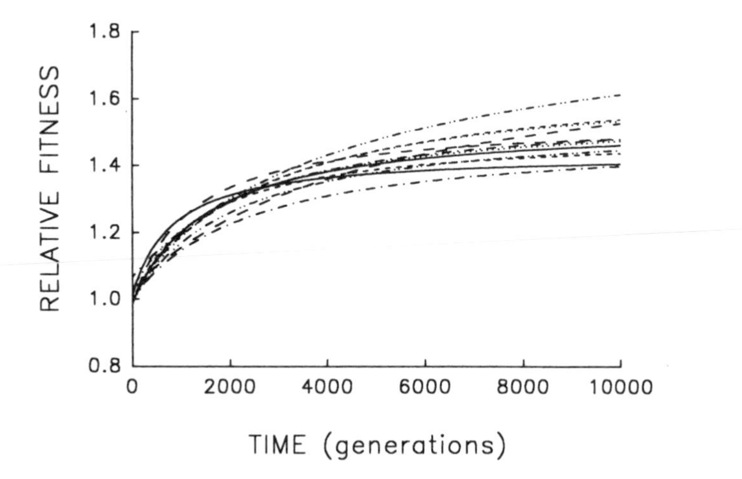

FIGURE 6 Trajectories for mean fitness relative to the ancestor in 12 replicate populations of *E. coli* during 10,000 generations. Each curve represents the best fit of a hyperbolic model to data obtained for one population every 500 generations.

directed) opposes passage through a "valley" of maladapted intermediate states, even though a better solution may exist across the way. Theoreticians have identified processes that might facilitate peak shifts, but empiricists know very little about the structure of adaptive landscapes. A key question is, how often are there nearby fitness peaks of unequal height?

To address this question, we examine whether the evolving populations diverged from one another in mean fitness, as they did in morphology. Figure 6 shows the estimated trajectories for mean fitness in the replicate populations during 10,000 generations. All 12 adapted much more rapidly soon after their introduction into the experimental environment than they did subsequently, when their environment had been constant for several thousand generations. We performed analyses of variance to estimate the among-population variance component for mean fitness at each time. If the populations were approaching different fitness peaks from the same initial state, then this variance component should increase from zero to some plateau. Figure 7 shows the fit of the hyperbolic model to the trajectory for the among-population standard deviation for mean fitness. The fit of the model is poor ($r = 0.286$). *However*, the twenty separate analyses of variance (after $t = 0$) yielded estimated variance components > 0 in 18 cases. The associated significance levels were <0.05 in 7 cases and between 0.05 and 0.25 in 8 others.

The joint probability of obtaining, by chance, so many low-probability outcomes is very remote [$P < 0.001$, based on Fisher's technique for combining probabilities from independent tests of significance (Sokal and Rohlf, 1981)]. Therefore, although we cannot discern any clear trend in the among-population variance for mean fitness, we know it was initially equal to 0 and that significant variation arose early in the experiment and persisted throughout the 10,000 generations.

Nonetheless, we cannot absolutely exclude the possibility that the populations might eventually converge in mean fitness. In fact, when similar analyses were performed after 2000 generations, we suggested that the populations might have diverged in mean fitness only transiently, owing to stochastic variation in the timing of a series of substitutions leading to the same fitness peak (Lenski *et al.*, 1991). The reason for our earlier caution in claiming sustained divergence was not because the among-population genetic variance was much smaller in the first 2000 generations (Figure 7), but rather because the rate of adaptation continued at a much more rapid pace (Figure 6), leaving open the possibility that the less fit populations would soon catch up to their better-adapted cousins. But with an additional 8000 generations, the rate of fitness improvement has become so slow that eventual convergence now seems very unlikely. Whereas the average rate of improvement between generations 1000 and 2000 was 0.108 (±0.020 SEM) per 1000

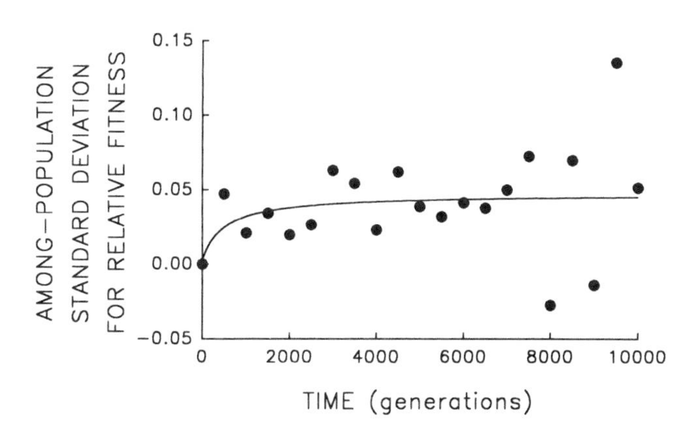

FIGURE 7 Trajectory for diversification of the 12 populations in their mean fitness. Analyses of variance were performed at 500-generation intervals to partition the observed variation in mean fitness into its components. Each point represents the among-population SD for mean fitness (i.e., the square root of the among-population variance component); negative values indicate that the estimated variance component was negative. Curve shows the best fit of a hyperbolic model.

generations (Lenski *et al.*, 1991), the rate of further improvement had fallen to only 0.008 (±0.004 SEM) per 1000 generations between generations 5000 and 10,000. Also, using the hyperbolic models for each population, there is a positive correlation between fitness and the rate of continued improvement at 10,000 generations (r = 0.848; n = 12; P < 0.001), suggesting that the replicate populations will become more divergent in fitness, not less.

Thus, the replicate populations have approached distinct fitness peaks of unequal height (or, just maybe, they are climbing slowly along different ridges of unequal elevation toward the same peak). After 10,000 generations, the among-population standard deviation for fitness was ≈0.05, while the average fitness gain from the ancestral state was ≈0.48.

The Relationship of Size and Fitness. One cannot help but notice the similarity of the trajectories for cell size and fitness. Does this correspondence imply that cell size was a "target" of selection? Or might size merely be correlated with other traits responsible for the improved fitness? Is the relationship between cell size and fitness rigidly fixed or is it evolutionarily malleable?

Even with this powerful experimental system, these questions are vexing and difficult to resolve. Ideally, one would like to manipulate cell size, holding all other traits constant. But this proposition presumes an atomization of traits that is implausible, given the pleiotropic action of alleles and the functional interconnections inherent in any organism. Thus, for the time being, we are forced to rely on statistics to describe the relationship between size and fitness.

To describe the *functional* relationship between two traits, it is necessary to use a regression coefficient (the slope in a linear model) rather than a correlation coefficient. A problem arises, however, because standard regression methods require that one of the variables either be manipulated experimentally or measured without error (Sokal and Rohlf, 1981), whereas both average cell size and mean fitness of a population are measured with error. Although there is no general solution to this problem of "model II" regression, a precise solution exists when the error variances associated with measurement of each variable are known (Mandel, 1964). Fortunately, we can estimate the error variances associated with both variables, because each size/fitness datum in our analysis is the mean of two (size) and three (fitness) independent assays.

Figure 8*A* shows the correspondence between mean fitness and average cell size over all populations and time points. Figure 8*B* illustrates the hypothesis that size and fitness are tightly coupled, as

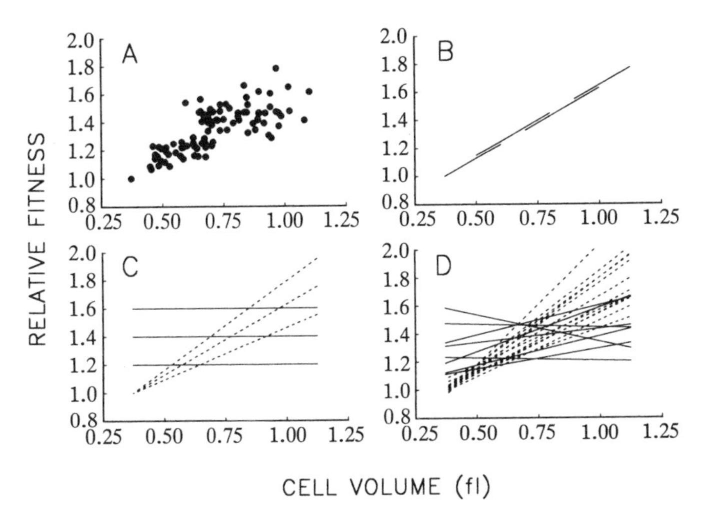

FIGURE 8 Relationship between mean fitness and average cell size. (*A*) Data from all 12 populations and 9 time points for which both fitness and size were measured. (*B*) Hypothesis I: There exists a rigid functional relationship between average cell size and mean fitness, such that the latter can be improved only by increasing the former. Longitudinal (one population over all time points) and cross-sectional (all populations at one time point) regressions would yield the same slope. (*C*) Hypothesis II: The functional relationship between size and fitness is malleable, so that replicate populations may diverge in the relationship between size and fitness. Although longitudinal regressions (dashed lines) may be strong, cross-sectional regressions (solid lines) need not show any systematic coupling between size and fitness. (*D*) Dashed and solid lines show actual longitudinal and cross-sectional regressions, respectively, using the data shown in *A*. All regressions were performed according to model II procedures that are applicable when both variables are measured with error, but the corresponding error variances are known (Mandel, 1964). From analyses of variance performed on repeated measures, the ratio of the error variances for average cell size and mean fitness was 0.138 (adjusted for sample sizes of two and three, respectively, and averaged over all populations and generations).

though size were the actual target of natural selection. According to this hypothesis, one should obtain the same regression line whether the data are analyzed *longitudinally* (i.e., using a single population over all time points) or *cross-sectionally* (i.e., using all populations at a single time point). Put another way, any variation among populations in fitness is *because* some have achieved larger cells than others.

Figure 8C illustrates an alternative hypothesis, which states that traits other than cell size are the actual targets of selection; larger size is correlated with some of the selected traits, but there is no rigid coupling between size and fitness. Thus, the replicate populations may diverge not only in size and fitness, but also in the functional relationship between size and fitness. In that case, there need not be any systematic relationship between size and fitness in cross-sectional regressions, even if the longitudinal regressions are strong.

Figure 8*D* shows the results of the 12 longitudinal (1 for each population) and 8 cross-sectional (1 for each time point) regressions. The mean of the longitudinal slopes is 1.063 fl^{-1}. (That is, relative fitness, which is a dimensionless quantity, increases by 1.063 per fl increase in cell volume.) This mean is significantly greater than 0 ($t = 11.72$; 11 df; $P < 0.001$). In contrast, the mean of the cross-sectional slopes is only 0.187 fl^{-1}, which is not significantly different from 0 ($t = 1.61$; 7 df; $P = 0.150$) but is significantly less than the mean of the longitudinal slopes ($t = 6.01$; 18 df; $P < 0.001$). Therefore, these analyses do not support the hypothesis that the functional relationship between size and fitness is causal and rigidly fixed (Figure 8*B*) but suggest instead that the replicate populations have diverged in this relationship (Figure 8*C*). These results therefore also challenge the controversial assumption of certain evolutionary analyses that genetic covariances between traits are constant over long periods (Lande, 1975, 1979; Kohn and Atchley, 1988; Turelli, 1988).

Conclusions and Discussion

Chance and Necessity. The 12 bacterial populations had similar trajectories for both cell size and fitness (Figures 2 and 6). It is perhaps surprising that the populations evolved in such a parallel fashion, given that their evolution depended on mutations that arose independently in each population. Of course, a critical factor promoting parallel evolution was the simple fact that populations evolved in identical environments. A second factor promoting parallelism may have been large population sizes, which would give rise to identical mutations in the replicate populations. Each population underwent $\approx 7.5 \times 10^{11}$ cell replications (5×10^7 cell replications per ml per day $\times$ 10 ml $\times$ 1500 days). The estimated rate of mutation in *E. coli* (Drake, 1991) is $\approx 2.5 \times 10^{-3}$ mutation per genome replication (5×10^{-10} mutation per bp replication $\times\ 5 \times 10^6$ bp per genome). Thus, each population experienced $\approx 2 \times 10^9$ mutations. With 5×10^6 bp per genome and three alternative point mutations at each bp (ignoring more complex mutations), this translates to >100 occurrences, on average, for every point mutation in the whole genome! (Of course, drift eliminates many mutations shortly after they occur, but even so this figure suggests redundancy.)

However, it is also important that the replicate populations diverged *somewhat* in both morphology and mean fitness (Figures 3 and 7). After 10,000 generations, the standard deviation for mean cell size among the derived populations was ≈30% of the average difference between the derived populations and their common ancestor. For mean fitness, the standard deviation among the derived populations was ≈10% of the average improvement from the common ancestor. Moreover, the populations also diverged in the functional relationship between cell size and fitness (Figure 8*D*).

Evolutionary biologists usually regard diversification as being caused by either (*i*) adaptation to different environments, which often produces conspicuous phenotypic variation, or (*ii*) random genetic drift, which is usually seen in molecular genetic variation. Yet our experiments demonstrate diversification, in identical environments and with very large populations, of no less selected a trait than fitness itself. Someone confronted with the variability among our derived populations (and unaware of the experimental design) might attribute this diversity to environmental heterogeneity or phylogenetic constraints, but any such "just-so story" would clearly be misguided in this case. Instead, our experiment demonstrates the crucial role of chance events (historical accidents) in adaptive evolution.

In a previous analysis of the first 2000 generations of this experiment (Lenski *et al.*, 1991), it was not possible to reject the hypothesis that the populations had diverged only transiently in mean fitness but would soon converge on the same mean fitness. It was proposed that this hypothesis could be rejected, in favor of sustained divergence, "if the level of between-population variance of mean fitness remains significant indefinitely, even in the absence of further increases in mean fitness" (Lenski *et al.*, 1991, p. 1337). We have shown here that variation among populations in mean fitness does persist for thousands of generations, even after improvement in mean fitness has slowed to an almost imperceptible rate.

Sustained divergence in mean fitness supports a Wrightian model of evolution (Wright, 1932, 1982, 1988; Barton and Hewitt, 1989; Mani and Clarke, 1990; Wade and Goodnight, 1991), in which replicate populations found their way onto different fitness peaks. Although the experimental populations were so large that the same mutations occurred in all of them, the order in which various mutations arose would have been different (Mani and Clarke, 1990). As a consequence, some populations may have incorporated mutations that were beneficial over the short-term but led to evolutionary dead ends.

Beyond promoting the idea of fitness surfaces with adaptive peaks separated by maladapted intermediate states, Wright (1932, 1982, 1988) identified processes by which populations might move from one peak to another. One process that can produce a peak shift, without environmental change, is random genetic drift (including founder effects). Although this hypothesis has been highly influential [e.g., for models of speciation (Mayr, 1954; Templeton, 1980)], the mathematical conditions conducive to such peak shifts appear to be restrictive (Barton and Rouhani, 1987). In this respect, it is important that, in our experiment, populations were not on one adaptive peak and asked to "jump" to another peak; instead, they were thrown into an arbitrary environment and asked to climb any accessible peak.

Adaptation, Diversification, and Stasis. For about 2000 generations after their introduction into the experimental environment, all 12 populations underwent rapid changes in both morphology and fitness, whereas these properties were nearly static between generations 5000 and 10,000. The initially rapid evolution was presumably due to intense selection triggered by the sudden environmental changes imposed at the start of our experiment. Although the ancestors of the founding bacterium used in this study had been "in captivity" for several decades, they were not systematically propagated under the experimental conditions that we imposed (serial dilution in glucose-minimal medium). The experimental regime was thus a novel environment. Unfortunately, we cannot say anything quantitative about the evolutionary dynamics of the study organism before the start of our experiment. However, in two other studies, we have used derived bacteria from this study to found new populations, which were introduced into environments that differed either in temperature (Bennett *et al.*, 1992) or in limiting nutrient (Travisano, 1993). In both cases, these environmental changes led to more rapid adaptive evolution.

Our results also reveal the quasi-punctuated dynamics expected when selection depends on new mutations. However, we saw no compelling evidence for any more radical punctuation, such as when one adaptive change sets off a cascade of further changes [cf. "genetic revolutions" (Mayr, 1954; Templeton, 1980) and "epochal mutations" (Kubitschek, 1974)]. Such an effect might have been manifest by a period of renewed, rapid evolutionary change in a population that had previously been at or near stasis. Perhaps 12 populations and 10,000 generations were too few to see such rare events.

If environmental change and the dynamics of selection caused the initially rapid changes in morphology and fitness, then the eventual stasis (or near stasis) presumably resulted from the constancy of the environment and a paucity of mutations that would produce further improvements comparable to those seen earlier in the experiment. The trajectories for cell size and fitness are superficially similar to data from other experiments, wherein stasis results from a depletion of the genetic variation present in a base population (Falconer, 1983). However, this interpretation is not relevant to our experiment, because there was no initial variation and mutation is an ongoing process. One might also suggest that the eventual stasis in cell size was a consequence of stabilizing selection, although it is not clear why selection would stabilize the replicate populations at different sizes. And the idea that there might be stabilizing selection for intermediate fitness is an evolutionary oxymoron. Hence, the most reasonable interpretation for the eventual stasis in our experimental populations is that the organisms have "run out of ways" to become much better adapted to their environment. Either further major improvements (with fitness increments of more than a few percent in this environment) do not exist or else they are evolutionarily inaccessible (e.g., adaptations requiring multiple genetic changes in which the intermediate states are unfit).

A common pattern in the fossil record is that periods of rapid change in a lineage from its ancestral state (anagenesis) are also periods of rapid diversification (cladogenesis). This association is a central issue in the debate over the theory of punctuated equilibrium (Eldredge and Gould, 1972; Charlesworth *et al.*, 1982; Levinton, 1988; Gould and Eldredge, 1993). In our experiment, rates of anagenetic and cladogenetic evolution were tightly coupled, with the most rapid anagenetic change and diversification both occurring shortly after the populations were introduced into the experimental environment. Thus, our experiment recreates one of the major features taken as evidence for the theory of punctuated equilibrium. We believe the explanation for this concordance of anagenetic and cladogenetic rates is that environmental change (i.e., introduction of the study organism into the experimental environment) radically perturbed the adaptive landscape. This perturbation precipitated rapid adaptive evolution, while diversification resulted from the stochastic effects of mutation and drift that pushed replicate populations into the domains of attraction of different adaptive slopes and fitness peaks. [Diversification would presumably be even more pronounced if the populations were not only isolated but in different environments (Bennett *et al.*, 1992).] This explanation is similar to that put forward by Charlesworth *et al.* (1982, p. 482): "Ecological opportunities offered by the opening of new niches, either by changes in the environment or by the evolution of a key adaptation, will [in the] classical view, generate an association between rapid morphological evolution and the proliferation of species."

Coda

> [Experiments] may reveal what happens to a hundred rats in the course of ten years under fixed and simple conditions, but not what happened to a billion rats in the course of ten million years under the fluctuating conditions of earth history. (Simpson, 1944, p. xvii)

We acknowledge the severe limitations inherent in our study of evolutionary dynamics. Foremost among these are the short time span (even 10,000 generations is but a brief moment in evolution) and the simple environment (which ignores the complexity and changeability of nature). The former limitation reflects our lack of access to better machines for time travel, and the latter our desire as experimentalists to keep things simple enough that we may understand the results. In terms of these limitations, we are certainly studying the tempo and mode of microevolution.

But our studies begin to explore macroevolution, insofar as we address the repeatability of adaptation, the diversification of lineages, and thus the causes and consequences of the uniqueness of evolutionary history. Moreover, it is important that, in our experimental system (unlike other model systems more widely used in population genetics), all of the genetic variation available to selection is generated *de novo*, by mutation, during the course of the experiment. Thus, our study is concerned with the origin of novelties as well as their fate. And although our experiments are small in the evolutionary scheme of things, their duration and size are nonetheless such that each population explores literally millions of genetic changes and at least several of these changes eventually go to fixation (or near fixation). Therefore, our experiment reflects more complex and encompassing evolutionary dynamics than studies of responses to selection that depend either on quantitative variation already present in a population or on a single allele of major effect.

In fact, we observed several hallmarks of macroevolutionary dynamics, including periods of rapid evolution and stasis, altered functional relationships between traits, and concordance of anagenetic and cladogenetic trends. For now, the generality of our results remains an open question: one might well wonder what outcomes would be observed with a sexual organism, with larger or smaller population sizes, with different population structures, or with a more complex environmental regime. We believe that, with appropriate experimental systems and designs, all of these questions can be rigorously addressed.

The differences between microevolution and macroevolution are ones of spatial and temporal scale, of course, but they are also more than that. Microevolution deals primarily with the fundamental "laws" of evolution (the processes of selection, drift, mutation, recombination and migration), whereas macroevolution typically focuses on the uniqueness (the accidental nature) of evolutionary history. But we would argue that the uniqueness of evolutionary history is itself amenable to careful experimental analysis, and that this uniqueness may be an inevitable consequence of the "laws" of microevolution.

SUMMARY

We followed evolutionary change in 12 populations of *Escherichia coli* propagated for 10,000 generations in identical environments. Both morphology (cell size) and fitness (measured in competition with the ancestor) evolved rapidly for the first 2000 generations or so after the populations were introduced into the experimental environment, but both were nearly static for the last 5000 generations. Although evolving in identical environments, the replicate populations diverged significantly from one another in both morphology and mean fitness. The divergence in mean fitness was sustained and implies that the populations have approached different fitness peaks of unequal height in the adaptive landscape. Although the experimental time scale and environment were microevolutionary in scope, our experiments were designed to address questions concerning the origin as well as the fate of genetic and phenotypic novelties, the repeatability of adaptation, the diversification of lineages, and thus the causes and consequences of the uniqueness of evolutionary history. In fact, we observed several hallmarks of macroevolutionary dynamics, including periods of rapid evolution and stasis, altered functional relationships between traits, and concordance of anagenetic and cladogenetic trends. Our results support a Wrightian interpretation, in which chance events (mutation and drift) play an important role in adaptive evolution, as do the complex genetic interactions that underlie the structure of organisms.

This research is supported by a grant from the National Science Foundation to R.E.L. (DEB-9249916) and by the National Science Foundation Science and Technology Center for Microbial Ecology (BIR-9120006). We thank S. C. Simpson for excellent technical assistance; A. F. Bennett, D. E. Dykhuizen, D. L. Hartl, R. Korona, M. H. Lenski, M. R. Rose, P. D. Sniegowski, D. R. Taylor, and F. Vasi for helpful comments and discussion; and F. J. Ayala and W. M. Fitch for organizing a stimulating conference.

REFERENCES

Barton, N. H. and Hewitt, G. M. (1989) Adaptation, speciation and hybrid zones. *Nature (London)* **341,** 497–503.

Barton, N. H. and Rouhani, S. (1987) The frequency of shifts between alternative equilibria. *J. Theor. Biol.* **125,** 397–418.

Bennett, A. F., Lenski, R. E. and Mittler, J. E. (1992) Evolutionary adaptation to temperature. I. Fitness responses of *Escherichia coli* to changes in its thermal environment. *Evolution* **46,** 16–30.

Charlesworth, B., Lande, R. and Slatkin, M. (1982) A neo-Darwinian commentary on macroevolution. *Evolution* **36,** 474–498.

Drake, J. W. (1991) A constant rate of spontaneous mutation in DNA-based microbes. *Proc. Natl. Acad. Sci. USA* **88,** 7160–7164.

Eldredge, N. and Gould, S. J. (1972) Punctuated equilibria: an alternative to phyletic gradualism. In *Models in Paleobiology*, ed. Schopf, T. J. M. (Freeman, San Francisco), pp. 82–115.

Falconer, D. S. (1983) *Introduction to Quantitative Genetics* (Longman, London), 2nd Ed., pp. 194–206.

Gould, S. J. and Eldredge, N. (1993) Punctuated equilibrium comes of age. *Nature (London)* **366,** 223–227.

Kohn, L. A. P. and Atchley, W. R. (1988) How similar are genetic correlation structures? Data from mice and rats. *Evolution* **42,** 467–481.

Kubitschek, H. E. (1974) Operation of selective pressure on microbial populations. *Symp. Soc. Gen. Microbiol.* **24,** 105–130.

Lande, R. (1975) The maintenance of genetic variability by mutation in a polygenic character with linked loci. *Genet. Res.* **26,** 221–234.

Lande, R. (1979) Quantitative genetic analysis of multivariate evolution, applied to brain:body size allometry. *Evolution* **33,** 402–416.

Lenski, R. E. and Bennett, A. F. (1993) Evolutionary response of *Escherichia coli* to thermal stress. *Am. Nat.* **142,** S47–S64.

Lenski, R. E., Rose, M. R., Simpson, S. C. and Tadler, S C. (1991) Long-term experimental evolution in *Escherichia coli*. I. Adaptation and divergence during 2,000 generations. *Am. Nat.* **138,** 1315–1341.

Levinton, J. (1988) *Genetics, Paleontology and Macroevolution* (Cambridge Univ. Press, New York).

Mandel, J. (1964) *The Statistical Analysis of Experimental Data* (Wiley, New York), pp. 288–292.

Mani, G. S. and Clarke, B. C. (1990) Mutational order: a major stochastic process in evolution. *Proc. R. Soc. London B* **240,** 29–37.

Mayr, E. (1954) Change of genetic environment and evolution. In *Evolution as a Process*, eds. Huxley, J., Hardy, A. C. and Ford, E. B. (Allen and Unwin, London), pp. 157–180.

Monod, J. (1971) *Chance and Necessity* (Knopf, New York).

Muller, H. J. (1932) Some genetic aspects of sex. *Am. Nat.* **66,** 118–138.

Simpson, G. G. (1944) *Tempo and Mode in Evolution* (Columbia Univ. Press, New York).

Sokal, R. R. and Rohlf, F. J. (1981) *Biometry*, 2nd Ed. (Freeman, San Francisco).

Templeton, A. R. (1980) The theory of speciation via the founder principle. *Genetics* **94,** 1011–1038.

Travisano, M. (1993) Adaptation and divergence in experimental populations of the

bacterium *Escherichia coli*: the roles of environment, phylogeny and chance. Dissertation (Michigan State Univ., East Lansing, MI).
Turelli, M. (1988) Phenotypic evolution, constant covariances, and the maintenance of additive variance. *Evolution* **42,** 1342–1347.
Vasi, F., Travisano, M. and Lenski, R. E. (1994) Long-term experimental evolution in *Escherichia coli*. II. Changes in life history traits during adaptation to a seasonal environment. *Am. Nat.* **144,** 432–456.
Wade, M. J. and Goodnight, C. J. (1991) Wright's shifting balance theory: an experimental study. *Science* **253,** 1015–1018.
Wright, S. (1932) The roles of mutation, inbreeding, crossbreeding and selection in evolution. *Proc. 6th Int. Congr. Genet.* **1,** 356–366.
Wright, S. (1982) Character change, speciation, and the higher taxa. *Evolution* **36,** 427–443.
Wright, S. (1988) Surfaces of selective value revisited. *Am. Nat.* **131,** 115–123.

14

Explaining Low Levels of DNA Sequence Variation in Regions of the *Drosophila* Genome with Low Recombination Rates

RICHARD R. HUDSON

Recent developments in biotechnology have led to a huge burst in available DNA sequence information. DNA sequences from a large number of loci from a large number of taxa are currently available and more sequences are becoming available. With this information, much has been learned about the rates of molecular evolution in different taxa, at a variety of loci, and at different kinds of sites in the genome. Although the territory is large and much remains to be explored, some descriptive generalizations are now possible. In contrast, the population genetic processes underlying this molecular evolution remain almost entirely obscure. In other words, quite a bit is known about the tempo of molecular evolution, but very little is known about the mode. Understanding the population genetic forces that are most important is made exasperatingly difficult by the fact that very small selective effects, much too small to be directly measured, can be the determining factor in evolution. As a consequence, some effort has been devoted to making inferences about the evolutionary process by indirect means such as analyzing the patterns of molecular divergence and polymorphism in a variety of taxa and loci. Excellent recent reviews of empirical and theoretical aspects of these issues are available (Kimura, 1983; Gillespie, 1991). I will describe here some recently collected data and some efforts to make inferences from that data about underlying

Richard R. Hudson is professor of ecology and evolutionary biology at the University of California, Irvine.

population genetic processes. Several hypotheses and population genetic models that may account for the data will be discussed. Most of the discussion will concern a "background selection" model that may account for some important aspects of the data.

The Data

The data that will be considered here are primarily DNA polymorphism data from a number of loci primarily in *Drosophila melanogaster* and *Drosophila simulans*. The salient feature of these data, the focus of this paper, is the following: Regions of the *Drosophila* genome with low rates of recombination per base pair exhibit low levels of polymorphism within populations (Aguadé *et al.*, 1989; Stephan and Langley, 1989; Berry *et al.*, 1991; Begun and Aquadro, 1992; Martín-Campos *et al.*, 1992; Stephan and Mitchell, 1992; Langley *et al.*, 1993). A summary of these data that display the remarkable correlation between recombination rates and levels of polymorphism is given by Begun and Aquadro (1992) and Aquadro *et al.*, (1994). In the following paragraphs, three hypotheses to account for these data will be described. They are (*i*) a strictly neutral hypothesis, (*ii*) a hitchhiking with selective sweeps of advantageous mutations hypothesis, and (*iii*) a background selection of deleterious mutations hypothesis.

A Strictly Neutral Hypothesis

A very simple though interesting hypothesis to explain the correlation of recombination rates with polymorphism levels is a completely neutral one: Regions of low recombination might have low levels of polymorphism because they have low neutral mutation rates. These low neutral mutation rates in regions of low recombination might result from high average levels of constraint in those parts of the genome or because the spontaneous mutation rates are low there. Either of these possibilities would be interesting and surprising, since there is no *a priori* reason to suspect that mutation rates are lower or constraints are higher in regions of low recombination. Fortunately, there is a simple and powerful way to test this strict neutral interpretation: Examine the levels of divergence between species in these regions of high and low recombination. Under our neutral interpretation, regions of low recombination ought to have low levels of divergence compared to the levels of divergence in regions of high recombination, because the rate divergence under the neutral model is equal to the neutral mutation rate. The data in this regard are quite clear. Regions of low recombination are not diverging more slowly between species than are regions of high recombination (Berry *et al.*,

1991; Begun and Aquadro, 1992; Martín-Campos *et al.*, 1992; Langley *et al.*, 1993). We therefore reject our strict neutral hypothesis, concluding that regions of low recombination do not have low levels of variation because they have low neutral mutation rates.

Hitchhiking with Favorable Mutations Hypothesis

Another hypothesis that has been proposed to explain the pattern of sequence variation is the hitchhiking or "selective sweeps" model (Maynard Smith and Haigh, 1974; Kaplan *et al.*, 1989; Stephan *et al.*, 1992; Wiehe and Stephan, 1993). Under this model, the low levels of polymorphism in regions of low recombination are due to the hitchhiking effect of selectively advantageous mutants that sweep through the population and, in the process, eliminate variation at tightly linked sites. In regions of low recombination, large chunks of DNA are swept to fixation by such selection events, whereas in regions of high recombination, only small chunks are swept to fixation. If such selection events are steadily occurring in both high and low recombination regions, the result will be a lower steady-state level of variation in regions of low recombination. If this selective sweeps interpretation is correct, it should be possible to estimate some of the parameters of the population genetic process involved. Indeed, Wiehe and Stephan (1993) have developed a method to estimate an important rate parameter from the patterns of reduced variation. (They assume that most of the variation that is seen is neutral but that the levels of variation are, in some cases, strongly affected by occasional selective sweeps.) The development of this estimation method is quite significant, demonstrating an additional way in which inferences about the mode of molecular evolution can be made from patterns of polymorphism and divergence. Further work is clearly warranted to investigate other properties of this model and to assess the robustness of the rate estimates. Although there remains much to be investigated, this model can account for some important features of the data.

Background Selection Model

Recently, a quite different hypothesis, referred to as the background selection model, has been proposed to account, at least in part, for the low levels of variation in regions of low recombination (Charlesworth *et al.*, 1993). In this model, as with the selective sweeps model, one focuses on the level of neutral variation that will be maintained at a locus embedded in the midst of a large number of other loci at which mutations can occur that are not selectively neutral. Figure 1*A* illustrates

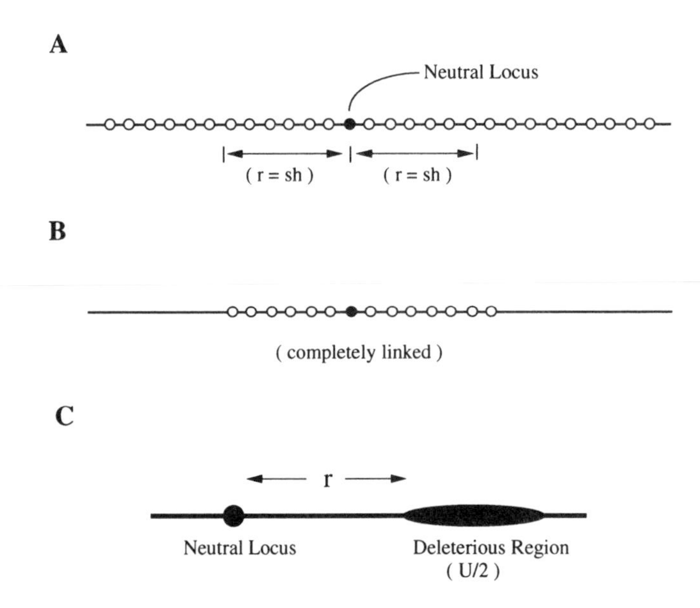

FIGURE 1 Three models considered in the analysis of the background selection. In each case, a neutral locus (indicated by a solid circle) is linked to other loci at which deleterious mutations can occur. In models A and B, the loci at which deleterious mutations can occur are indicated by the open circles. In model A, recombination can occur between any of the indicated loci. The rate of recombination (r) between the neutral locus and the site indicated by the arrows is assumed to equal sh. (sh is defined to be the decrement in fitness produced by a single heterozygous deleterious mutation; that is, the relative fitness of an individual heterozygous for a single deleterious mutation is assumed to be $1 - sh$.) In model B, all loci are completely linked. In model C, the rate of recombination between the neutral locus and the deleterious region is assumed to equal r. The total deleterious mutation rate in the deleterious region is assumed to equal $U/2$.

this model. (The parameters r and sh indicated on the figure will be defined later.) The locus at which neutral variation is being monitored will be referred to as the neutral locus. The central question, at least initially, will be, To what extent is the mean level of variation at the neutral locus affected by the natural selection that occurs at the linked loci. I will quantify this effect of selection at linked loci by the ratio $R = \pi/\pi_0$, where π is the expected nucleotide diversity at the neutral locus under the model with selection at linked loci and π_0 is the expected nucleotide diversity at the neutral locus in the absence of selection at linked loci. [Nucleotide diversity is a commonly employed measure of the amount of DNA polymorphism in a sample (Chapter 10 of Nei,

1987)]. Under the selective sweep model, the mutations occurring at the linked loci are favorable mutations that more or less regularly sweep through the population. In contrast, under the background selection model, the mutations that affect fitness are all assumed to be deleterious.

In a sufficiently large population, recurrent deleterious mutation and natural selection lead to an equilibrium, with mutation continually adding deleterious alleles and natural selection removing them. For a particular region of the genome, the population at this equilibrium can be characterized by a vector, $\mathbf{f} = (f_0, f_1, f_2, \ldots)$, in which f_i is the frequency of chromosomes in the population that carry i deleterious mutations in the region being considered. The equilibrium value of $\mathbf{f}$ will depend on the mutation rate and the fitness of genotypes with different numbers of mutations. For the case where all loci are completely linked, as indicated in Figure 1*B*, the ratio R is equal to f_0, the frequency of chromosomes that carry no deleterious mutations (Charlesworth *et al.*, 1993). For a wide range of mutation rates and selection coefficients, the effect of background selection is simply to reduce the effective population size from N to f_0N. For at least one model, an explicit expression for f_0 has been obtained. Namely, if the deleterious effect of a single mutation in heterozygous state is sh and, further, if an individual heterozygous at i loci has relative fitness $(1 - sh)^i$, then the equilibrium value of f_0 is $\exp(-U/2sh)$, where $U/2$ is the total (haploid) deleterious mutation rate at the completely linked loci (Kimura and Maruyama, 1966; Crow, 1970). Thus for this model and without recombination, the effect of background selection on levels of variation can be calculated.

With recombination, as in the model of Figure 1*A*, it is difficult to obtain analytical results, and a simpler model, shown in Figure 1*C*, is used to gain insight initially. This simpler model with a neutral locus linked to a deleterious region with recombination rate, r, between them has been analyzed using a coalescent approach (Hudson and Kaplan in press). Except in special cases, simple analytic expressions are not obtained, but numerical results are easily obtained. The effect of recombination is shown in Figure 2, for two cases. The important properties are depicted in this figure: (*i*) if $r \ll sh$, the model behaves as if the deleterious region and the neutral locus were completely linked, that is $R = f_0 = \exp(-U/2sh)$, and (*ii*) if $r \gg sh$, then it is as if the deleterious region were not there—that is, $R = 1$. In addition, the transition from one situation to the other occurs fairly quickly, as r is increased. This suggests an approximation. Perhaps the effect of background selection in the difficult model of Figure 1*A* is approximately the same as for the model shown in Figure 1*B*, in which all loci more than sh recombination units away have been removed, and all the other loci

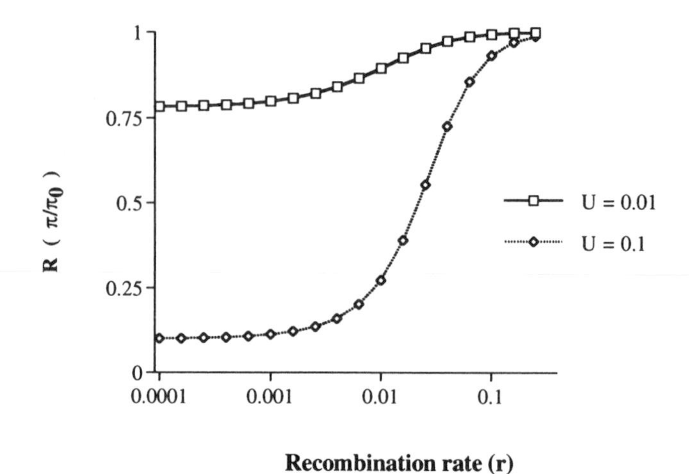

FIGURE 2 R, the ratio of nucleotide diversity with background selection to nucleotide diversity without background selection, is plotted as a function of r, the recombination rate between the neutral locus and the deleterious region. The model assumed is that depicted in Figure 1C. For these calculations, $N = 3200$ and $sh = 0.02$.

are assumed completely linked to the neutral locus. If this conjecture is correct, with the arrangement of loci depicted in Figure 1A, and if the multiplicative interaction model is assumed, the value of R would be

$$R = f_0 = \exp(-U_e/2sh), \tag{1}$$

where $U_e/2$ is the rate of deleterious mutation in the region within sh recombination units on either side of the neutral locus.

Now, some implications of this tentative approximation for patterns of variation seen in *Drosophila* can be explored. Since recombination usually occurs only in females in *Drosophila*, the effective recombination rate over many generations for autosomal genes is one-half the rate measured in females. Thus, in *Drosophila,* U_e would be the total deleterious mutation rate for loci within $2sh \times 100$ centimorgans of the neutral locus. If we denote the deleterious mutation rate of the entire diploid genome by U_T and the fraction of genome within z centimorgans of a locus by $F(z)$ and if the deleterious mutation rate is more or less constant per base pair across the genome, then U_e can be written as $F(200sh)U_T$, and substituting into equation **1**,

$$R = \exp[-F(200sh)U_T/2sh]. \tag{2}$$

U_T in *Drosophila* has been estimated as approximately 1.0 (Mukai *et al.*, 1972; Houle *et al.*, 1992) and sh has been estimated to be 0.02 (Crow and

Simmons, 1983). Hence, R can be written as $\exp[-F(4)/0.04]$. To predict R for any locus, one need only determine $F(4)$, the fraction of the genome within 4 centimorgans of the locus. From published genetic and cytological maps, $F(z)$ can be estimated for many loci. For example, consider the *cta* locus (located at cytological position 40F) near the base of the second chromosome. A region 4 centimorgans in each direction from this locus extends from cytological position 35C to about 44E, which is about 9% of the genome. That is, $F(4)$ for the *cta* locus is about 0.09. Therefore, we expect R for *cta* to be about $\exp(-0.09/0.04) = 0.11$. For *adh* located at position 35B, the region within 4 centimorgans of *adh* extends from about 33A to 37B, or about 4% of the genome. Therefore, we expect R for *adh* to be $\exp(-0.04/0.04) = 0.37$.

Note that in a region with uniform recombination rates, $F(200sh)$ will be proportional to sh, the constant of proportionality being the fraction of the genome per 400 centimorgans in the region. Consider a neutral locus in a large region with uniform recombination rate. Let r_{bp} denote the recombination rate per base pair in females in the neighborhood of the neutral locus. The number of base pairs within $400sh$ centimorgans of neutral locus is $4sh/r_{\text{bp}}$. By using the fact that the haploid genome is about 3×10^8 base pairs in size, $F(200sh) = 4sh/(r_{\text{bp}} \times 3 \times 10^8)$. By substituting this into equation **2** and assuming that $U_{\text{T}} = 1.0$, one obtains

$$R = \exp(-6.7 \times 10^{-9}/r_{\text{bp}}). \tag{3}$$

Note that the parameter sh has canceled out. It should be emphasized that this simplified expression is only applicable if the recombination rate per base pair is roughly constant throughout the region $200sh$ centimorgans on each side of the neutral locus. This will clearly not be true for loci near the tips and centromeres of chromosomes. If the recombination rate per base pair is 10^{-8} [a typical rate of recombination for regions away from tips and bases of chromosomes (Chovnick *et al.*, 1977)], then $R = \exp(-0.67) = 0.5$. Thus, even in regions of normal recombination, it appears that background selection could have a substantial effect on standing levels of neutral variation.

Further support for the idea that deleterious mutations are reducing variation in regions of normal recombination is obtained from the observation of transposable element insertion polymorphism from the *adh* region of *Drosophila melanogaster*. In a restriction site polymorphism survey of a 15-kb region including the *adh* locus (Aquadro *et al.*, 1986), it was found that 20% of their chromosomes carried one or more transposable-element insertions. Particular insertions were almost all rare and several arguments suggest that these insertions are destined to be lost due to some form of natural selection (Golding, 1987). If these transposable-element insertions are indeed deleterious, then the background selection operating on these transposable elements should by itself reduce the level of variation in the *adh* region by 20%. Other transposable elements at somewhat larger distances from *adh* should also have an effect unless the selection against them is very weak. Other forms of deleterious mutations are also presumably occurring in this region and would make R even smaller.

To analyze X chromosome-linked loci, slightly modified equations are required for f_0 (for the no-recombination case, see Charlesworth *et al.*, 1993). Using these modified equations and methods analogous to those described above, one can calculate R values for loci at the tip and base of the X chromosome for which data are available. The predicted R, obtained in this way, for the yellow-achaete-scute region is 0.7. This value is considerably above the very low levels of variation actually observed in this region. Similarly, the low levels of polymorphism observed on the fourth chromosome of *Drosophila melanogaster* are not predicted by the background selection model. These are the same conclusions reached by Charlesworth *et al.*, (1993) using slightly different methods.

In summary the background selection model predicts substantial reductions in polymorphism in some regions of the genome. However, the extreme reductions observed in the tip and base of the X chromosome and the fourth chromosome are not predicted with the versions of the model currently analyzed and the best available estimates of the parameters.

Discussion and Conclusion

The hitchhiking model appears to be able to account for the major features of the data. Current models of background selection do not appear to be able to account for the very large reductions in polymorphism levels observed in some regions of the *Drosophila* genome. However, it appears quite probable that background selection does have a substantial effect on some loci and analyses of observed patterns of variation need to take it into consideration. Other aspects of the data such as the frequency spectrum of the observed polymorphisms and the patterns of geographic variation are beginning to be analyzed and may shed additional light on the underlying process (Stephan and Mitchell, 1992; Begun and Aquadro, 1993).

These analyses of molecular divergence and polymorphism illustrate some of the difficulties and the promise of indirect analysis of molecular data. To make progress toward understanding the population genetic processes that produce the patterns of molecular polymorphism and the patterns of divergence observed requires a continual interplay between collection and analysis of informative data sets and the consideration of appropriate models. There is clearly a long way to go on the road to understanding the mode of molecular evolution.

SUMMARY

Different regions of the *Drosophila* genome have very different rates of recombination. For example, near centromeres and near the tips of chromosomes, the rates of recombination are much lower than in other regions. Several surveys of polymorphisms in *Drosophila* have now documented that levels of DNA polymorphism are positively correlated with rates of recombination; i.e., regions with low rates of recombination tend to have low levels of DNA polymorphism within populations of *Drosophila*. Three hypotheses are reviewed that might account for these observations. The first hypothesis is that regions of low recombination have low neutral mutation rates. Under this hypothesis between-species divergences should also be low in regions of low recombination. In fact, regions of low recombination have diverged at the same rate as other regions of the genome. On this basis, this strictly neutral hypothesis is rejected. The second hypothesis is that the process of fixation of favorable mutations leads to the observed correlation between polymorphism and recombination. This occurs via genetic hitchhiking, in which linked regions of the genome are swept along with the selectively favored mutant as it increases in frequency and eventually fixes in the population. This hitchhiking model with fixation of favorable mutations is compatible with major features of the data. By assuming this model is correct, one can estimate the rate of fixation of favorable mutations. The third hypothesis is that selection against continually arising deleterious mutations results in reduced levels of polymorphism at linked loci. Analysis of this background selection model shows that it can produce some reduction in levels of polymorphism but cannot explain some extreme cases that have been observed. Thus, it appears that hitchhiking of favorable mutations and background selection against deleterious mutations must be considered together to correctly account for the patterns of polymorphism that are observed in *Drosophila*.

REFERENCES

Aguadé, M., Miyashita, N. & Langley, C. H. (1989) Reduced variation in the *yellow-achaete-scute* region in natural populations of *Drosophila melanogaster*. *Genetics* **122,** 607–615.

Aquadro, C. F., Begun, D. J. & Kindahl, E. C. (1994) Selection, recombination, and DNA polymorphism in *Drosophila*. In *Alternatives to the Neutral Model,* ed. Golding, G. B. (Chapman & Hall, New York), pp. 46–56.

Aquadro, C. F., Deese, M. M., Bland, C. H., Langley, C. H. & Laurie-Ahlberg, C. C. (1986) Molecular population genetics of the alcohol dehydrogenase gene region of *Drosophila melanogaster*. *Genetics* **114,** 1165–1190.

Begun, D. J. & Aquadro, C. F. (1992) Levels of naturally occurring DNA polymorphism correlate with recombination rates in *D. melanogaster*. *Nature (London)* **356,** 519–520.

Begun, D. J. & Aquadro, C. F. (1993) African and North American populations of *Drosophila melanogaster* are very different at the DNA level. *Nature (London)* **365,** 548–550.

Berry, A. J., Ajioka, J. W. & Kreitman, M. (1991) Lack of polymorphism on the Drosophila fourth chromosome resulting from selection. *Genetics* **129,** 1111–1117.

Charlesworth, B., Morgan, M. T. & Charlesworth, D. (1993) The effect of deleterious mutations on neutral molecular variation. *Genetics* **134,** 1289–1303.

Chovnick, A., Gelbart, W. & McCarron, M. (1977) Organization of the *Rosy* locus in *Drosophila melanogaster*. *Cell* **11,** 1–10.

Crow, J. F. (1970) in *Mathematical Topics in Population Genetics,* ed. Kojima, K.-I. (Springer, Berlin), pp. 128–177.

Crow, J. F. & Simmons, M. J. (1983) in *The Genetics and Biology of Drosophila,* eds. Ashburner, M., Carson, H. L. & Thompson, J. N. (Academic, London), pp. 1–35.

Gillespie, J. H. (1991) *The Causes of Molecular Evolution* (Oxford Univ. Press, New York).

Golding, G. B. (1987) The detection of deleterious selection using ancestors inferred from a phylogenetic history. *Genet. Res.* **49,** 71–82.

Houle, D., Hoffmaster, D. K., Assimacopoulos, S. & Charlesworth, B. (1992) The genomic mutation rate for fitness in Drosophila. *Nature (London)* **359,** 58–60.

Hudson, R. R. & Kaplan, N. L. (1994) Gene trees with background selection. In *Alternatives to the Neutral Model,* ed. Golding, G. B. (Chapman & Hall, New York), pp. 140–153.

Kaplan, N. L., Hudson, R. R. & Langley, C. H. (1989) The "hitchhiking effect" revisited. *Genetics* **123,** 887–899.

Kimura, M. (1983) *The Neutral Theory of Molecular Evolution* (Cambridge Univ. Press, Cambridge, U.K.).

Kimura, M. & Maruyama, T. (1966) The mutational load with epistatic gene interactions in fitness. *Genetics* **54,** 1337–1351.

Langley, C. H., MacDonald, J., Miyashita, N. & Aguadé, M. (1993) Lack of correlation between interspecific divergence and intraspecific polymorphism at the *suppressor of forked* region in *Drosophila melanogaster* and *Drosophila simulans*. *Proc. Natl. Acad. Sci. USA* **90,** 1800–1803.

Martín-Campos, J. M., Comeron, J. P., Miyashita, N. & Aguadé, M. (1992) Intraspecific and interspecific variation at the *y-ac-sc* region of *Drosophila simulans* and *Drosophila melanogaster*. *Genetics* **130,** 805–816.

Maynard Smith, J. & Haigh, J. (1974) The hitchhiking effect of a favorable gene. *Genet. Res.* **23,** 23–35.
Mukai, T., Chigusa, S. I., Mettler, L. E. & Crow, J. F. (1972) Mutation rate and dominance of genes affecting viability in *Drosophila melanogaster*. *Genetics* **72,** 335–355.
Nei, M. (1987) *Molecular Evolutionary Genetics* (Columbia Univ. Press, New York).
Stephan, W. & Langley, C. H. (1989) Molecular genetic variation in the centromeric region of the X chromosome in three *Drosophila ananassae* populations. I. Contrasts between the *vermillion* and *forked* loci. *Genetics* **121,** 89–99.
Stephan, W. & Mitchell, S. J. (1992) Reduced levels of DNA polymorphism and fixed between-population differences in the centromeric region of *Drosophila ananassae*. *Genetics* **132,** 1039–1045.
Stephan, W., Wiehe, T. H. E. & Lenz, M. W. (1992) The effect of strongly selected substitutions on neutral polymorphism—analytical results based on diffusion theory. *Theor. Popul. Biol.* **41,** 237–254.
Wiehe, T. H. E. & Stephan, W. (1993) Analysis of a genetic hitchhiking model, and its application to DNA polymorphism data from *Drosophila melanogaster*. *Mol. Biol. Evol.* **10,** 842–854.

15

The History of a Genetic System

ALEKSANDAR POPADIĆ AND WYATT W. ANDERSON

Nineteen forty-four was a special year in the history of evolutionary biology. George Gaylord Simpson published *Tempo and Mode in Evolution* (1944), bringing paleontology and the concepts of macroevolution into the modern synthesis of evolutionary theory. Nineteen forty-four also saw the publication of *Contributions to the Genetics, Taxonomy, and Ecology of Drosophila pseudoobscura and Its Relatives* by Theodosius Dobzhansky and Carl Epling (Dobzhansky and Epling, 1944). This monograph presented a detailed analysis of chromosomal polymorphism for inversions in *D. pseudoobscura* and *Drosophila persimilis,* summarizing earlier studies and presenting extensive new data. These inversions, often referred to as gene arrangements, constitute a genetic system that has played a prominent role in studies of population genetics and evolutionary biology during the past 50 years. In one of the three papers making up the 1944 monograph, Epling reasoned on biogeographical grounds that the distribution pattern of these gene arrangements was ancient, dating from perhaps the Miocene. It follows from Epling's hypothesis that the gene arrangements themselves are ancient. Simpson was one of the principals in a lively correspondence that ensued between Epling and other evolutionary biologists about the age of this genetic system. Simpson, Mayr, and Stebbins published their views as a *Symposium on the Age of the Distribution Pattern of the Gene Arrangements in Drosophila pseudoobscura* in 1945 (Mayr *et al.*, 1945).

Aleksandar Popadić is a Ph.D. candidate in genetics and Wyatt W. Anderson is professor of genetics at the University of Georgia, Athens.

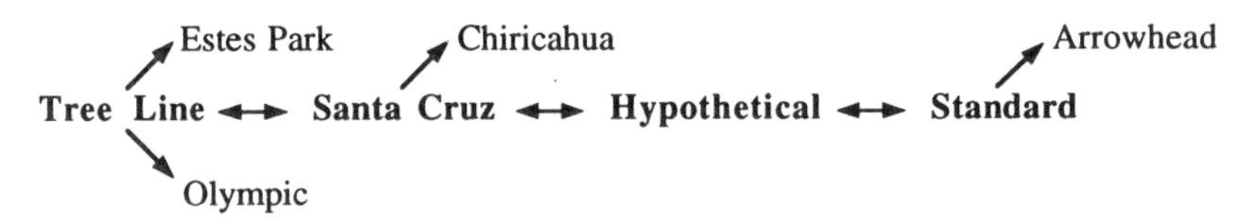

FIGURE 1 Cytogenetic phylogeny of the *D. pseudoobscura* gene arrangements examined in the present study.

Stebbins supported Epling's hypothesis and argued that the inversion system was quite old, while Mayr disagreed with Epling and favored a more recent origin of the distribution pattern. Simpson concluded that the age of the system could not be determined from the available evidence. It thus seems appropriate in this colloquium honoring the 50th anniversary of Simpson's seminal work to address a question about the history of the *D. pseudoobscura* inversions, not quite the question of age that Simpson considered, but rather the related question of which gene arrangement was the ancestral one.

In *D. pseudoobscura*, the third chromosome is polymorphic for more than 40 gene arrangements resulting from overlapping paracentric inversions (Olvera *et al.*, 1979), which can be ordered in a phylogeny based on the breakpoints of inversions under the parsimonious assumption that each inversion arose only once (Figure 1). With the single exception of Hypothetical, all of the gene arrangements necessary to reconstruct the complete phylogeny have been observed in nature. Four of these arrangements—Standard (ST), Hypothetical (HY), Santa Cruz (SC), and Tree Line (TL)—are central to the phylogeny, with all the others being their one- or two-step derivatives. The tree in Figure 1 is unrooted, and the question of which of these four gene arrangements is ancestral has remained unanswered for more than 50 years, inasmuch as cytogenetic, biogeographic, and electrophoretic data have not consistently supported a single hypothesis.

ST was the first arrangement proposed as ancestral (Sturtevant and Dobzhansky, 1936), because it is the only one shared by *D. pseudoobscura* and its sibling species, *Drosophila persimilis*. HY was suggested (Dobzhansky and Sturtevant, 1938) because its inverted region resembles the banding pattern of the homologous chromosome in *Drosophila miranda*, a related species more distant from *D. pseudoobscura* than is *D. persimilis*. Historically, SC has received less attention than the other arrangements, although it too has been suggested (Dobzhansky and Epling, 1944, Brown, 1989). More recently, TL has been considered a favorite candidate for the ancestral gene arrangement on the basis of its distribution pattern (Wallace, 1966; Olvera *et al.*, 1979; Wallace, 1988) and comparison of alleles at protein loci in *D. pseudoobscura*, *D. persimilis*, and *D. miranda* (Olvera *et al.*, 1979) and because it pairs more fully with the *miranda* homolog of the *pseudoobscura* third chromosome in interspecies hybrids than does either SC or ST (Wallace, 1966; Morrow, 1970; Wallace, 1988). The only informative molecular data on this topic come from a recent analysis of restriction site polymorphism (RSP) within these arrangements (Aquadro *et al.*, 1991) that produced two important results. First, the phylogeny based on RSP data corroborates the cytogenetic phylogeny. Second, it was estimated that the TL branch diverged from the SC–ST group about 1.7 million years ago. The greater depth of the TL branch and its early splitting have been used to support the ancestral status of the TL arrangement, which represents the consensus view today (Aquadro *et al.*, 1991; Powell, 1992).

We decided to approach the question of the ancestral arrangement in the following way. First, an independent assessment of the phylogenetic relationships among the central gene arrangements was made on the basis of nucleotide sequences flanking the amylase 1 (*Amy1*) gene, which is located within the inversions. Second, a framework of competing hypotheses regarding the ancestral type was devised on the basis of the known relationships among central arrangements, and each of these hypotheses was compared with the empirically derived phylogenies. Third, several additional chromosomes were added to our earlier RSP data set (Aquadro *et al.*, 1991) and analyzed by several different methods to determine whether this more extensive phylogenetic analysis still supported an ancestral status for TL.

MATERIALS AND METHODS

Lines of *D. pseudoobscura* stocks homozygous for the third chromosome were constructed using balancer stocks (Pavlovsky and Dobzhansky, 1966). Salivary glands were dissected from third-instar larvae, 30 per line, and gene arrangements were diagnosed from squash preparations of the polytene chromosomes. The six strains used for determining DNA sequences flanking *Amy1* are as follows: ST, Ayala reference strain, from northern California; SC, strain BAJA 859#3, from Baja California, Mexico; Chiricahua (CH), strain AH 87#2, from northern California; TL, strain AH 73#2, from northern California; Estes Park (EP), strain BC p430#4, from British Columbia, Canada; Olympic (OL), strain s14AR-D; from British Columbia, Canada. Three SC strains from Michoacan, Mexico, were added to the original RSP data set: strain MEX z67w, strain MEX z53y, and strain MEX z13w. Restriction mapping of these strains was carried out as described previously (Aquadro *et al.*, 1991). In addition to the inversions already identified, the RSP data set includes Arrowhead (AR).

Isolation of total genomic DNA was accomplished by extraction from freshly ground flies and purification by CsCl density gradient centrifugation (Bingham *et al.*, 1981). Genomic DNA was digested with *Hin*dIII and *Eco*RI restriction enzymes and then loaded onto a 5–30% sucrose step gradient (Ausubel *et al.*, 1987). The fraction containing 5- to 6-kb fragments was cloned into the vector pBluescript SK− (Stratagene) and transformed by high-voltage electroporation (Maniatis *et al.*, 1982, Dower *et al.*, 1988). *D. pseudoobscura* clones containing *Amy* homologous sequences were isolated from a genomic library by colony hybridization (Maniatis *et al.*, 1982) using the plasmid pFA4 (Brown *et al.*, 1990) containing the *D. pseudoobscura Amy1* coding sequence as probe. All sequences were determined using an automatic sequencer (Applied Biosystems 373A) following the manufacturer's protocol. In all cases both strands were sequenced. Some regions were sequenced again manually (Sanger *et al.*, 1977) using the Sequenase DNA sequencing kit (United States Biochemical). Sequenced regions include 667 nucleotides (from bases 701 to 35) upstream of the start codon and 391 nucleotides (from bases 62 to 452) downstream of the stop codon. These sequences have been deposited in GenBank (accession numbers U09746–U09757). Sequences were aligned with each other using the GENALIGN program (IntelliGenetics) and checked again by eye. Sequence divergence estimates were calculated as direct counts of nucleotide sequence differences, since no correction is needed for differences as small as those in our study (Nei, 1987).

The phylogenetic analysis was carried out using the neighbor-joining (NJ) and maximum likelihood (ML) methods in the PHYLIP package (Felsenstein, 1993), the NJ method in the MEGA package (Kumar *et al.*, 1993), and the maximum parsimony (MP) method in PAUP (Swofford, 1991). For the RSP data, we excluded strains with a redundant restriction pattern, which reduced the number of strains that were phylogenetically analyzed from 33 to 21. To bootstrap the NJ tree derived from RSP data, we followed the advice kindly given to us by Walter Fitch. First, "1" and "0" in the data set were replaced with "A" and "T," respectively. Second, the SEQBOOT program (PHYLIP) was used to produce 100 bootstrapped data sets. Third, the DNADIST, NEIGHBOR, and CONSENSE programs (PHYLIP) were used in succession to produce the bootstrap values.

RESULTS AND DISCUSSION

A Test of Hypotheses Regarding the Ancestral Gene Arrangement. Amylase in *D. pseudoobscura* is a family of three genes, located within the inverted region in most gene arrangements (Aquadro *et al.*, 1991). The *Amy1* gene

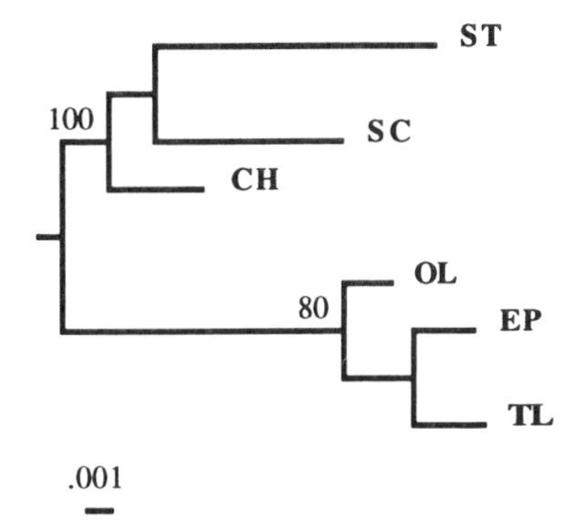

FIGURE 2 Dendrogram based on the combined flanking sequences of the *Amy1* gene, obtained by the NJ algorithm. A separate analysis of the two (5′ and 3′) flanking regions generated the same basic topology. Numbers refer to the percentage of times a node was supported in 200 bootstrap replications.

is the only copy present in all arrangements, and it has been suggested that the *Amy2* and *Amy3* copies arose by duplication of *Amy1* (Brown *et al.*, 1990). In this study, we used only 5′ and 3′ flanking regions specific to the *Amy1* gene.

A phylogenetic analysis of each flanking region generated the same branching topology, so we combined the two regions (Figure 2). In the deduced phylogeny, the TL arrangement splits off early from the SC–ST group, a finding concordant with the results of the previous RSP data analysis (Aquadro *et al.*, 1991). An intuitive interpretation of this phylogeny is that the TL arrangement is ancestral to both SC and ST arrangements. An estimate of sequence divergence (3.2%) between the TL arrangement and the SC–ST group calculated from our sequence data agrees well with that based on RSP data (2.9%). These findings seem to support the ancestral status of TL. What they do not provide is an exclusion of other gene arrangements as potential ancestors. Thus, we are still left with the following questions: Are our results consistent with another arrangement than TL being ancestral? And what about the HY arrangement, which has never been found in nature? Since no data are available for it, the phylogenies in Figure 2 cannot be used to rule out the possibility that HY was in fact ancestral.

To settle these questions, we developed a framework of competing hypotheses based on the assumption that each inversion type is monophyletic in origin, which accords with the RSP data (Aquadro *et al.*, 1991), and the assumption that the relationships among the four central arrangements (Figure 1) are derived parsimoniously, with each inversion arising by two breakpoints from its parental arrangement. Under these assumptions, we successively chose each of the four possible arrangements to represent the ancestral type and asked in each case what the branching pattern of the resulting phylogenetic tree would be.

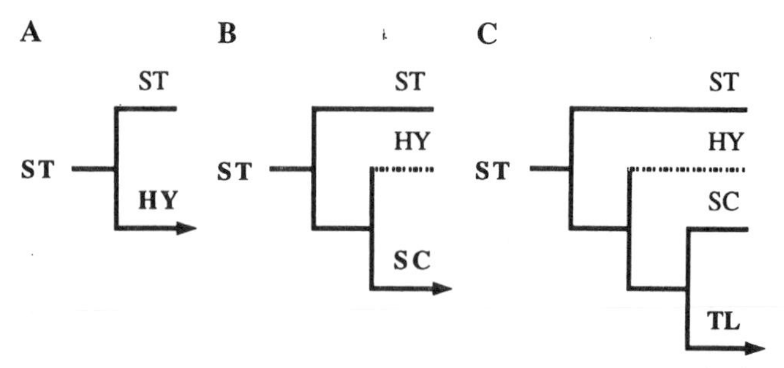

FIGURE 3 An illustration of the phylogeny that would result from three successive inversion events, beginning with the ST gene arrangement. In *A*, an inversion of the ST gene arrangement gives HY; in *B*, an inversion of HY gives SC; and in *C*, an inversion of SC gives TL.

For the reader not familiar with the construction of phylogenies, this reasoning is shown in Figure 3. By applying this approach to each of the four central arrangements, we generated the six phylogenies displayed in Figure 4. Since the ST and TL arrangements are at the ends of the phylogeny, each produces only one tree (Figure 4 *A* and *B*). Because they are located in the middle of the phylogeny, the HY and SC arrangements have two possibilities for the first node, thus producing two possible trees each (Figure 4 *C1*, *C2*, *D1*, and *D2*).

Of the six possible trees, four are incompatible with the empirically obtained tree in Figure 2, which rules out the scenarios based on the ancestral status of ST and HY (Figure 4 *A*, *C1*, and *C2*), as well as the one assuming both that SC is ancestral and that the first node leads to HY (Figure 4*D2*). Although the HY arrangement has never been found in nature, we are still able to exclude it as a potential ancestor. Each of the two remaining trees agrees with the empirical tree. One of them (Figure 4*B*) follows from the assumption that TL is ancestral, while the other (Figure 4*D1*) represents the case where SC is the ancestor and the first node leads to TL. On the basis of this analysis, then, either the TL or the SC arrangement could be ancestral.

Could SC be the Ancestral Arrangement? In the previous section, we used *Amy* flanking regions to generate the empirical phylogeny. These regions appear to be functionally unconstrained, with sequence divergence levels comparable to that of the RSP data. This makes them useful for phylogenetic analysis. At the present moment, however, the lack of information from the outgroup species, *D. miranda*, limits our ability to distinguish between the two candidates (TL and SC) on the basis of sequence data. This leaves the RSP data as the only source of further insight into the ancestral arrangement. Therefore, we decided to reanalyze the data set from the original RSP study with the following additions. First, we added three more SC strains to assure that SC was adequately represented, since the original data set contained only two SC strains. Second, the data were analyzed with the NJ, MP, and ML methods. All three methods have been shown (Jin and Nei, 1991) to be superior in finding the correct tree topology over the unweighted pair-group method of averages (UPGMA) algorithm used previously. It

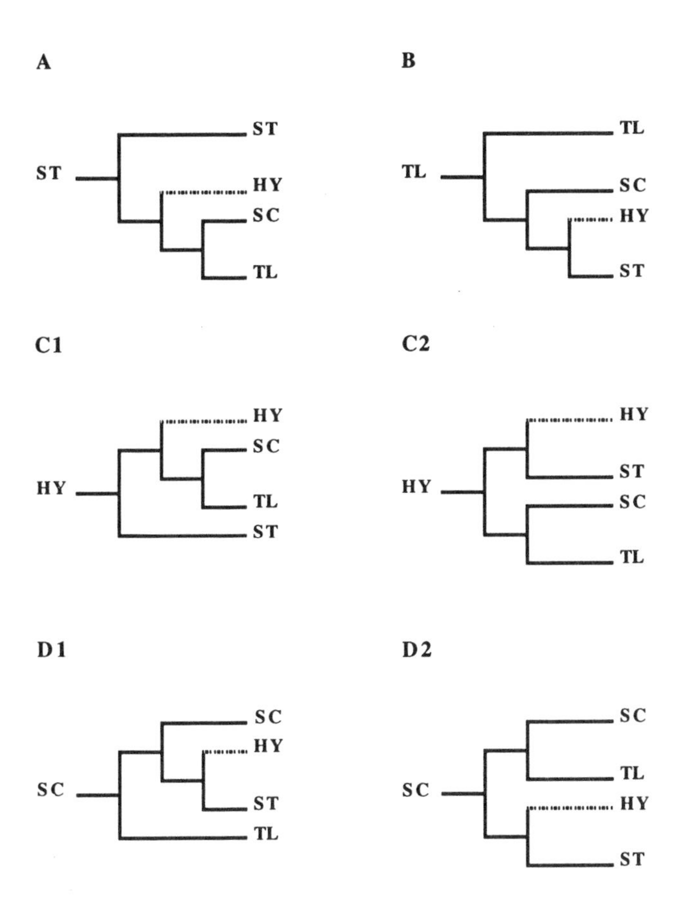

FIGURE 4 The six possible scenarios showing relationships among ST, HY, SC, and TL gene arrangements. In *A*, ST is ancestral, whereas in *B*, TL occupies that position. In *C1* and *C2*, HY is the ancestor and the first bifurcation leads to ST and SC. In *D1* and *D2*, SC is the ancestral type with the first bifurcation leading to TL and HY. Branches leading to HY are indicated by dotted lines, since this gene arrangement has not been found in nature.

is especially instructive to see if the phylogenetic analysis of the enlarged RSP data set also supports an ancestral status for TL.

The NJ tree (Figure 5*A*) shows one SC strain and all of CH strains as being closest to the outgroup (*D. miranda*), with very short branch lengths. The rest of the SC strains join either the TL or the ST arrangements. In contrast, the ST and TL arrangements form separate clades (along with their derivatives). Except for the short branches that are replaced with a multifurcation at the ancestral node, the same tree topology was generated with the ML method (Figure 5*B*). The parsimony analysis was performed both on the complete data set (results not shown) and on selected subsets of the data, two of which are shown (Figure 6*A*, *B*). Again, all of the MP trees show no difference from the previously generated NJ and ML trees. Therefore, the same basic topology was generated with the three different methods, although each method is based on a different set of assumptions. The major characteristic of this common topology is that while the TL and ST arrangements (and their derivatives) form separate clades, the SC arrangements are spread throughout the phylogeny. One SC arrangement is closest to the outgroup

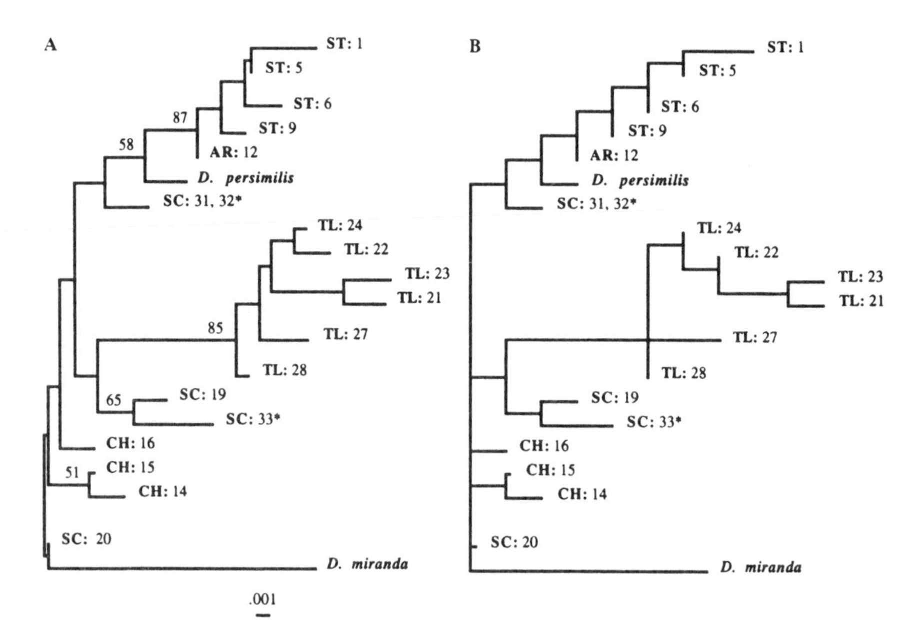

FIGURE 5 Phylogenetic trees inferred from the RSP data. Taxa are labeled as in the previous study (Aquadro *et al.*, 1990), with three new SC strains indicated by an asterisk. (*A*) NJ tree. Numbers refer to the bootstrap values as percentages for 100 replicates. (*B*) Maximum likelihood tree.

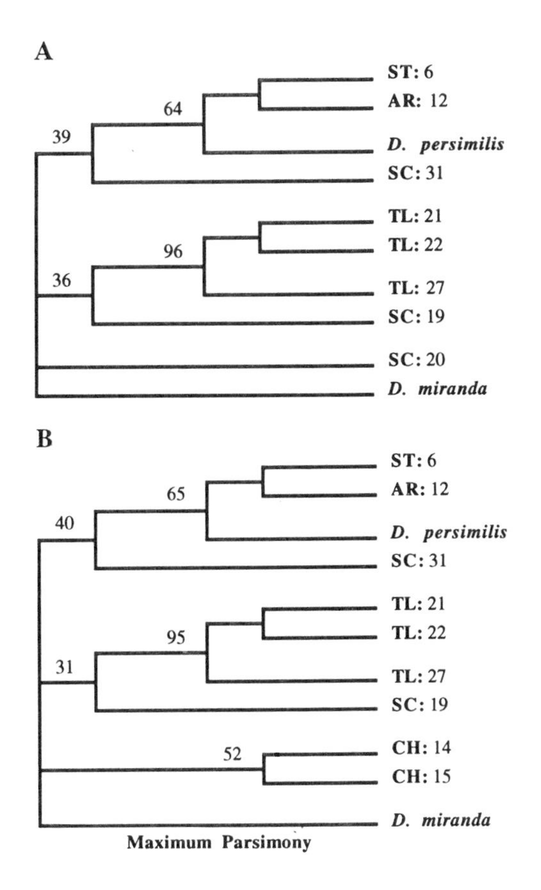

FIGURE 6 MP trees. Numbers refer to parsimony bootstrap values in percentages (out of 100 replicates). Branch lengths are not proportional to the number of nucleotide changes. (*A*) This tree is a strict consensus of 10 trees of length 25. The consistency index was 0.80. (*B*) Strict consensus of 10 trees of length 27, with a consistency index of 0.78.

(according to the NJ and ML methods), while the rest join either the TL or ST clades. Also, the CH arrangements split off from the ancestral node, without joining any of the SC strains. These results are somewhat different than those obtained with the UPGMA method in the previous study (Aquadro *et al.*, 1991). This discrepancy is due partially to the addition of three more SC strains, after which the scattering of SC arrangements throughout the phylogeny becomes obvious, and partially to the varying rate of evolution among different arrangements, combined with the low level of observed divergence (at most 4%). Recently, it has been shown that the UPGMA method performs inadequately under either circumstance, especially for RSP data (Jin and Nei, 1991).

We decided to investigate this matter further by modifying our framework of competing hypotheses so it would take into account all arrangements represented in the RSP data. Two main changes were inclusion of both the outgroup (*D. miranda*) as well as derivatives of the

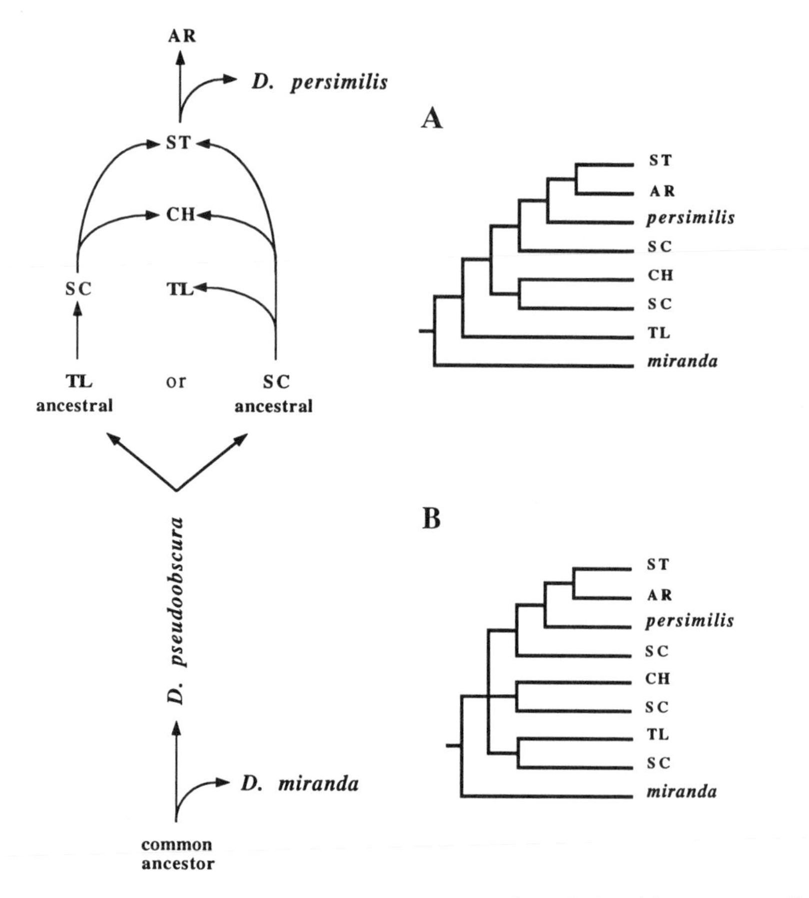

FIGURE 7 The two possible scenarios showing the relationships among all gene arrangements represented in the RSP data set. In *A*, TL is assumed to be ancestral; in *B*, SC is assumed to be ancestral.

SC and ST arrangements. Also, for simplicity, the HY inversion was not included. By following the same sort of reasoning depicted in Figure 4, we derived two trees that assume that either TL or SC was the ancestral gene arrangement (Figure 7*A*, *B*). The main difference between them is the branching pattern at the ancestral node. If TL is the ancestral arrangement, then TL and *D. miranda* should share a common ancestor, whereas SC and *D. miranda* should not. Phylogenetically, this scenario would require TL to branch off immediately after the outgroup. In contrast, the ancestry of SC predicts an unresolved branching pattern, represented with a trifurcation at the ancestral node. This same lack of resolution can be observed in all empirical trees, combined with the dispersal of SC throughout the phylogeny.

Therefore, comparison of the two competing hypotheses with the empirical results is more supportive of an ancestral status for SC. George Gaylord Simpson would probably have been pleased to know that 50 years after he considered the history of gene arrangements in *D. pseudoobscura*, a molecular evolutionary approach has brought us close to an understanding of the origin of this genetic system.

SUMMARY

Although the chromosomal polymorphism for inversions in *Drosophila pseudoobscura* is one of the best studied systems in population genetics, the identity of the ancestral gene arrangement has remained unresolved for more than 50 years. There are more than 40 gene arrangements, and 4 of them (Standard, Hypothetical, Santa Cruz, and Tree Line) have been considered as candidates for the ancestral type. We propose a framework of competing hypotheses to distinguish among the alternatives. Two conclusions come from contrasting each hypothesis with the results from DNA sequencing and restriction mapping. First, not only Standard but also Hypothetical can be excluded as the ancestral gene arrangement. Second, although either Tree Line or Santa Cruz could be the ancestral type, the available data provide greater support for Santa Cruz.

We thank Danijela Popadić for technical assistance and J. C. Avise, F. Ayala, M. Arnold, M. Ball, Y. Fu, J. Hamrick, E. McCarthy, J. McDonald, R. Meagher, and B. Wallace for comments on the manuscript; B. Bowen provided an especially helpful review. We also thank M. T. Clegg, W. M. Fitch, and E. Mayr for helpful suggestions about our analysis. This work was supported in part by a grant from the National Science Foundation (BSR-8516188 to W.W.A.).

REFERENCES

Aquadro, C. F., Weaver, A. L., Schaeffer, S. W. & Anderson, W. W. (1991) Molecular evolution of inversions in *Drosophila pseudoobscura*: the amylase gene region. *Proc. Natl. Acad. Sci. USA* **88,** 305–309.

Ausubel, F. M., Brent R., Kingston, R. E., Moore, D. D., Seidman, J. G., Smith, J. A. & Struhl, K. (1987) *Current Protocols in Molecular Biology* (Wiley/Green, New York).

Bingham, P. M., Levis, R. & Rubin, G. M. (1981) Cloning of DNA sequences from the *white* locus of *D. melanogaster* by a novel and general method. *Cell* **25,** 693–704.

Brown, C. J. (1989) Ph.D. thesis (Univ. of Georgia, Athens).

Brown, C. J., Aquadro, C. F. & Anderson, W. W. (1990) DNA sequence evolution of the amylase multigene family in *Drosophila pseudoobscura*. *Genetics* **126,** 131–138.

Dobzhansky, T. & Sturtevant, A. H. (1938) Inversions in the chromosomes of *Drosophila pseudoobscura*. *Genetics* **23,** 28–64.

Dobzhansky, T. & Epling, C. (1944) Contributions to the genetics, taxonomy, and ecology of *Drosophila pseudoobscura* and its relatives. *Carnegie Inst. Washington Publ.* **554**.

Dower, W. J., Miller, J. F. & Ragsdale, C. W. (1988) High efficiency transformation of *E. coli* cells by high voltage electroporation. *Nucleic Acids Res.* **16,** 6127.

Felsenstein, J. (1993) PHYLIP, Phylogeny Inference Package (Univ. of Washington, Seattle), Version 3.5c.

Jin, L. & Nei, M. (1991) Relative efficiencies of the maximum-parsimony and distance-matrix methods of phylogeny construction for restriction data. *Mol. Biol. Evol.* **8,** 356–365.

Kumar, S., Tamura, K. & Nei, M. (1993) MEGA, Molecular Evolutionary Genetics Analysis (Pennsylvania State Univ., University Park, PA), Version 1.

Maniatis, T., Fritsch, E. F. & Sambrook, J. (1982) *Molecular Cloning: A Laboratory Manual* (Cold Spring Harbor Lab. Press, Plainview, NY).

Mayr, E., Stebbins, G. L. & Simpson, G. G. (1945) *in* Symposium on age of the distribution pattern of the gene arrangements in *Drosophila pseudoobscura*. *Lloydia* **8,** 69–108.

Morrow, D. (1970) M.S. thesis (Cornell Univ., Ithaca, NY).

Nei, M. (1987) *Molecular Evolutionary Genetics* (Columbia Univ. Press, New York).

Olvera, O., Powell, J. R., de la Rosa, M. E., Salceda, V. M., Gaso, M. I., Guzman, J., Anderson, W. W. & Levine, L. (1979) Population genetics of Mexican *Drosophila* VI. Cytogenetic aspects of the inversion polymorphism in *Drosophila pseudoobscura*. *Evolution* **33,** 381–395.

Pavlovsky, O. & Dobzhansky, T. (1966) Genetics of natural populations. XXXVII. The coadapted system of chromosomal variants in a population of *Drosophila pseudoobscura*. *Genetics* **53,** 843–854.

Powell, J. (1992) Inversion polymorphism in *Drosophila pseudoobscura* and *Drosophila persimilis*. In *Drosophila Inversion Polymorphism*, eds. Krimbas, C. & Powell, J. R. (CRC, London), pp. 73–125.

Sanger, F., Nicklen, S. & Coulson, A. R. (1977) DNA sequencing with chain terminating inhibitors. *Proc. Natl. Acad. Sci. USA* **74,** 5463–5467.

Simpson, G. G. (1944) *Tempo and Mode in Evolution* (Columbia Univ. Press, New York).

Sturtevant, A. H. & Dobzhansky, T. (1936) Inversions in the third chromosome of wild races of *Drosophila pseudoobscura*, and their use in the study of the history of the species. *Proc. Natl. Acad. Sci. USA* **22,** 448–450.

Swofford, D. L. (1991) PAUP, Phylogenetic Analysis Using Parsimony (Illinois Nat. Hist. Surv., Champaign, IL), Version 3.0s.

Wallace, B. (1966) *Chromosomes, Giant Molecules, and Evolution* (Norton, New York).

Wallace, B. (1988) In defense of verbal arguments. *Perspect. Biol. Med.* **31,** 201–211.

16

Genome Structure and Evolution in *Drosophila*: Applications of the Framework P1 Map

DANIEL L. HARTL, DMITRY I. NURMINSKY,
ROBERT W. JONES, AND ELENA R. LOZOVSKAYA

A clone-based physical map consists of a set of ordered, overlapping inserts of cloned genomic DNA. Such a map affords a unique resource for studying the structure and function of the genome. The map facilitates the molecular identification of mutant genes, and the clones in the map provide ready access to substrates for genomic sequencing. A clone-based physical map also opens up new opportunities for studies of genome evolution.

Clone-based physical maps have been assembled for several species chosen as model organisms in the Human Genome Project (Collins and Galas, 1993). These include *Escherichia coli* (Kohara *et al.*, 1987), *Saccharomyces cerevisiae* (Olson *et al.*, 1986), *Caenorhabditis elegans* (Coulson *et al.*, 1986), and *Drosophila melanogaster* (Hartl, 1992). There is also a first-generation physical map of the human genome (Cohen *et al.*, 1993). In most organisms, sets of overlapping clones covering an uninterrupted stretch of the genome (contigs) are assembled by detecting overlaps by means of shared restriction fragments in fingerprints or shared sequence-tagged sites (STS). *Drosophila* is unique among model organisms in presenting giant polytene chro-

Daniel L. Hartl is professor of biology and Dmitry I. Nurminsky and Elena R. Lozovskaya are research associates in the Department of Organismic and Evolutionary Biology at Harvard University, Cambridge, Massachusetts. Robert W. Jones is a senior associate in the Department of Biochemistry and Biophysics at Washington University Medical School, St. Louis, Missouri.

mosomes in the larval salivary glands so that clones can be assigned positions in the genome by means of *in situ* hybridization. There are about 5000 polytene bands, most ranging in DNA content from 5 to 50 kilobase pairs (kb), with an average DNA content of ≈20 kb (Heino *et al.*, 1994). The limit of cytological resolution with *in situ* hybridization is approximately 20 kb (Merriam *et al.*, 1991). One advantage of the hybridization approach is that the approximate locations of clones covering much of the genome can be assembled relatively rapidly (yielding a "framework map"), although clones that appear to be adjacent in the framework map need not necessarily overlap at the molecular level. The *in situ* mapping strategy therefore requires that molecular overlaps be determined after the framework map is completed rather than during assembly.

A physical map of the *Drosophila* genome based on yeast artificial chromosomes (YACs) ordered by *in situ* hybridization has been reported previously (Ajioka *et al.*, 1991; Merriam *et al.*, 1991; Hartl, 1992; Hartl and Lozovskaya, 1992; Lozovskaya *et al.*, 1993; Cai *et al.*, 1994). The YAC map includes ≈1200 clones with inserts averaging 200 kb that cover 90% of the euchromatic part of the genome. These clones have been grouped into ≈150 "cytological contigs," averaging 650 kb in extent, based on apparent overlaps detected by means of *in situ* hybridization; the gaps between cytological contigs average ≈50 kb (Hartl, 1992). The YAC clones provide molecular access to much of the *Drosophila* genome, but the vector has a low copy number and it is difficult to separate large quantities of YAC from contaminating yeast genome.

A second-level framework map based on bacteriophage P1 clones has now been assembled and is reported here. The map is based on 2461 clones with insert sizes averaging ≈80 kb that have been ordered by *in situ* hybridization. The P1 map includes an estimated 85% of the sites in the euchromatic genome with an average depth of coverage of 1.8. The mapped P1 clones, as well as 6755 additional clones, are being screened with STS markers in order to be arrayed in molecular contigs. The localizations of P1 clones and other information, as well as sources from which P1 clones may be obtained, are available to the scientific community by anonymous file-transfer protocol from ftp.bio.indiana.edu in the directory flybase or by electronic mail requests to flybase@morgan.harvard.edu.

MATERIALS AND METHODS

Drosophila Strains and Procedures. Drosophila P1 clones were produced from DNA extracted from nuclei of adult *D. melanogaster* flies from an

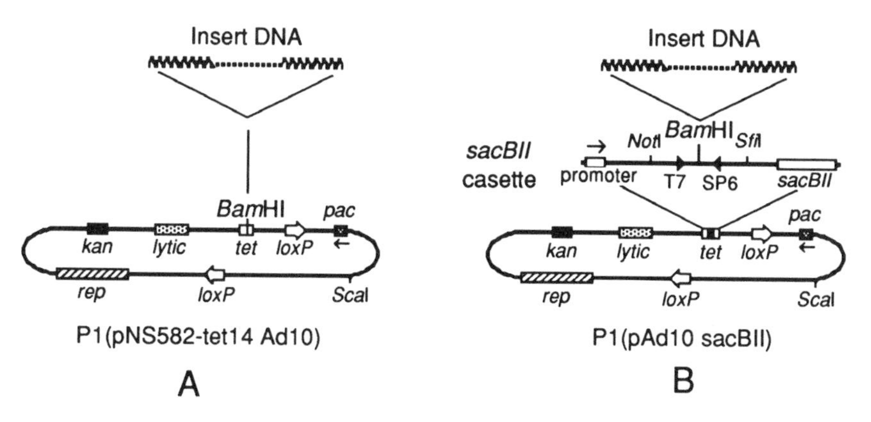

FIGURE 1 Bacteriophage P1 vectors.

isogenic strain of genotype *y; cn bw sp* according to the methods described in Smoller *et al.*, (1991) and Lozovskaya *et al.*, (1993). High molecular weight DNA was partially digested with *Sau*3A1 and fractionated according to molecular weight in sucrose gradients (10–40% sucrose) in order to isolate fragments in the size range 75–100 kb. These fractions were dialyzed, concentrated with 2-butanol, and precipitated in ethanol in preparation for ligation to the P1 vectors.

P1 Cloning Vectors. P1 clones were produced by using the P1 vectors pNS582-tet14 Ad10 (Figure 1*A*) and pAd10 sacBII (Figure 1*B*), which differ in the region around the cloning site. In pNS582-tet14 Ad10 (Sternberg, 1990), the *Bam*HI cloning site interrupts the tetracycline-resistance gene; in pAd10 sacBII (Pierce *et al.*, 1992), the *Bam*HI cloning site is flanked on one side by a T7 promoter, a *Not* I site, and the promoter of the *sacBII* gene of *Bacillus amyloliquefaciens*; and it is flanked on the other side by an SP6 promoter, an *Sfi* I site, and the *sacBII* structural gene for levansucrase. Large fragments of DNA inserted into the cloning site disrupt expression of the *sacBII* gene and thereby allow cells of *E. coli* to survive in medium containing 5% sucrose (Pierce *et al.*, 1992).

Vector arms resulting from digestion of either vector with *Bam*HI and *Sca* I were treated with calf intestinal alkaline phosphatase, and an equimolar ratio of vector-arm DNA and genomic DNA was ligated in the presence of T4 DNA ligase as described (Smoller *et al.*, 1991). Packaging of the ligated DNA was carried out as described in Sternberg (1990). During packaging, molecules are packaged stepwise in the counterclockwise direction from the *pac* site (as Figure 1 is drawn) until the phage head has been filled (100–115 kb), after which cleavage occurs.

Following packaging of the vector pNS582-tet14 Ad10, *E. coli* strain NS3145 (Sternberg, 1990) was infected and plated on LB plates (Miller, 1972) with kanamycin (25 μg/ml). Kanamycin-resistant colonies were tested for tetracycline resistance, and tetracycline-sensitive bacterial colonies were isolated. Following packaging of the vector pAd10 sacBII, *E. coli* strain NS3529 (Pierce *et al.*, 1992) was infected and plated on LB agar containing kanamycin (25 μg/ml) and 5% sucrose in order to select for insert-containing vectors. Both NS3145 and NS3529 produce a site-specific recombinase that targets the *loxP* sites (Figure 1) and circularizes the vector by recombination. After isolation, a sample of clones from each vector was isolated and the size of the insert was determined by contour-clamped homogeneous electric field (CHEF) gel electrophoresis (Vollrath and Davis, 1987).

DNA Preparation from P1 Clones. Bacterial isolates containing single P1 clones were inoculated into LB medium containing kanamycin (25 μg/ml) and isopropyl β-D-thiogalactopyranoside (1 mM), which induces the lytic replicon (*rep* in Figure 1) to amplify the plasmid copy number (Sternberg, 1990), and grown overnight at 37°C. Plasmid DNA was extracted by the alkaline lysis method (Birnboim and Doly, 1979).

Cytological Analysis. Localization of clones was carried out with laboratory strain Oregon RC. Polytene chromosomes were prepared as in Atherton and Gall (1972). Chromosomes were pretreated in 2× SSC at 65°C for 30 min (1× SSC is 0.15 M NaCl/0.015 M sodium citrate, pH 7), dehydrated in 70% and 95% (vol/vol) ethanol, denatured in 0.07 M NaOH for 2.5 min, washed twice in 2× SSC, dehydrated again, and dried in air. DNA from P1 clones was labeled with biotin derivatives of dNTPs (GIBCO/BRL) by means of the random hexamer method (Feinberg and Vogelstein, 1984). Hybridization of labeled DNA to polytene chromosome squashes *in situ* was carried out overnight at 37°C in 1.4× SSC/7% (wt/vol) dextran sulfate/35% (vol/vol) *N,N*-dimethylformamide containing sonicated denatured salmon sperm DNA (0.6 mg/ml). Hybridization was detected with the Detek I horseradish peroxidase signal-generation system (Enzo Diagnostics) and 3,3′-diaminobenzidine (Sigma). Chromosomes were stained with Giemsa stain and embedded in Permount.

PCR Amplification of Insert–Vector Junctions. STS markers were determined from sequences at the termini of the genomic fragments present in P1 clones after amplification by the polymerase chain reaction (PCR) as described in Nurminsky and Hartl (1993); in this method, a partially double-stranded anchor–adapter oligonucleotide is ligated onto the genomic fragment near the junction of vector and insert, and a template suitable for PCR amplification is produced by primer extension from a site in the flanking vector sequence. After amplification of the genomic fragment, DNA sequencing was performed with an Applied Biosystems model 373A DNA sequencing system and the *Taq* DyeDeoxy terminator cycle-sequencing kit.

STS Markers from Known Genes. STS markers were also determined from known *Drosophila* sequences present in GenBank release 79.0. Oligonucleotides suitable as PCR primers were chosen with an algorithm (Rychlik and Rhoads, 1989) implemented in the program OLIGO (National Biosciences, Plymouth, MN), version 4.0, for the Macintosh. The primer oligonucleotides, ranging from 18-mers to 21-mers, were either synthesized with an Applied Biosystems model 392 DNA Synthesizer or supplied by Research Genetics (Huntsville, AL) or Genset (La Jolla, CA).

PCR Screening of the P1 Library. For PCR screening, the P1 library was organized in 96-well microtiter dishes (arrays of 8 rows × 12 columns). The screening was carried out in two stages. First, a set of 96 reactions was carried out on DNA pools from all the clones present in each 8 × 12 array; each array containing a clone able to support amplification was subjected to a second set of 20 reactions carried out on DNA pools from clones in the 8 rows and 12 columns of the array. The DNA pools were prepared and generously provided by Bill Kimmerly and collaborators at Lawrence Berkeley Laboratories as part of the *Drosophila* Genome Center described in the acknowledgments. The individual PCRs were carried out in MJ Research (Watertown, MA) PTC-100 thermal cyclers in 96-well, V-bottomed, polycarbonate microtiter plates. PCR was carried out in 20-μl reaction mixtures containing 1–3 mM $MgCl_2$ (optimized for each pair of oligonucleotides) overlaid with mineral oil and subjected to 25 cycles of 3 sec at 96°C, 45 sec at 92°C, 90 sec at the annealing temperature (optimized for each pair of oligonucleotides), and 90 sec at 72°C. The duration of time at the annealing temperature was increased by 1 sec in each cycle, and the duration of time at 72°C was increased by 3 sec in each cycle. The reactions were terminated by holding at 72°C for 5 min and stored at 4°C. PCR products were fractionated by electrophoresis in 1.2% agarose gels.

RESULTS

The *Drosophila* P1 library presently in use consists of 9216 clones. Approximately 40% of the clones are in pNS582-tet14 Ad10 (Figure 1*A*);

the remainder, ≈60%, are in pAd10 sacBII (Figure 1*B*). Inserts in pNS582-tet14 Ad10 cluster in the size range 70–100 kb with a mean and standard deviation of 83.0 ± 6.2 kb (n = 25) (Smoller *et al.*, 1991); inserts in pAd10 sacBII cluster in the same size range and have a mean and standard deviation of 82.5 ± 5.8 kb (n = 20) (data not shown).

Distribution of Clones by Chromosome. A total of 3104 clones were localized by *in situ* hybridization with the polytene salivary-gland chromosomes. Among the localized clones, 388 hybridized with the chromocenter, the underreplicated mass consisting largely of the pericentromeric heterochromatin and the Y chromosome, and/or with multiple euchromatic sites (typically, 10–100) without any apparent major euchromatic site of hybridization. A total of 64 clones yielded dual hybridizations (strong signals in two distinct euchromatic sites); these clones have not been investigated further, but some of them may represent chimeric clones containing ligated fragments from two different parts of the genome. An additional 191 clones were deliberate duplicates introduced into the workstream as blind controls in order to verify the accuracy of the procedures and reproducibility of the cytological localizations.

The remaining 2461 localized clones yielded single major sites of euchromatic hybridization. Approximately 10% of these clones also exhibited multiple (typically, 10–100) secondary sites of hybridization in the euchromatin; about half of this class also hybridized with the chromocenter. The multiple sites of hybridization are interpreted as resulting from transposable elements or other types of moderately repetitive, dispersed DNA contained in the cloned insert. With such clones, it is usually not difficult to identify a principal site of hybridization in the euchromatin in which the signal is intense and encompasses several bands, compared to which the multiple sites of hybridization are usually much weaker and present in single bands; the principal site of hybridization is the site to which the clone is assigned. (The distinction between the major site and secondary sites of hybridization can be difficult if the probe is excessive in amount or too heavily labeled.) A final class of clones, ≈2% of those localized, hybridized with the chromocenter and also to one principal site of hybridization in the euchromatin, without detectable secondary sites of euchromatic hybridization.

Because the average insert size of the clones is ≈80 kb, the mapped clones with unique major sites of hybridization include ≈200 megabase pairs (Mb) of DNA, or the equivalent of ≈1.8 copies of the haploid euchromatic genome. If we assume that all euchromatic sites are equally likely to be present in the clones, the proportion of euchromatic sites

TABLE 1 Distribution of P1 clones in the euchromatic genome

Chromosome arm	No. of P1 clones	No. of P1 clones per lettered subdivision (mean ± SEM)	No. of P1 clones per 80 kb
X[a]	317	2.7 ± 0.2	1.5
2L	490	4.2 ± 0.3	2.1
2R	463	3.9 ± 0.3	1.6
3L	531	4.6 ± 0.4	2.0
3R	653	5.7 ± 0.4	1.9

[a]Applying the correction factor of 4/3 for X-linked segments yields an average of 3.6 P1 clones per lettered subdivision on the X chromosome and an average of 2.0 P1 clones per 80 kb.

expected to be represented at least once among the clones is ≈85%. The distribution of clones localized in the arms of the large euchromatic chromosomes is given in Table 1. (The X chromosome is acrocentric, chromosomes 2 and 3 are metacentric.) The ≈25% deficiency of clones localized to the X chromosome is expected because the libraries were constructed from a mixture of male and female DNA (Ajioka *et al.*, 1991); however, the relative underrepresentation of clones from 2R is statistically significant ($P < 0.01$). Although a small percentage of the clones are duplicates arising from inadvertent transfer of the same bacterial colony during clone isolation, these are not sufficient in number to account for the nonrandomness. An additional 7 clones were localized to chromosome 4 which, based on its DNA content, would be expected to have ≈25; the underrepresentation of chromosome 4 is also statistically significant ($P < 0.01$).

The large autosome arms are each divided into 20 numbered divisions, and each of these (except for those at the base) is divided into six lettered subdivisions (A–F) with an average DNA content of ≈200 kb. The distribution of P1 clones by lettered subdivision is given in Figure 2. Overall, >90% of the lettered subdivisions have at least one P1 clone, and the average number of clones per lettered subdivision is 4.2.

Framework Map. The P1 framework map is summarized in Figure 3. Sections 1–20 comprise the X chromosome, 21–40 and 41–60 the respective left and right arms of chromosome 2, and 61–80 and 81–100 the respective left and right arms of chromosome 3. In each numbered section, the short vertical tick marks, from left to right, set off the lettered subdivisions A–F (unlabeled). The histograms give the number of P1 clones in each lettered subdivision. A clone overlapping two

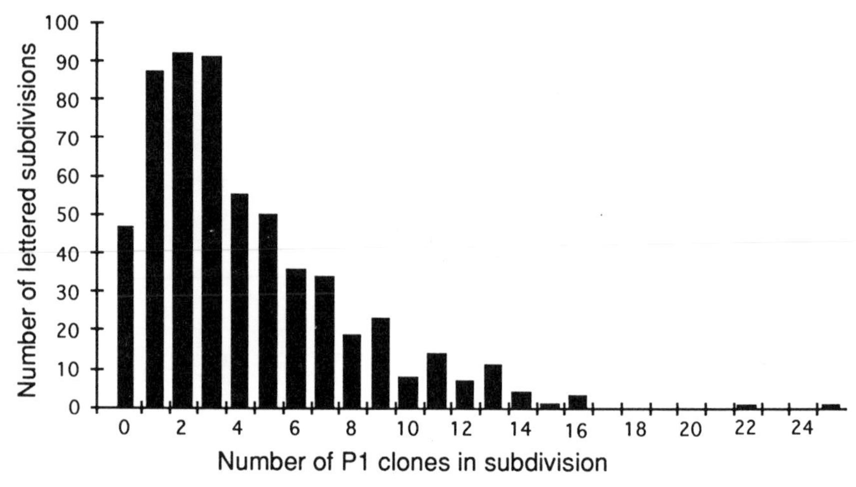

FIGURE 2 Distribution of P1 clones by lettered subdivision. See text for explanation.

subdivisions is counted as one-half clone in each subdivision. At the level resolution of Figure 3, the alignment between the chromosome bands and the delineators is approximate. Detailed information on the clones is accessible electronically from the flybase database and in printed form in Hartl and Lozovskaya (1994).

Contig Assembly. While the framework map in Figure 3 was being completed, contig assembly began, using the strategy of STS-content mapping, in which overlapping P1 clones are identified by virtue of their ability to support the PCR amplification of single-copy genomic sequences. The STS markers used for contig assembly are derived from (*i*) known genes with sequences available in GenBank, (*ii*) the termini of the genomic fragments present in P1 clones, and (*iii*) sites of insertion of the transposable element *P* (Hartl and Palazzolo, 1993). Although genome-wide contig assembly is still in its early stages, progress in one of the targeted regions near the tip of the X chromosome is shown in Figure 4 and illustrates the general strategy.

In Figure 4, the heavy black bars across the top represent the bands in the polytene chromosome. The positions of the STS markers are indicated by vertical dashed lines intersecting with horizontal black bars below that represent P1 clones; the ability of a P1 clone to support amplification of a STS is indicated with a cross at the site of intersection. The P1 contigs are indicated by long horizontal black bars across the bottom. In this region, the STS markers include 7 known genes and 11 termini of P1 inserts (hyphenated names) over ≈650 kb. There is a contig of 10 clones extending ≈240 kb from bands 2E2 to 3A1 and another contig of 15 clones extending ≈370 kb from 3A1 to 3B3. Between these contigs there may be a small gap of ≤40 kb, which we have not, as yet, attempted to bridge with additional P1 clones identified by screening the entire library with the ends of the inserts in the flanking P1 clones. The average density of STS markers per P1 clone in this region is 2.5.

DISCUSSION

The framework map reported here and summarized in Figure 3 affords a unique resource for research in the genetics of *Drosophila*. The framework map includes ≈85% of the euchromatic genome with an average redundancy of coverage of 1.8. Although there is some nonrandomness in the distribution of clones, the coverage of the euchromatic genome is very broad, and DNA from most euchromatic regions of interest should be accessible from the clones in the framework map. Less well represented are clones from the meshlike region of the polytene chromosomes denoted β heterochromatin, constituting the base of each chromosome arm and most of chromosome 4. Unrepresented in the framework map are clones from the chromocenter, comprising the α heterochromatin, which is grossly underreplicated in salivary gland nuclei and which includes the pericentromeric heterochromatin and the entire Y chromosome. Some of this material may be present in the ≈15% of P1 clones that hybridize with the chromocenter and that have no major sites of hybridization in the euchromatin.

Applications of the *Drosophila* physical map in studies of genome structure and function have been stressed elsewhere (Kafatos *et al.*, 1991; Merriam *et al.*, 1991; Hartl *et al.*, 1992; Hartl and Lozovskaya, 1992). Less consideration has been given to the utility of clone-based physical maps in studies of genome evolution. *Drosophila* has a long history of studies of genome evolution because, within many species groups, the banding patterns of the polytene chromosomes are sufficiently similar that phylogenetic relationships can be inferred. The principal limitation of cytological analysis is that, between species groups, the differences in banding patterns are usually too great for homologous chromosomal regions to be identified reliably, even between species whose morphological similarity implies virtual certainty of close relationship (Stone, 1962). The result is a very incomplete understanding of the patterns, processes, and functional significance of genome evolution in *Drosophila*. A case in point is the obscura species group, long an object of evolutionary studies, which includes such species as *D. pseudoobscura*,

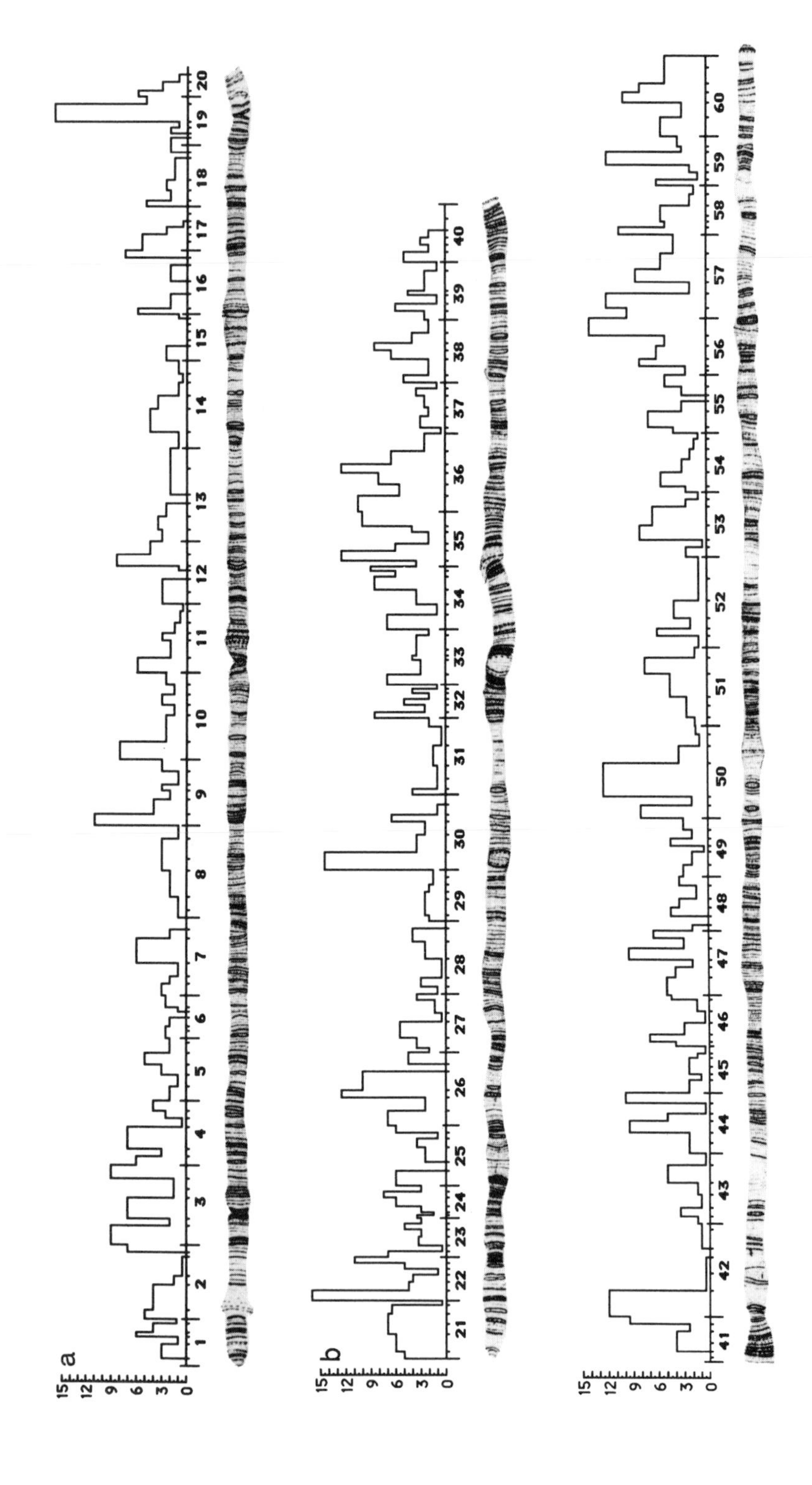
a
b
15
12
9
6
3
0

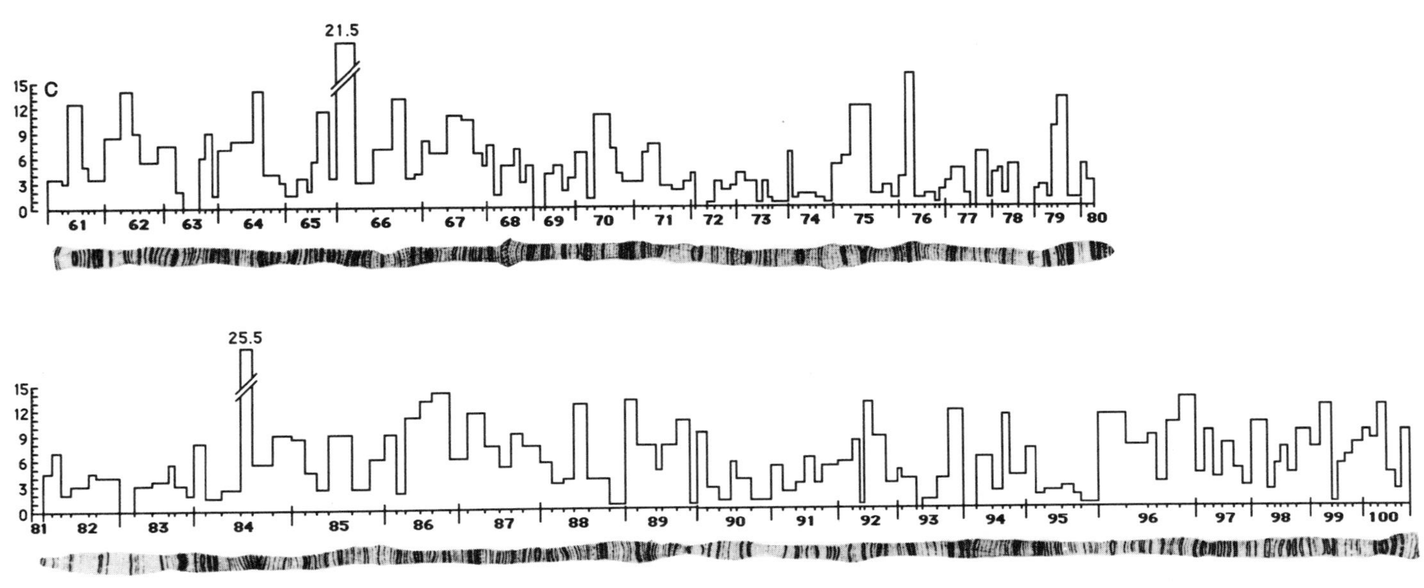

FIGURE 3 Framework P1 map. The tick marks in each numbered subdivision delineate, from left to right, the lettered subdivisions A–F. (*a*) X chromosome. (*b*) Chromosome arms 2L (*Upper*) and 2R (*Lower*). (*c*) Chromosome arms 3L (*Upper*) and 3R (*Lower*).

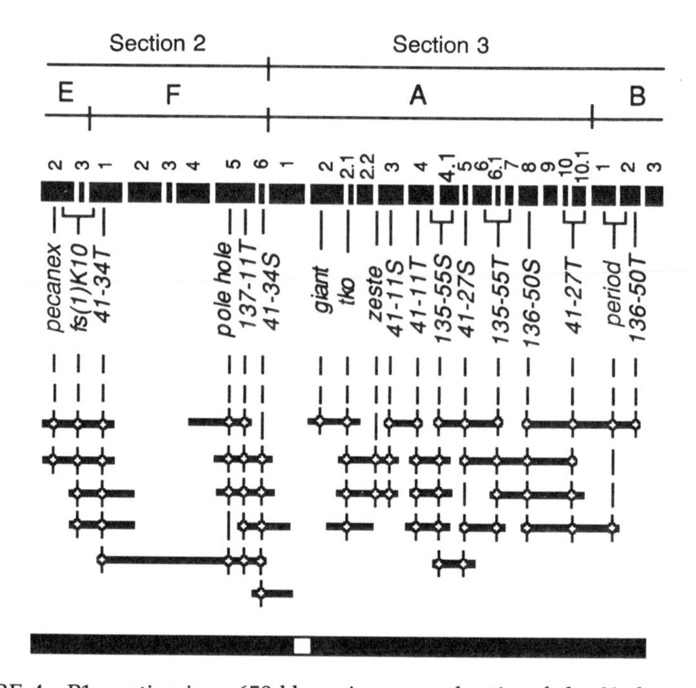

FIGURE 4 P1 contigs in a 650-kb region near the tip of the X chromosome. Numbered black bars across the top represent the polytene chromosome bands; narrow black bars below represent P1 clones. Positions of STS markers are denoted by dashed vertical lines intersecting P1 clones able to support PCR amplification. STS markers with hyphenated designations are the termini of P1 inserts; others are known genes. The thick horizontal bars across the bottom are two contigs.

D. miranda, *D. subobscura*, *D. guanche*, and *D. affinis* (Lakovarra and Saura, 1982). Although the correspondences between the chromosome arms of *D. melanogaster* and various species in the obscura group have been ascertained by *in situ* hybridization with probes from *D. melanogaster* (Segarra and Aguadé, 1992), no detailed, point-by-point cytological comparison between the genomes is possible because the polytene chromosome band morphologies, after an estimated 30–40 million years of evolutionary divergence (Beverley and Wilson, 1984), are too dissimilar. However, in studies with P1 clones from the framework map of *D. melanogaster*, >80% yielded strong, single hybridization patterns with the salivary gland chromosomes of *D. pseudoobscura* (Segarra and Lozovskaya, unpublished observations). No ambiguities resulting from hybridization with transposable elements have been noted, probably because, relative to single-copy genes, the transposable elements are sufficiently divergent in sequence (de Frutos *et al.*, 1992). Hence, the framework map in Figure 3 provides the material for a detailed alignment of the *D. melanogaster* and *D. pseudoobscura* euchromatic genomes with a density of markers of approximately one every 100 kb. Undoubtedly, many of the 80-kb P1 clones will also contain sites of rearrangement breakpoints between the species and thereby provide the material for determining the molecular mechanisms, as well as the patterns, of genome evolution.

In contrast to the situation with *D. pseudoobscura*, no reliable hybridization signals with the salivary-gland chromosomes of *D. virilis* have been possible with P1 clones from *D. melanogaster* (Lozovskaya *et al.*, 1993). The simplest explanation is that there has been too much DNA sequence divergence in the estimated 60 million years since the existence of a common ancestor of *D. virilis* and *D. melanogaster* (Beverley and Wilson, 1984). However, a bacteriophage P1 library of *D. virilis* has been constructed and initial progress toward a framework P1 map described; the framework map of *D. virilis* can be put into correspondence with that of *D. melanogaster* by means of DNA hybridization with suitable single-copy probes (Lozovskaya *et al.*, 1993). Similar to the manner in which the framework map of *D. melanogaster* affords access to genome organization in *D. pseudoobscura* and other species in the subgenus *Sophophora*, it is anticipated that a framework map of *D. virilis* will afford access to genome organization and evolution in the subgenus *Drosophila*, including the Hawaiian drosophila. The P1 clones in the *D. virilis* framework map also provide the materials for large-scale sequencing of selected regions of the genome for comparison with the corresponding regions in *D. melanogaster*.

SUMMARY

Physical maps showing the relative locations of cloned DNA fragments in the genome are important resources for research in molecular genetics, genome analysis, and evolutionary biology. In addition to affording a common frame of reference for organizing diverse types of genetic data, physical maps also provide ready access to clones containing DNA sequences from any defined region of the genome. In this paper, we present a physical map of the genome of *Drosophila melanogaster* based on *in situ* hybridization with 2461 DNA fragments, averaging ≈80 kilobase pairs each, cloned in bacteriophage P1. The map is a framework map in the sense that most putative overlaps between clones have not yet been demonstrated at the molecular level. Nevertheless, the framework map includes ≈85% of all genes in the euchromatic genome. A continuous physical map composed of sets of overlapping P1 clones (contigs), which together span most of the euchromatic genome, is currently being assembled by screening a library of 9216 P1 clones with single-copy genetic markers as well as with the ends of the P1 clones already assigned positions in the framework map. Because most P1 clones from *D. melanogaster* hybridize *in situ* with chromosomes from related species, the framework map also makes it possible to determine the genome maps of *D. pseudoobscura* and other species in the subgenus *Sophophora*. Likewise, a P1 framework map of *D. virilis* affords potential access to genome organization and evolution in the subgenus *Drosophila*.

We thank Lara Brilla for administrative support, Yaping Xu for data base management, and Jie Wei for technical help. The work reported here was supported in part by the National Center for Human Genome Research, *Drosophila* Genome Center (HG00750), Gerald M. Rubin, Principal Investigator. The *Drosophila* Genome Center includes investigators at the University of California at Berkeley, Lawrence Berkeley Laboratories, Harvard University, and the Carnegie Institution of Washington in Baltimore. We are grateful to G. M. Rubin, A. C. Spradling, and other participating investigators in the *Drosophila* Genome Center, especially Bill Kimmerly, Mike Palazzolo, and Chris Martin. David Smoller participated in making the *Drosophila* P1 libraries, and we are grateful to Nat Sternberg, James Pierce, and Phil Moen for their advice.

REFERENCES

Ajioka, J. W., Smoller, D. A., Jones, R. W., Carulli, J. P., Vellek, A. E. C., Garza, D., Link, A. J., Duncan, I. W. & Hartl, D. L. (1991) *Drosophila* genome project: one-hit coverage in yeast artificial chromosomes. *Chromosoma* **100,** 495–509.

Atherton, D. & Gall, J. (1972) Salivary gland squashes for in situ nucleic acid hybridization studies. *Drosophila Info. Serv.* **49,** 131–133.

Beverley, S. M. & Wilson, A. C. (1984) Molecular evolution in *Drosophila* and the higher diptera II. Time scale for fly evolution. *J. Mol. Evol.* **21,** 1–13.

Birnboim, H. C. & Doly, J. (1979) A rapid alkaline extraction procedure for screening recombinant plasmid DNA. *Nucleic Acids Res.* **7,** 1513–1523.

Cai, H., Kiefel, P., Yee, J. & Duncan, I. (1994) A yeast artificial chromosome clone map of the *Drosophila* genome. Genetics **136,** 1385–1401.

Cohen, D., Chumakov, I. & Weissenbach, J. (1993) A first-generation physical map of the human genome. *Nature (London)* **366,** 698–701.

Collins, F. & Galas, D. (1993) A new five-year plan for the U.S. human genome project. *Science* **262,** 43–49.

Coulson, A., Sulston, J., Brenner, S. & Karn, J. (1986) Toward a physical map of the genome of the nematode *Caenorhabditis elegans*. *Proc. Natl. Acad. Sci. USA* **83,** 7821–7825.

de Frutos, R., Peterson, K. R. & Kidwell, M. G. (1992) Distribution of *Drosophila melanogaster* transpoable element sequences in species of the obscura group. *Chromosoma* **101,** 293–300.

Feinberg, A. P. & Vogelstein, B. (1984) A technique for radiolabeling DNA restriction endonuclease fragments to high specific activity: Addendum. *Anal. Biochem.* **137,** 266–267.

Hartl, D. L. (1992) Genome map of *Drosophila melanogaster* based on yeast artificial chromosomes. In *Genome Analysis: Strategies for Physical Mapping,* eds. Davies, K. E. & Tilghman, S. M. (Cold Spring Harbor Lab. Press, Plainview, NY), Vol. 4, pp. 39–69.

Hartl, D. L., Ajioka, J. W., Cai, H., Lohe, A. R., Lozovskaya, E. R., Smoller, D. A. & Duncan, I. W. (1992) Towards a *Drosophila* genome map. *Trends Genet.* **8,** 70–75.

Hartl, D. L. & Lozovskaya, E. R. (1992) The *Drosophila* genome project: Current status of the physical map. *Comp. Biochem. Physiol. B* **103,** 1–8.

Hartl, D. L., and E. R. Lozovskaya. (1994) *The Drosophila Genome Map: A Practical Guide.* R. G. Landes, Austin, TX

Hartl, D. L. & Palazzolo, M. J. (1993) Drosophila as a model organism in genome analysis. In *Genome Research in Molecular Medicine and Virology,* ed. Adolph, K. W. (Academic, Orlando, FL), pp. 115–129.

Heino, T. I., A. O. Sura, and V. Sorsa (1994) *Salivary chromosome maps of Drosophila melanogaster.* Drosophila Information Service, Vol. **73.**

Kafatos, F. C., Louis, C., Savakis, C., Glover, D. M., Ashburner, M., Link, A. J., Siden-Kiamos, I. & Saunders, R. D. C. (1991) Integrated maps of the *Drosophila* genome: progress and prospects. *Trends Genet.* **7,** 155–161.

Kohara, Y., Akiyama, K. & Isono, K. (1987) The physical map of the whole *E. coli* chromosome: Application of a new strategy for rapid analysis and sorting of a large genomic library. *Cell* **50,** 495–508.

Lakovarra, S. & Saura, A. (1982) Evolution and speciation in the *Drosophila obscura* group. In *The Genetics and Biology of Drosophila,* eds. Ashburner, M., Carson, H. L. & Thompson, J. N., Jr. (Academic, New York), pp. 1–59.

Lozovskaya, E. R., Petrov, D. A. & Hartl, D. L. (1993) A combined molecular and cytogenetic approach to genome evolution in *Drosophila* using large-fragment DNA cloning. *Chromosoma* **102,** 253–266.

Merriam, J., Ashburner, M., Hartl, D. L. & Kafatos, F. C. (1991) Toward cloning and mapping the genome of *Drosophila. Science* **254,** 221–225.

Nurminsky, D. I. & Hartl, D. L. (1993) Amplification of the ends of DNA fragments cloned in bacteriophage P1. *BioTechniques* **15,** 201–208.

Olson, M. V., Dutchik, J. E., Graham, M. Y., Brodeur, G. M., Helms, C., Frank, M., MacCollin, M., Scheinman, R. & Frank, T. (1986) Random-clone strategy for genomic restriction mapping in yeast. *Proc. Natl. Acad. Sci. USA* **83,** 7826–7830.

Pierce, J. C., Sauer, B. & Sternberg, N. (1992) A positive selection vector for cloning high molecular weight DNA by the bacteriophage P1 system: Improved cloning efficacy. *Proc. Natl. Acad. Sci. USA* **89,** 2056–2060.

Rychlik, W. & Rhoads, R. E. (1989) A computer program for choosing optimal oligonucleotides for filter hybridization, sequencing and in vitro amplification of DNA. *Nucleic Acids Res.* **17,** 8543–8551.

Segarra, C. & Aguadé, M. (1992) Molecular organization of the X chromosome in different species of the obscura group of *Drosophila. Genetics* **130,** 513–521.

Smoller, D. A., Petrov, D. & Hartl, D. L. (1991) Characterization of bacteriophage P1 library containing inserts of *Drosophila* DNA of 75–100 kilobase pairs. *Chromosoma* **100,** 487–494.

Sorsa, V. (1988) *Chromosome Maps of Drosophila* (CRC, Boca Raton, FL), Vols. 1 and 2.

Sternberg, N. (1990) Bacteriophage P1 cloning system for the isolation, amplification, and recovery of DNA fragments as large as 100 kilobase pairs. *Proc. Natl. Acad. Sci. USA* **87,** 103–107.

Stone, W. S. (1962) The dominance of natural selection and the reality of superspecies (species groups) in the evolution of *Drosophila*. *Univ. Texas Publ.* **6205,** 507–537.

Vollrath, D. & Davis, R. W. (1987) Resolution of DNA molecules greater than 5 megabases by contour-clamped homogeneous electric fields. *Nucleic Acids Res.* **15,** 7865–7876.

Index

A

F

N

O

P

R

T

U

V

W